Artificial Intelligence for Engineers

Zhen "Leo" Liu

Artificial Intelligence for Engineers

Basics and Implementations

Zhen "Leo" Liu
Department of Civil and Environmental Engineering
University of Virginia
Charlottesville, VA, USA

ISBN 978-3-031-75955-0 ISBN 978-3-031-75953-6 (eBook)
https://doi.org/10.1007/978-3-031-75953-6

Cover illustration: As a book for AI, this book adopts an image generated using generative AI as the cover image, in order to provide a snapshot of the fast-evolving AI realm. The image exhibits the power of contemporary AI, though it will become "obsolete" as soon as new AI tools are available. Such generative AI can craft high-quality images with a text description or "prompt" from users. If this AI wave lasts longer, more such AI innovations may transform more aspects of human society, such as arts and entertainment, more than ever before.

This Springer imprint is published by the registered company Springer Nature Switzerland AG
The registered company address is: Gewerbestrasse 11, 6330 Cham, Switzerland

This book is dedicated to people

who do not limit themselves,
who believe in miracles and the future, and
who trust and are cautious about the
power of science and technology.

Also, for my wife Sarah, son Brent, and
daughter Leah.

Preface

This book presents basic knowledge and essential toolsets needed for people who want to step into artificial intelligence (AI). The book is especially suitable for those college students, graduate students, instructors, and IT hobbyists who have an engineering mindset. That is, it serves the idea of getting the job done quickly and neatly with an adequate understanding of why and how. It is designed to allow one to obtain big pictures for both AI and essential AI topics within the shortest amount of time. Based on the picture(s), suitable amounts of theoretical knowledge are contextualized to help the learner gain information about the most essential concepts and algorithms. These algorithms are introduced and formulated in a way that the learner can easily implement them for real-world applications with a small amount of effort. In short, you read it, you understand it, you try it, and you can solve it.

The book is prepared by a learner for learners. Available learning materials like textbooks are diverse due to the broad scope and fast development of AI. There are inconsistent terminologies, distinct focuses of people from different application areas, inaccurate/misleading information from the Internet, and fast evolvement of available tools. As a result, a major problem of learners is not the lack of information but the opposite: there is too much information. A common frustration is the numerous gaps and conflicts that we can easily run into when we try to organize such knowledge to form a consistent and efficient literacy. We may need to read multiple books, take several actual or online courses, go through a lot of blogs or tutorials, and learn several tools via their user guides/manuals before we finally get what we want. Eventually, we may sadly find that much of the effort is unnecessary, while a lot of time is used to address the inconsistencies, gaps, and confusion arising during the above process. This book is prepared by a learner who has suffered from all the above issues for other learners who want to avoid such issues.

This book, though titled Artificial Intelligence, is mostly devoted to numeric AI represented by machine learning algorithms, which predominate the so-called third wave/tide of AI. The most common and useful machine learning topics are selected and introduced. This includes introductions to topics that are needed to gain a basic understanding of machine learning, such as linear models, decision trees,

Bayesian algorithms, and clustering algorithms, as well as more advanced topics like deep learning and reinforcement learning. It attempts to cover the essential terms, basic/common algorithms, and useful tools that one may encounter in a typical journey for learning and performing contemporary AI.

The book aims to strike a balance between being pragmatic and theoretical. Many AI learners, especially those in engineering applications, tend to solve a problem as quickly as possible, for example, using some AI code or libraries from the Internet. This usually works well considering the maturity of many AI tools. However, it may lead to unnecessary, inappropriate use of such tools and hinder further learning of the topics. On the contrary, some other learners try to start with intricate math and bottom-level computer science knowledge, which can easily discourage them and eventually turn out to be not needed in many cases. Each of the topics (or chapters) in this book adopts its own storyline, which may share a commonality with other topics while still maintaining uniqueness that stems from its own historical development and algorithmic nature.

The book is designed for a typical undergraduate, graduate, or dual-listed course with a semester-based calendar. Accordingly, its 16 chapters present 16 essential topics for the 15 to 16 weeks in a typical university semester. The book has been tested in a dual-listed course containing both undergraduate and graduate students from engineering, in which each week had one lecture and one computer lab for each topic. Thus, additional teaching and learning materials, including lecture videos, notes, and code, are available. The book and associated materials have been proven to be able to assist a student with limited or even no knowledge in coding, data analytics, and machine learning to develop a basic and competing ability to understand, communicate, and implement AI. Due to the same reason, the book can also be considered by learners who are not an expert on or major in computer science but want to take 15 to 16 weeks or 45 cumulative hours (1.5 hours per session $\times$ 2 sessions per week $\times$ 15 weeks) to learn AI. This book can also be used as a reference book for more advanced AI practitioners who want to check the basic algorithms for reviewing a learned algorithm or learning new unfamiliar topics.

This book starts with a chapter for the basics of AI, which both outlines a big AI picture by discussing why to look into AI, what is AI, the history of AI, AI vs. traditional engineering methods, as well as AI applications in engineering and other sectors, and presents the AI basics including the basic concepts, common algorithms, and challenges and issues. Next, in Chap. 2, common tools needed for implementing AI, including coding language and environment, data manipulation and visualization tools, machine learning and data analysis libraries, and deep learning packages, will be introduced. Following that, typical supervised learning topics, including linear models, decision trees, support vector machines, Bayesian algorithms, neural networks, and deep learning, will be presented in Chaps. 3–7. Ensemble learning will also be introduced in Chap. 9 based on them. After that, four typical unsupervised learning topics, i.e., clustering, dimensionality reduction, anomaly detection, and association rule learning, will be explored in Chaps. 10–13. Reinforcement learning will be introduced using two chapters: Chap. 14 for the basics of reinforcement learning and value-based reinforcement learning, and

Chap. 15 for policy-based reinforcement learning. Finally, Appendix A delivers appendices for more detailed information on AI's foundational and ancillary knowledge in math, optimization (solvers), and evaluation metrics.

Charlottesville, VA, USA Zhen "Leo" Liu

Acknowledgments

Much obliged for students who worked on AI-relevant research with me, including but not limited to Behnam, Aynaz, Hossein, and Muchun, and those students who have taken and will take my AI course.

Special thanks to the Department of Civil, Environmental, and Geospatial Engineering at Michigan Technological University and the Department of Civil and Environmental Engineering at the University of Virginia. Their strongest and long-lasting support is critical to the production of this work.

The support from the U.S. National Science Foundation (NSF) via Project NSF1742656 and the U.S. Federal Highway Administration (FHWA) via Project FHWA 693JJ320C000022 is sincerely appreciated. The support enabled this book.

Contents

Chapter 1
Preparation Knowledge: Basics of AI

1.1 Overview

This chapter begins with an introduction to artificial intelligence (AI) to discuss the first things that most AI learners want to know: why look into AI, what is AI, the history of AI, AI versus traditional engineering methods, and AI applications in engineering and other sectors. Next, the basics of AI are laid out, including the basic concepts, common algorithms, and challenges and issues.

Mathematics takes different roles in different AI topics and is thus essential in the understanding and implementation of many of them. Considering the possibly diverse backgrounds of the readers, such knowledge is provided systematically at the end of the book as Appendices. Mathematics knowledge that is needed for AI, i.e., statistics, information theory, and array operations, can be reviewed there as needed. Some math knowledge that is essential to specific AI topics will be provided in the corresponding chapters.

1.2 Introduction to Artificial Intelligence

1.2.1 Why Look into AI?

Deemed the core of the fourth industrial revolution or Industry 4.0, AI has been reshaping our lives in many ways towards what we have long pictured in science fiction. From chatbots like ChatGPT to autonomous cars, more widespread AI applications have been transforming our society, bringing benefits such as increased efficiencies, more intelligent products, and fewer repetitive tasks. AI is projected to boost the corporate profitability of 16 industries in 12 economies by an average of 38% by 2035 [1].

Z. "L." Liu, *Artificial Intelligence for Engineers*,
https://doi.org/10.1007/978-3-031-75953-6_1

The popularity and impacts of AI can be seen in related intellectual products. First, the number of AI-related publications almost tripled in one decade, e.g., from 88,000 in 2010 to 240,000 in 2022, in AI categories like pattern recognition, machine learning, and algorithms [2, 3]. Besides publications, the number of AI-related patent filings multiplied by a factor of 30 times from 2015 to 2021. The global market size of AI software is expected to grow at a rate of 22.97% from 2023 and surpass $1094.52 billion by 2032 [4].

This so-called third wave/tide of AI has enabled world-changing applications represented by deep learning and reinforcement learning. Deep learning gained success in application areas like computer vision and natural language processing. Computer vision is an AI subfield that teaches machines to understand images and videos for the purposes of image classification, object detection, mapping of position and movement of objects, and so on. In addition to understanding existing images, AI can also generate "fake" images and videos that are nearly indistinguishable from real ones. Natural language processing, a subfield that focuses on summarizing contents, inferring outcomes, identifying emotional context, speech recognition and transcription, and translation, is impacting our lives again via many language-related innovations and tools, attributed to its cumulative development since the 1950s and recent breakthroughs via deep learning especially large language models. Besides deep learning, reinforcement learning is another major area of AI innovations, which helps us advance the cutting edge of AI. Its triumphs have swept from video games to complicated board games like chess and Go and, more recently, in engineering decision-making and control tasks like robotics.

AI grows and impacts the world via strong synergy with the blooming of data and improvements in computing hardware. The third wave of AI is propelled by big data, improvements in computing software (GPU), and advances in machine learning algorithms. In engineering, the explosion of data is partly attributed to the widespread use of low-cost information-sensing mobile devices, remote sensing, software logs, cameras, microphones, radio-frequency identification (RFID) readers, and wireless sensor networks. The vast amounts of data are enabling us to explore AI tools with more complicated architectures, higher capacity, and better generalization, such as neural networks that are deeper, wider, and more intricate inside. In return, AI breakthroughs driven by data promoted tech giants like Google, Microsoft, Meta, IBM, and Amazon to improve AI tools, algorithms, applications, and data. Owning to the development of AI, a system that would have taken 6 minutes to train in 2018 would only need about 13 seconds in 2021 [2]. In terms of cost, the costs of AI training drop at a rate of 70% per year as of 2024 [5], outpacing the famous Moore's Law.

The dramatic development of AI has also been obviously affecting the future of the workforce. More AI applications across industries create higher demands for AI education and jobs. Taking the United States, for example, California, Texas, New York, and Virginia have exhibited high demands for AI-related occupations. In computer science, the most popular specialties among PhD students have been secured by AI and machine learning in the past decade. In addition to the trends in the workforce demand and development, corporate investments in AI are at an all-time high, totaling $189 billion in 2023, representing a 1300% increase from 2013, and could amount to $200 billion globally by 2025 [3]. AI companies in the "healthcare" sector led the champion, followed by "data management, processing, and cloud," and financial technology (or called "fintech" for short).

AI also generates far-reaching impacts on the economy, politics, mobility, healthcare, security, and the environment. Such influence on the economy can take place via disruptions to the labor market, alternation of the nature of long-established roles, and changes in political thinking and opinions. For mobility, AI is estimated to be capable of helping reduce the number of road accidents by as much as 90% and boosting multimodal transportation through better transportation options and operations, while also bringing about new challenges in liability, ethics, and management [6]. The further development of AI for healthcare can possibly eradicate many incurable diseases and help deliver care to remote areas and groups with difficulties. In security and defense sectors, AI-powered software is dramatically altering the digital security threat landscape via innovative cyberattack detection, prevention, and risk control, which could easily save economic loss in excess of $50 billion in one major global cyberattack [7].

Therefore, AI is a must-know for the new generations. For engineers, we may need to know the basics as engineering is being further impacted by and fused with AI. Some engineers who will be more exposed to or deal with AI may need to know the common AI techniques, from the entry level of having common and useful AI techniques in their toolbox to a more advanced level of assessing, modifying, extending, and coding some newer and complicated AI algorithms. This book is proposed to help engineers quickly bridge these gaps.

1.2.2 What Is AI?

AI has been defined from different perspectives by people from many distinct areas and thus encompasses a wide variety of techniques. In this book, AI is defined as a method or the use of such a method for making a computer or a computer-controlled agent, either hardware like a robot or software like a computer program, to think intelligently like the human mind. Thus, AI is accomplished by studying the working mechanisms of human brains and by analyzing the cognitive processes. Such AI studies develop products like data, algorithms, intelligent software and systems, and paradigms for specific applications. All of such AI products enable the computer or any agent controlled by it to exhibit some types of human beings' intelligence to some extent.

AI can be classified in different ways, which is a topic that can trigger many inconsistencies, conflicts, and debates. This fact is attributed to numerous reasons. In particular, as an area contributed by people from different disciplines, the history, convention, and backgrounds of these contributors in AI could lead to the adoption of the same terms for different meanings, names, and usages and different terms with

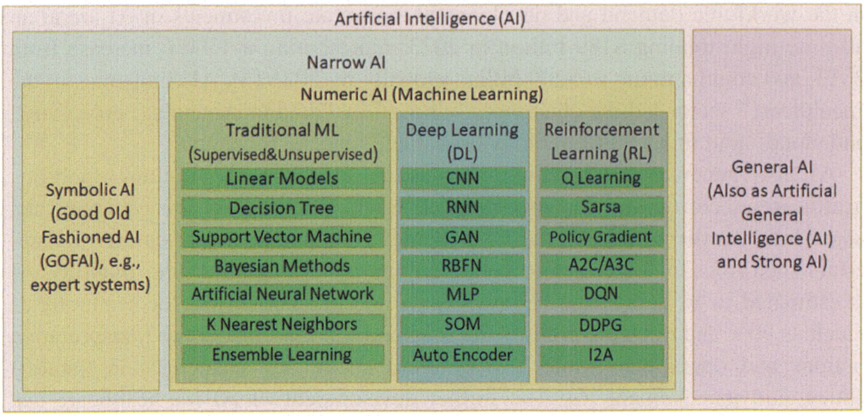

Artificial Intelligence (AI)

Narrow AI

Symbolic AI (Good Old Fashioned AI (GOFAI), e.g., expert systems)

Numeric AI (Machine Learning)

Traditional ML (Supervised&Unsupervised)	Deep Learning (DL)	Reinforcement Learning (RL)
Linear Models	CNN	Q Learning
Decision Tree	RNN	Sarsa
Support Vector Machine	GAN	Policy Gradient
Bayesian Methods	RBFN	A2C/A3C
Artificial Neural Network	MLP	DQN
K Nearest Neighbors	SOM	DDPG
Ensemble Learning	Auto Encoder	I2A

General AI (Also as Artificial General Intelligence (AI) and Strong AI)

Fig. 1.1 Subareas in AI

the same meaning. In addition, the evolution of the AI field including the technical development and other incidents such as rebranding of AI topics for promoting publications and fundraising further created gaps, overlaps, conflicts, and confusion when talking about the types of AI as well as the categorization of its subareas.

One classification constructed based on the consensuses in the literature will be adopted throughout this book for consistency. As shown in Fig. 1.1, AI can be roughly categorized into general AI and narrow AI. General AI is what we see in science fiction, in which AI can enable intelligent agents like supermen, e.g., Terminator and Wall-E, for handling much different tasks like combating and flying. Though general AI is always a dream and has been repeatedly discussed in the history of AI including the recent artificial general intelligence, it may still be far from us. By contrast, narrow AI is what we have been mostly working on. Narrow AI is set for a lower goal and thus more feasible. More interestingly, the successes in narrow AI in different phases of the AI history promoted people to pursue general AI.

Narrow AI can be divided into two groups: symbolic AI and nonsymbolic AI (or called numeric AI less frequently). Symbolic AI is represented by logic reasoning that was extensively studied in the early stages of AI and by more successful applications in expert systems. Nonsymbolic AI mostly refers to machine learning. What differentiates these two major AI streams is not the use of symbols or languages that can be easily understood by human beings. Although symbolic AI is usually associated with the use of symbols or languages, while nonsymbolic AI is not, there are exceptions. Instead, they essentially represent two ways of learning or gaining intelligence/knowledge: deduction and induction. Symbolic AI features deduction (or deduction reasoning), in which we make inferences based on widely accepted factors or premises, whereas nonsymbolic AI boils down to induction (or induction reasoning), in which we extract general knowledge from observations on specific cases. From another perspective, deduction goes from general to special, whereas induction moves from special to general. As a result, symbolic AI features the use of reasoning, usually performed with languages that humans could understand, while nonsymbolic AI, e.g., machine learning, is characterized by learning from or via data.

This book is mostly devoted to machine learning, which is predominant in contemporary AI studies. This arrangement was made considering the fact that many people including AI researchers use AI and machine learning interchangeably these days. Despite this fact, it is still worthwhile to mention that symbolic AI might not disappear or go out of date. In fact, in some people's opinions, they have more likely dissolved into our lives. Symbolic AI, which is called "good old-fashioned artificial intelligence (GOFAI)," is believed by some other people to be the classical and most successful AI approach till now. AI techniques or efforts along this line make computers more "intelligent" by using logic, e.g., mathematically provable logical methods, to manipulate "terms" that were specific to the target task. Humans could define "rules" for the concrete manipulation of such "terms" and create rule-based systems. Thus, computers and smartphones that can fulfill tasks based on predefined rules built on reasoning like "if statements" can also be viewed as symbolic AI, though we no longer view them as AI in most of the modern AI contexts.

In machine learning, it is common to classify algorithms into supervised learning and unsupervised learning. In addition, semi-supervised learning, as something between and still distinctly different from supervised and unsupervised learning, has also caught lots of attention, especially in recent years. Reinforcement learning is usually deemed as another major category of machine learning in addition to supervised, unsupervised, and semi-supervised learning. This viewpoint is made based on the thought that supervised learning requires a labeled dataset for training, and unsupervised learning identifies hidden data patterns from an unlabeled dataset, while reinforcement learning does not require data as it learns by interacting with the environment—generating data in the learning process. Notwithstanding, reinforcement learning can be understood using the frameworks of supervised and unsupervised learning in some way, which makes some people view it as a special case of supervised or unsupervised learning, depending on how labels are defined in the consideration of the rewards. This book primarily covers supervised learning, unsupervised learning, and reinforcement learning.

Another way of classifying machine learning, which is based on the major characteristics or underlying mechanisms, groups machine learning algorithms into symbolism, connectionism, Bayesianism, evolutionism, and analogism. This classification method is much less common. However, some terms from it such as connectionism may frequently appear in the context of AI and may confuse people without prior knowledge. Symbolism (investigated by symbolists) here includes symbolic AI in a broad sense. But within machine learning, the narrow definition of it refers to methods characterized by the use of symbols and some types of logic reasoning processes together with learning from data. Examples include decision trees and random forests. Connectionists or neuroscientists create models based on the brain and thus employ artificial neural networks and their variations including the most recent deep learning, which is a rebrand of the "machine learning with deep neural networks." Bayesians, or more broadly as Bayesian methods, treat machine learning as a form of probabilistic inference. Examples of this category include naive Bayesian, Bayesian network, hidden Markov chains, and graphical models. Evolutionaries or biologists use genetic algorithms and evolutionary programming, respectively. It is worthwhile to mention that statistical (machine) learning usually represents a much broader concept, in which statistics is used to reinterpret the most popular machine learning algorithms. Thus, it overlaps with machine learning and is distinct from the above Bayesian methods (as a subcategory of machine learning). The use of genetic algorithms appears to be more frequently discussed in the area of optimization than in AI. Analogizers or psychologists fulfill machine learning tasks based on the similarity between samples or groups. Machine learning models that analogizers frequently use are K-nearest neighbor algorithms, SVMs, and unsupervised machine learning methods.

1.2.3 *History of AI*

The history of AI is primarily about technical innovations. However, an in-depth understanding of this history also needs the consideration of many financial, political, and cultural factors.

Spawning (1930–1952)

As marked in Fig. 1.2, the fertilization of AI started in myths, stories, and rumors in which artificial beings were endowed with consciousness or intelligence by master craftsmen. Next, the implantation occurred with further developments in science fiction, e.g., during the Golden Age of science fiction between 1938 and 1946 [8].

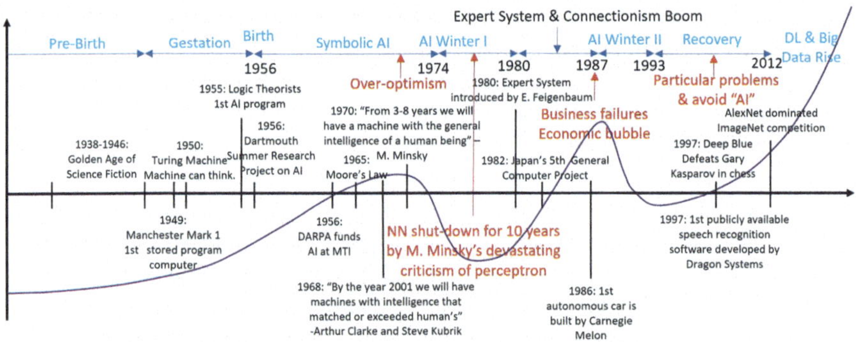

Fig. 1.2 Historical development of AI

Then, AI's fast maturation into a fetus was driven and marked by breakthroughs like philosophers' effort at describing the process of human thinking and materialized as mechanical manipulation of symbols was realized in the invention of programmable digital computers in the 1940s.

During this time, the confluence of several closely related ideas from different areas provided theoretical support for constructing the electronic brain, which cast the foundation for modern AI. These include research in neurology that showed the brain is a network of neurons fired in all-or-nothing electrical pulses. In particular, Walter Pitts and Warren McCulloch reported the use of networks of idealized artificial neurons for performing simple logical functions in 1943 [9], which opened the door to artificial neural networks as well as the ups and downs of connectionism in the later AI history. In 1950, Alan Turing made the first serious proposal in the philosophy of AI by presenting his famous Turing test, in which a machine is thought to be "thinking" if it could conduct a conversation that was indistinguishable from a conversation with a human being [10]. This study allowed Turing to convincingly argue that a "thinking machine" was plausible and answered the most common objections to the proposition.

Other breakthroughs included the first stored program computer in 1948, i.e., the Ferranti Mark 1 machine [11], the use of this machine to write a checkers program, and Arthur Samuel's checkers program developed in the mid-1950s and early 1960s, which gained skills comparable to respectable amateur players [12]. When access to digital computers became possible in the mid-1950s, a few scientists identified a new approach to creating thinking machines: a machine that manipulates numbers could also manipulate symbols, and such a manipulation of symbols could be the essence of human thoughts. As a result, the first AI program, i.e., Logic Theorists [13], was created in 1955.

Birth (1952 and 1956)

Between 1952 and 1956, the advent of computers inspired a handful of scientists to seriously discuss the possibility of building an electronic brain. Usually, it is believed that the milestone where AI becomes a field of study (or an academic discipline) was marked by a workshop held on the campus of Dartmouth College, USA, in 1956 [14]. Most event attendees later led AI research with millions of dollars of financial support, and many of them predicted machines as intelligent as human beings would be created in no more than one generation.

Symbolic AI (1956–1974)

After the Dartmouth Workshop, fast developments in AI programs, especially symbolic AI, were achieved for applications such as solving algebra problems, proving theorems in geometry, and learning to speak English. There were many efforts at maze-alike games in the paradigm of "reasoning as search." Attempts at enabling computers to communicate in natural languages yielded programs like "Student," AI programs written using a semantic net (conceptual dependency theory), and chatterbots (later clipped to chatbots) like ELIZA [15, 16]. The MIT AI Laboratory proposed to focus on artificially simple situations known as micro-worlds based on a perception that, in successful sciences like physics, basic principles were often best understood using simplified models like frictionless planes or perfectly rigid bodies. In Japan, the world's first full-scale "intelligent" humanoid robot or android was created via the WABOT (WAseda roBOT 1) project.

Such successes led to overoptimism among the first generation of AI researchers as well as some funding agencies. Many researchers believed that, within ten to twenty years, the problem of creating AI and generating machines that can handle most of human beings' work would become possible. Meanwhile, the optimism prompted major research funding, such as that from the Advanced Research Projects Agency in the United States (later known as DARPA).

First AI Winter (1974–1980)

In the 1970s, AI started receiving major critiques and financial setbacks. In particular, AI researchers failed to appreciate the difficulty of the AI problems they faced. Meanwhile, the overoptimism had raised expectations impossibly high, and when the promised results failed to be delivered, funding for AI was withdrawn. During this time, the field of connectionism (or neural nets) was shut down almost completely for 10 years partially due to Marvin Minsky's devastating criticism of perceptrons [17]. Despite the recession and criticisms against AI in the late 1970s, new ideas were explored in logic programming, commonsense reasoning, and many other areas.

Expert System and Connectionism Bloom (1980–1987)

In the 1980s, a form of AI program called "expert systems" gained popularity in the industry worldwide [18], and knowledge became the focus of mainstream AI research. In the same period, the government of Japan aggressively funded AI through its fifth-generation computer project [19]. Another remarkable advance in the early 1980s was the revival of connectionism represented by the work of John Hopfield and David Rumelhart [20]. Once again, AI gained success in a variety of ways during this relatively short blooming period.

Second AI Winter (1987–1993)

The rise and drop of AI in the 1980s, especially the involvement of industries and governments, exhibited a clear correlation with the economy. The AI collapse in the later 1980s was partially due to the failure of commercial vendors to develop a wide variety of workable solutions and the burst of the economic bubble during that time. The AI technology was deemed not viable as the public was also discouraged by the failures of dozens of companies. Despite the setbacks and pessimism, the AI field continued to advance in multiple ways. For example, many researchers, including robotics developers Rodney Brooks and Hans Moravec [21], advocated approaching AI in other ways.

Recovery (1993–2011)

After more than half a century of development, the field of AI finally achieved some of its oldest goals. In particular, widespread use of many AI techniques in industries finally became realistic. Some of the successes were due to the increasing computer power, while some others were achieved by focusing on specific problems and pursuing them with the highest standards of scientific accountability. Despite the progress, the reputation of AI, at least in the business world, was less than pristine. Within the field, there was little agreement on the reasons why AI failed to fulfill the dream of human-level intelligence in the 1960s. All these factors led to the evolvement of AI into competing subfields focused on particular problems or approaches, sometimes even under new names to blur, rebrand, or dissociate their AI tattoo. This is a time when AI became both more cautious and more successful than it had ever been.

Deep Learning and Big Data Rise (2011–present)

The third wave marked the explosive development of deep learning, which may be attributed to three factors: improvements in deep learning architectures, especially for addressing the vanishing gradient issue, increases in computational power represented by the use of GPU, and growth of data in a "big data" era especially image data. Important milestones in these three aspects include the publication of the deep belief networks [22], the advocacy for the use of GPUs for training deep neural networks [23], the launch of the ImageNet database (14 million labeled images) that led to later ImageNet annual competitions (ILSVRC) [24], and the development of ReLU and other techniques for the vanishing gradient problem [25]. Then in 2012, AlexNet's victory in the ImageNet competition triggered a new deep learning bloom globally and attracted industry giants' attention [26]. Deep learning started gaining more momentum and making impacts in or even sweeping many disciplines such as computer vision and natural language processing. Advancements in algorithms such as generative adversarial network (GAN) [27] and the continuous development of deep learning architectures for various purposes, platforms, and tasks have been continuing. This led to the dramatic growth of the market for AI-related products, which was called an AI "freeze" by the New York Times. During this time, the rise of reinforcement learning, especially its integration with deep learning, also generated astonishing breakthroughs, and further developments along this direction gained more momentum and popularity. This freeze continues as large language models and generative AI impact and reshape many business areas [28].

1.2.4 AI Versus Traditional Engineering Methods

From the mathematical perspective, most engineering problems boil down to the problem of finding a mapping, f, from the input, x, to the output, y:

$$y = f(x) \tag{1.1}$$

Traditional physics-based engineering methods and data-driven methods like numerical AI (machine learning) are just two different ways of finding such a mapping. Coupled physics-based analysis represented by multiphysics, multiscale, and multi-fidelity and AI analysis represented by deep and reinforcement learning reflect the trends of collaboration and automation in modern industries, respectively, and both address the needs for more complicated and accurate analysis. Thus, both types of methods represent the future directions of modern industries and deserve close attention regardless of the rises and falls of AI compared with traditional engineering analysis. Traditional physics-based methods use mathematical equations derived on the basis of physics rules to construct models. These models are employed to explain or predict phenomena, e.g. states and processes.

By contrast, machine learning aims to find a model that approximates the solution to a real-world problem by analyzing the data. Such models need to be constructed with tools that computers can conveniently process and human beings can easily understand. Thus, mathematical models like basic mathematical functions (like polynomials) or linear algebra equations (formulated using arrays) are intuitive choices. In fact, mathematical models are what have been predominantly adopted. These mathematical functions are also referred to as "mathematical models" or just models. As a result, machine learning models are usually viewed as mathematical equations/functions that represent or model real-world problems/scenarios. Terms such as "mathematical form," "mathematical being," and "machine" are also used in the replace of "models" in some literature. In some cases, machine learning models are also called function approximations. This is because it is usually difficult to find exact functions to represent real-world problems.

Therefore, data-driven AI, like machine learning, also uses models. However, distinct from physics-based methods, machine learning derives the model from data during the analysis instead of constructing the model based on physics before the analysis.

As illustrated in Fig. 1.3, in traditional engineering analysis, we first build the model, e.g., a mathematical model that derives the analytical solution and a numerical model like a finite element analysis model, for a problem based on its underlying physics. Then this model, together with some data that informs the

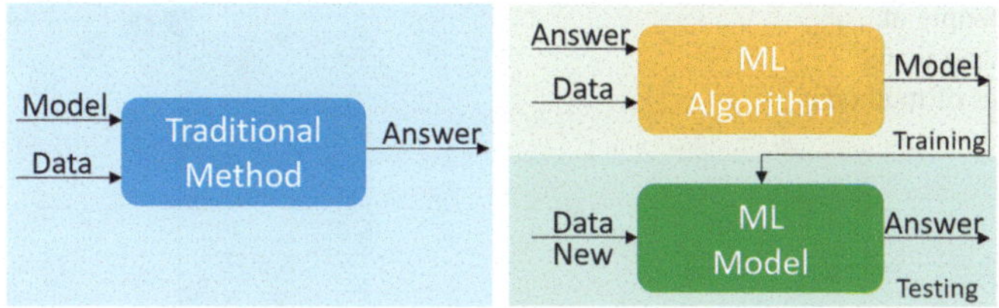

Fig. 1.3 Workflows of physics-based methods (left) and data-driven methods (right)

initial/boundary conditions and material properties, will be fed into the model to obtain an answer, which can be predictions of some state variables or others.

In machine learning analysis, the construction of the model occurs during the analysis instead of before the analysis. Also, the construction of the model is a significant or even a major part of machine learning analysis in many cases. The word "training" is used to refer to this process of generating the model. Due to this reason, this word is used everywhere in data-driven methods. Notwithstanding, it may be totally new to people who are not familiar with machine learning. To generate a model in machine learning, we need to get the basic construction materials:

(1) Data and some types of predefined knowledge about the data (like labels in supervised learning, metrics to assess data in unsupervised learning, and reward functions in reinforcement learning)
(2) A way of constructing the model, or more commonly, called an algorithm

In particular, algorithms dictate how data can be used to generate models. Therefore, algorithms form a major body of the knowledge for machine learning. That is, machine learning is usually introduced in terms of different types of algorithms. By contrast, "models" represent the knowledge extracted from the data based on the employed algorithms. In a simple way, we can understand data as cooking materials, algorithms as recipes, and models as cooked food or dishes. Thus, a typical machine learning cooking book like this book does not talk about "models" specifically in detail, but instead, it focuses more on common recipes, typical ingredients, and their treatments, as well as tricks for cooking and evaluating dishes.

In machine learning, as illustrated in Fig. 1.3, after a model is obtained or "trained," this trained model will be used to obtain answers to new problems based on new data for these problems. This "testing" process is similar to the traditional engineering analysis.

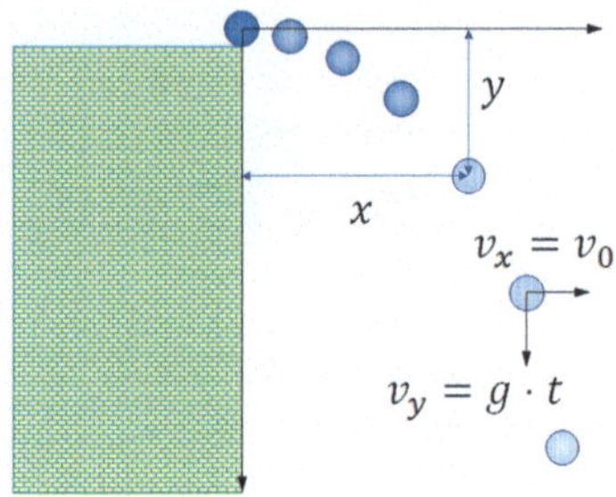

Fig. 1.4 Example of problem-solving using different types of methods

Practice: Prediction of Object Flying Trajectory (Physics Methods Versus Data Method)

A simple engineering problem is employed here to show how data-driven methods and traditional physics-based engineering methods can be used to solve the sample problem. The problem to be considered is to analyze and predict the trajectory of a ball. As shown in Fig. 1.4, the ball moves off the edge of an object with a horizontal velocity of $v_0 = 5$ m/s. The goal is to find out the trajectory of the moving ball from 0 s to 10 s in terms of a function $y = f(x)$ with a gravity constant of $\mathrm{g} = 9.81\ \mathrm{m/s^2}$.

Let us first take a look at two common traditional engineering methods: analytical solution and numerical analysis. Both methods are performed based on the physics underlying the process and a mathematical formulation of the process. The analytical method involves framing the problem in a well-understood form and calculating the exact solution. Numerical analysis is based on a numerical procedure that approaches the solution to the problem in a continuous world using a numeric approximation.

To solve the above problem using the ***analytical method***, we first need to recall the major physics: Newton's second law ($f = m \cdot a$). Along the x (horizontal) direction, the initial velocity will remain as a constant, because no force along the horizontal direction implies no acceleration. Thus, the traveling distance is $x = v_0 \cdot t$, in which v_0 is the initial velocity of the ball along the horizontal direction and t is time. Along the y (vertical) direction, $m \cdot \mathrm{g} = m \cdot a$, so the acceleration along the vertical direction is g. Accordingly, the vertical velocity is $\mathrm{g} \cdot t$ and the traveling distance is $y = \frac{1}{2}\mathrm{g}t^2$. Based on the above formulations, we can easily derive the function $y = f(x)$ as follows:

$$y = \frac{1}{2}\mathrm{g}t^2 = \frac{1}{2}\mathrm{g}\left(\frac{x}{v_0}\right)^2 = \frac{\mathrm{g}}{2v_0^2}x^2 \tag{1.2}$$

Based on the above deduction, the analytical solution is obtained by substituting the known constants into the equation: $y = 0.01962x^2$.

Numerical Analysis is usually preferred for complicated problems, especially those that cannot be easily addressed with an analytical solution, for example, problems with high nonlinearity and high dimensionality. Thus, it is usually not adopted for the above simple problem. Here, we just use it to show the idea of numerical analysis. For this purpose, we discretize the time, e.g., using a timestep of 0.1 s for the 10 s. Then, the numerical solution to the above problem can be obtained via the following iterative process.

Numerical process for predicting the trajectory:

Initialize distance and velocity: $x_i = 0$, $v_{xi} = v_0$, $y_i = 0$, $v_{yi} = 0$, $i = 0$, and $\Delta_t = 0.1$
Repeat until $i = 100$ (i.e., 10/0.1):
  $x_i = v_{xi} \cdot \Delta_t$
  $v_{yi} \leftarrow v_{yi} + \Delta_t$, and $y_i \leftarrow y_i + v_{yi} \cdot \Delta_t$
  $i = i + 1$

Data-Driven Methods seek the solution from data. Thus, distinct from the above physics-based methods, data-driven methods need to obtain some data first. Such data can be obtained from experiments or computer simulations. To illustrate this process, let us generate some data from the analytical solution and add random noise to represent "experimental data" with different sources of errors in real-world systems and measurements. Then, if we use a second-order polynomial function to fit the "experimental data," we can also obtain a model that describes the data. Such a model can also be used for predicting future behavior, for example, the trajectory beyond 10 s.

The results of the three methods and the "experimental data" are shown in Fig. 1.5. In this simple example, we can see that these three methods can help us achieve the same goal with comparable performance. In real-world engineering practices, the selection of methods usually relies on the characteristics of the problem, such as the nature and complexity of the problem, the knowledge about the physics and material properties, the available computational power, and the expectation for computing time and accuracy. Such a selection can be complicated and may require a certain level of expertise.

1.2.5 AI Applications

AI Applications in All Sectors

Aiming at enabling a machine to think itself, AI has been impacting the development of many sectors since its pre-birth, especially during the high tides of its

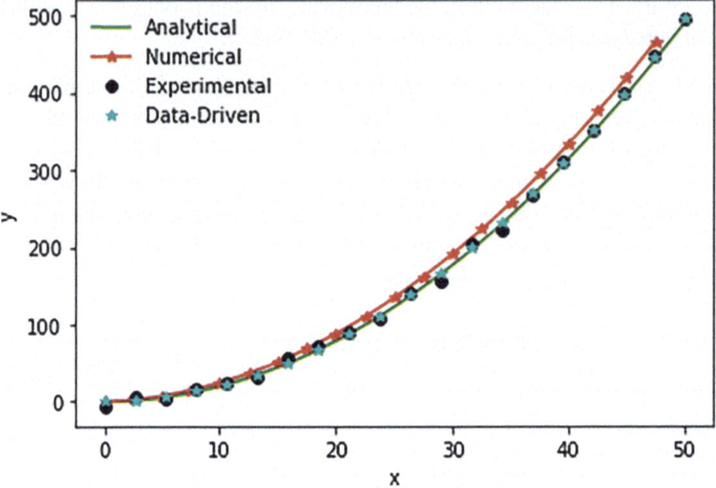

Fig. 1.5 Results of problem-solving using different types of methods

development. In the fourth industrial revolution, AI, automation, and big data are transforming virtually every sector. In particular, machine learning is being widely used in approximately all the sectors, including information technology, healthcare, finance, materials, communication services, etc. Some representative and impactful real-world examples of machine learning applications are listed below.

Autonomous Cars

Autopilot represented by autonomous cars is one of the most exciting applications of machine learning in today's world. The use of various machine learning techniques with the data from different sensors enabled different levels of vehicle autonomy. The integration of AI with electric cars has been revolutionizing the automotive industry and incubating many tech unicorns and new industry leaders in advanced driver assistance systems (ADAS) and autonomous driving like Tesla and BYD. The demands for technical innovations in this area also drive the development of technologies in computer vision, planning, decision-making, and control.

Smart Assistants

Speech recognition and face recognition are among the earliest and most popular applications of machine learning in the third wave of AI. Nowadays, most smartphones, tablets, and desktops provide voice search functions or voice-controlled personal assistants, which can communicate with the device users by speaking human languages. The core of such AI tools includes speech recognition as well as other AI applications for voice recognition, natural language processing, and personalized recommendations. There are many well-known examples including Google Assistant, SIRI, Alexa, and Cortana. Together with face recognition, speech recognition has become the interface for many other AI applications such as smart home.

Healthcare

AI tools based on machine learning algorithms have gained considerable acceptance in healthcare industries. For example, deep learning has become a useful tool for helping specialists analyze external medical data on patients' conditions such as CT, MRI, ultrasound, and various screening tests. In addition, machine learning has been explored as a way to assist disease diagnosis, find hidden knowledge from medical data, and recommend treatment methods. Other than treatment, machine learning has also been explored for automatic billing, clinical decision support, development of clinical care guidelines, and so on.

Recommendations AI

Most of us should have had some experience with AI-powered advertising. For example, after you perform a search for an item on Amazon, you may soon find that many web pages and apps that you browse later contain advertisements about this item. This is a typical example of recommendations AI, which reshaped our lives silently and significantly. Another example is the use of similar tools in content-based online social networks like YouTube, Flicky, and TikTok. Such apps can quickly detect your tastes and preferences and then deliver the most related contents to you.

Robots

Robots were studied and adopted in industries even before the new wave of AI. However, the advances in AI have been helping these machines with different levels of autonomy to acquire higher intelligence. Such intelligence can be gained via computer vision, natural language processing, decision-making ability via reinforcement learning, and general machine learning for prediction and diagnosis. The consequent innovations can bring forth increased uptime (e.g., identify issues and predict maintenance time), reduced programming time (e.g., less programming), and higher productivity (e.g., multitask, collaboration with humans and other robots) to industrial robots.

Finance

In the sector of finance and marketing, machine learning has also made remarkable progress in reshaping this area. First, the use of marketing chatbots and automated financial investing has been widely accepted. Besides, machine learning can now help marketers create various hypotheses, perform tests and evaluations, analyze financial data, and make predictions for future trends or events. Quantitative finance has become mainstream as stock trading bots are handling most of the trading based on calculations from machine learning algorithms. Deep learning models like convolutional neural network, recurrent neural network, and other deep learning architectures helped build such trading models.

Traffic Prediction

Transportation is another area where AI has gained good ground for development and application. In addition to connected and autonomous vehicles (CAVs), the prediction and management of traffic have also been benefiting from AI. Navigation tools like Google Maps now rely on machine learning to help us find the shortest route and predict traffic conditions to make such results in real time. In the future, such efforts will be further integrated with vehicle-based sensors, vehicle-to-everything (V2X, including infrastructure), social media, and other data sources to create more intelligent transportation systems.

AI Applications in Engineering

The applications of AI in the engineering sector is not as widespread as it should be. Notwithstanding, AI has been widely accepted as a significant future direction and an essential component of engineering. In particular, it has been widely recognized that the sooner AI is adopted, the sooner the engineering will reap its benefits. Also, the sector can stay competitive and sustain its global leadership by embracing the most advanced AI applications. Though possibly still in the early stage, AI is now reconstructing the ways of engineering practices. The following is a summary of some typical influential examples of AI applications in engineering. Engineers need to prepare themselves to be literate in the widespread AI tools and learn to work in collaboration with software and machines equipped with AI.

Data Processing
Autonomous processing and analysis of engineering data may be one of the first benefits that AI brings to engineers. This is because engineering is on the front line of data generation and utilization due to the ubiquitous use of sensing and testing equipment and the extensive adoption of data for various monitoring, analysis, design, and management tasks. Examples of such data include sensor readings, drawings, documents, 3D models, measurements, simulation results, and image and video data. AI applications ranging from conventional machine learning methods to more advanced deep learning approaches for image processing and text/visual recognition can all assist or replace engineers in processing and analyzing such data even in their raw formats. This can free engineers from many labor-intensive data processing and analysis work for improved productivity, objectivity, and accuracy.

System Monitoring
AI can be applied to monitor the performance of many engineering systems for various benefits, such as determining the time for maintenance, identifying system errors as anomalies, and recommending system operations like downtime selection. Such efforts have been attempted in a variety of engineering systems such as energy, civil infrastructure, and mechanical systems. For example, agencies like the Electrical Power Research Institute (EPRI) and the US Department of Transportation have explored the way of using machine learning algorithms to analyze images taken by drones to detect malfunctions and distress in the infrastructure.

Prediction
Many engineering tasks involve the prediction of time- or event-dependent factors. Typical examples include the service life of structural and mechanical components, the cost and time needed for specific engineering jobs like construction, the traffic

energy use in regular scenarios and special events, the behavior of many systems under external loadings, and so on. Many of such tasks may suffer from the lack of information and the existence of high uncertainty due to unknown factors. Under such conditions, traditional engineering methods may be hard to apply or yield results with needed accuracy and timeliness. Fortunately, these are where AI tools can step in and prevail.

Design
Design is a core practice in many engineering disciplines. Taking civil engineering for example, CAD (computer-aided design) has been widely adopted to help designers propose, visualize, compare, evaluate, and communicate designs. Now at the cutting edge, 3D CAD has been integrated with the building information modeling (BIM) to generate a tool more than just a 3D model to design building systems. With the assistance of BIM, engineers can design multidimensional models of the projects in a simulation before their execution in the field. Also, AI tools can bring autonomous data synthesis and analysis, estimates of parameters, time and costs, and decision-making for selecting methods and materials to the BIM. This will possibly further engineering design to a more efficient, powerful, automated, and accurate practice.

Management
The management of human resources, facilities and equipment, data and other digital assets, monitoring and testing programs, and materials is sometimes the major part of many engineering applications and determines the efficiency, cost, and outcomes of these applications. Many engineering management practices still rely on subjective and manual operations. Taking asset management in transportation agencies, for example, the prevalent management practices still heavily depend on engineers' presence, documentation, and judgment. These are exactly where AI can easily outperform human beings. Therefore, AI applications have great potential in management.

Optimization
AI is heavily coupled with optimization because AI is supposed to help us get the best mapping from the input to output. This fact may bring forth one benefit: AI can easily explore a great amount of existing data or possible conditions to obtain the possibility that best suits our needs. For discovery, AI has been studied for its potential of identifying better materials, manufacturing/construction procedures, testing schemes, and system operation approaches. AI also has the potential of serving as a higher-level decision-making tool to guide the maneuver of other tools, such as the use of reinforcement learning to guide basic controllers in robotics and other automation applications.

Automation
AI has been a popular option for implementing cyber-physical systems. This is because AI can enable many systems to handle complex tasks in addition to self-management and self-healing. In addition, AI helps many systems gain the ability to learn and upgrade themselves. Due to these reasons, many tasks that were performed

by human engineers can now be carried out by robots. For example, the use of advanced robots in automobile manufacturing has been steadily increasing. AI has been explored to serve multiple roles in automation from being in charge of some suitable parts on selected levels to controlling all parts on all levels.

Summary
In fact, many world-changing engineering applications of AI involve more than one capability of AI tools. Taking autonomous driving, for example, the autonomous processing of image data for road condition detection, road planning via route discovery and optimization, autonomous control of the mechanical components, power management, object detection, and emergency judgment and treatment all take significant roles. Such engineering applications may further expand and deepen their influences via the incorporation of big data and convergence with the Internet of Things (IoT).

1.3 Basics of AI

The representation, optimization, and evaluation, as well as underlying math, constitute the basics of AI. In this section, the basic concepts will be introduced first. What follows will be a quick overview of the common machine learning algorithms. The challenges and issues in machine learning will be discussed next. The math knowledge needed for machine learning, e.g., more details about optimization and evaluation, will be presented in the appendices.

1.3.1 Basic Concepts

This subsection covers the basic concepts including key machine learning elements, data format, and typical machine learning workflows.

Key Machine Learning Elements

Machine learning, or most machine learning algorithms, can be conceptually divided into three main elements:

(1) Representation: What does the model look like? ↔ How is knowledge represented?
(2) Solution (optimization): How are (optimal) models generated? ↔ How is knowledge extracted?
(3) Evaluation: How is the performance of models evaluated? ↔ How is the obtained knowledge measured?

There are eight key concepts in the above three main elements.

Representation
Concept 1: Data. Data is the food or nutrients for machine learning and the generation of machine learning models. Thus, data is where the experience or knowledge is embedded. In supervised learning, data is divided into input and output to represent the unlabeled data and labels, respectively. In unsupervised machine, there is only unlabeled data that can be processed to find specific patterns in such data. In reinforcement learning, data is generated as the learning agent interacts with the environment and "labeled' using a reward function. More explanation about the format/structure of data will be provided in a later subsection.
Concept 1.1: Input. Input is what we feed into machine learning models for learning. Such input data represents the observations or measurements excluding the labels or target values. Each observation is a sample or an instance composed of values for different aspects of the observation, which are called attributes, features, (random) variables, independent variables, and predictors (predictor variables) in different literature.
Concept 1.2: Output. This is what we want the models to predict or estimate. Output is also called labels, targets, dependent variables, and response variables in different places.
Concept 2: Algorithms. Algorithms or learning algorithms are different ways of constructing machine learning models based on data. We can understand an algorithm as a procedure that may contain one way of conceptualizing the model, mathematical equations, and logic statements. Algorithms usually can be outlined using pseudo-code to illustrate how to implement the procedure. We can understand algorithms as methods in a more general sense. Different algorithms can be used to address different machine learning tasks, e.g., regression, classification, and clustering.
Concept 3: Model. A model or a machine learning model is a math function or a more complicated entity that can be mathematically formulated. As the product of running an algorithm on data, a model can be a fitting function in the simplest case or neural networks with fixed weights. We may encounter untrained models, e.g., an artificial neural network with randomly generated weights, which have not gone through the training process and thus contain model parameters that are irrelevant to data. By contrast, trained or pre-trained models contain parameters that have been determined with data and thus represent some knowledge from such data.
Concept 4: Parameters. Parameters are also called coefficients and weights, depending on the algorithms and contexts. Models can be understood to be formed by two parts: a general template (or architecture) and detailed parameters that fix the template into a specific object. The number of parameters can range from a few, e.g., 2 in a linear model, and to millions, e.g., 138 million in the VGG12 deep neural network.
Concept 5: Hyperparameters. Hyperparameters are distinct from the (model) parameters in that they are not part of the model. Instead, hyperparameters are those numbers that we set in the initial configuration or setting before training machine

learning models. Such numbers are needed to determine how the learning will be performed and mostly determined by the selected algorithm including the solver. Typical hyperparameters include the coefficients determining the loss function, solvers, visualization, and the way that the data and model are processed. Hyper-parameters are critical to the success of learning tasks in addition to the selected algorithm, model architecture, and initial model parameters (if any).

Solution
Solution is the search for the model that best extracts the knowledge from the data with the selected machine learning algorithm. We can perform the solution by both deriving an analytical solution, i.e., equation(s) for calculating the model parameters directly, and using an optimizer or a solver to help us find a model. The former can be done within one or a couple of steps, while the latter will involve an iterative optimization process. Also, the former usually can secure the exact or best model, while the latter, in most cases, can only help us find a local optimum as a relatively good (approximate) model instead of the best model. While both approaches can be adopted for many machine learning models, the former is mostly adopted for algorithms with simple models like linear models, whereas the latter is usually adopted for algorithms with complicated models like deep neural networks.
Concept 6: Loss/cost/objective function. Loss functions, cost functions, and objective functions are usually used interchangeably, though the objective function can be minimized or maximized while the loss/cost functions need to be minimized. Such a function is an essential part of the algorithm especially when approximate optimization is needed for solution. This is because it measures the performance of machine learning models, which can determine the optimization direction during the solution process. The idea is to minimize the loss function so that the most appropriate parameters for the machine learning model can be obtained.
Concept 7: Optimization methods (solvers and optimizer). Optimization methods, or called solvers and optimizers, dictate how the solution process is performed to optimize the loss function. The optimization method is usually not fixed when developing an algorithm, and it entails knowledge that is much different from and independent of the machine learning algorithms and models. Therefore, optimization methods can be separate from the algorithms. Common solvers are borrowed from the optimization realm. Such methods can work as long as a loss function, parameters to be optimized (e.g., model parameters in machine learning), and constraints (if needed) are given.

Evaluation
Evaluation is usually implemented during testing or cross-validation to assess the performance of the model. Evaluation needs to be discussed with respect to different machine learning tasks (or problems). This is because different evaluation metrics have been proposed for different types of tasks, e.g., classification, regression, and clustering. A key in machine learning model evaluation is the selection of the evaluation metrics.

Concept 8: Evaluation metrics present different methods for measuring the performance of a model including but way beyond accuracy and error. A complete list and description of the common metrics will be provided in the appendices.

Data Format

This subsection discusses how to organize data, or, in other words, the structure of data. However, the term "data structure" has a connotation of the way of organizing and storing data in a computer's memory or storage. Thus, we here use data format in the title to avoid confusion. Despite this arrangement, data format and data structure are interchangeable in this book, and both refer to the structure or organization of data on a higher level that human beings can easily understand or visualize.

Data is a collection of numbers and symbols that we use to describe things. In machine learning, data is usually used to quantitatively describe measurements (or called observations) and their assessments. Therefore, an intuitive structure of data is to organize different measurements as different data points, which are the elements in the data structure. Due to the same reason, a data point can also be called a sample, an instance, or an observation (a record occasionally). As shown in Fig. 1.6, every observation may contain information for multiple aspects; thus, every data point may also have different values termed as attributes or features (less frequently as property). Data points are usually grouped together as a dataset for a purpose, e.g., training and testing a model, leading to training and testing datasets, respectively. Usually, all the data points in the same dataset have the same number of attribute values. The assessments of data are stored as labels (or called targets) in the dataset. In a labeled dataset, each data point has one to multiple label values. Every data point with its label(s) is called a labeled sample, a labeled instance, or an example.

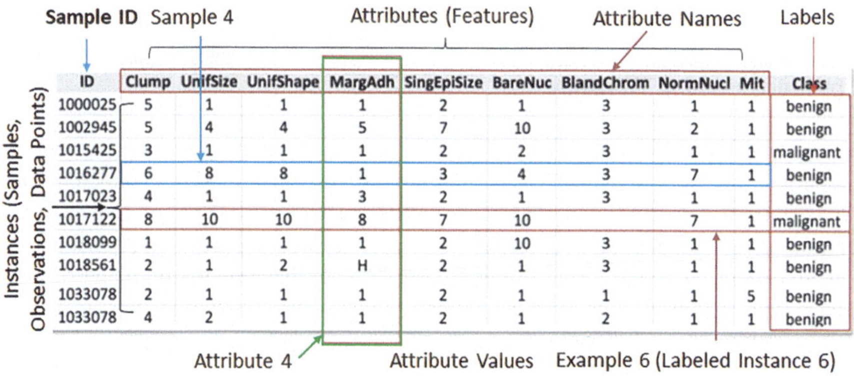

Sample ID Sample 4 Attributes (Features) Attribute Names Labels

Instances (Samples, Observations, Data Points)

ID	Clump	UnifSize	UnifShape	MargAdh	SingEpiSize	BareNuc	BlandChrom	NormNucl	Mit	Class
1000025	5	1	1	1	2	1	3	1	1	benign
1002945	5	4	4	5	7	10	3	2	1	benign
1015425	3	1	1	1	2	2	3	1	1	malignant
1016277	6	8	8	1	3	4	3	7	1	benign
1017023	4	1	1	3	2	1	3	1	1	benign
1017122	8	10	10	8	7	10		7	1	malignant
1018099	1	1	1	1	2	10	3	1	1	benign
1018561	2	1	2	H	2	1	3	1	1	benign
1033078	2	1	1	1	2	1	1	1	5	benign
1033078	4	2	1	1	2	1	2	1	1	benign

Attribute 4 Attribute Values Example 6 (Labeled Instance 6)

Fig. 1.6 Structure of data for (supervised) machine learning

The realizations of data structure for different machine learning purposes share a lot in common in spite of minor differences between such realizations in different packages. Figure 1.6 presents a representative way of organizing data. One item that was not mentioned above is the sample ID. In some AI tools, the sample ID is explicitly used for referring to different data points, while in some other tools, ID is not explicitly defined, and parameters such as the location, row number, and order are implicitly used as the ID. As can be seen, it is common to use different rows for different data points and different columns for different attributes. Label values are usually stored either as a separate array or column(s) to the right of the unlabeled data.

As can be seen in the above introduction, many terms are used interchangeably, though there may be some negligible differences. In this book, though these terms are used interchangeably most of the time, we try to follow the convention of their use in individual machine learning topics. In addition to the convention, though not always true, the following rules will be followed. "Data point" is preferred when talking about plotting, space, and distributions; "sample" is used when an experimental context is emphasized; "instance" is used in a pure machine learning context. "Attribute" is preferred for samples whose individual aspect can be described with simple math entities such as a real number; "feature" is used for samples with more complicated characteristics, such as those that need to be represented by a collection of numbers. "Label" is used when emphasizing the application or experimental flavor; "target" is used in contexts with heavy math content.

Finally, we need to point out that data has been a core component in both data mining and machine learning. However, the role of data can still be slightly different due to the different objectives of these two areas: data mining aims to uncover patterns in the data, while machine learning is intended for reproducing known patterns and making predictions based on them. That is, data is being explored in data mining to gain knowledge; by contrast, data is used in machine learning for creating models that can handle future prediction tasks.

Machine Learning Workflow

The general workflow of machine learning can be summarized as a process of preparing data and feeding it into a model so that the model parameters can be optimized to reach a model with satisfactory performance in the analysis of new data. Despite the common traits, the detailed workflow of performing machine learning varies across different categories of machine learning algorithms: supervised, unsupervised, semi-supervised, and reinforcement machine learning. Typical differences are listed below.

- How is data prepared? For example, collected before training (supervised, unsupervised) versus during training (reinforcement learning)

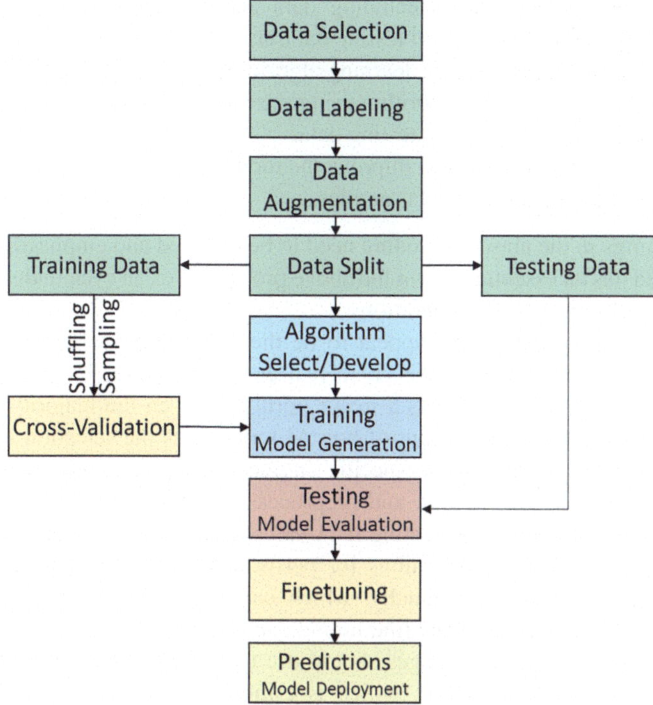

Fig. 1.7 Machine learning (supervised) workflow

- What does the data look like? For example, labeled (supervised) versus unlabeled (unsupervised)
- How is the best performance defined? For example, labels (supervised) versus data characteristics (unsupervised) versus rewards (reinforcement)

In fact, the process can be slightly different even between algorithms in the same category.

Despite the differences, some terms referring to specific stages of the workflow are widely used everywhere. In particular, the process and some terms associated with supervised learning are more familiar to us due to the predominant role of supervised learning in traditional machine learning. Here, we will introduce the machine learning process in common supervised learning tasks. The workflows of other categories of machine learning algorithms can be understood with deviations from this one. Those workflows, together with the deviations, can be seen in later chapters for other categories of machine learning algorithms.

Let us first take a look at how (supervised) machine learning is adopted for addressing typical engineering problems. As shown in Fig. 1.7, the following steps are what we typically adopt in supervised machine learning:

(1) Collect and prepare data (including data augmentation, labeling the data, dividing data into training and testing datasets).
(2) Choose or develop a machine learning algorithm.
(3) Train the model with the algorithm using the training data.
(4) Evaluate the model using the testing data.
(5) Fine-tune hyperparameters to improve the model.
(6) Make predictions for new data samples.

A few things in the above procedure need to be clarified and emphasized.

First, data has an essential role in the above process. We can even understand this process as a sequence of data operations.

Second, training and testing appear to be the core stages, whose significance usually needs no emphasis. However, in real engineering practice, it is common to find that data preparation takes a major portion or even the majority of effort. In particular, the availability of out-of-box AI tools eases the selection and use of algorithms/models, while data in specific engineering applications usually needs to be collected, cleaned, labeled, and preprocessed before use. Work needed for generating, cleaning, and augmenting data varies case by case, depending on the data quality, algorithms, expectations for performance, etc. In many cases, data labeling needs to be done manually, which can be trivial and labor-intensive. Sometimes, model evaluation and fine-tuning can also take a lot of time and care.

Third, training and testing stages should use two datasets that are independent and identically distributed (IID), which is critical to the success of the above process. Independence and distribution are two major characteristics of datasets. The independence implies that the two datasets share no common samples. Otherwise, testing will be redundant with training in some way, depending on the overlap. The identical distribution denotes that the random variables (attributes of the data samples) should have the same distributions in both datasets. Otherwise, the two datasets will have different natures, leading to two different problems for training and testing. Particularly, a model trained to work well for one problem very likely works poorly for another problem with a totally different nature. In some cases, we also have a substage called cross-validation in training. This cross-validation is different from testing in that testing usually involves exclusively IID data, while cross-validation shares data with training. Cross-validation typically involves shuffling and sampling steps and runs alternately with the training process.

Finally, the above process may be simpler than many actual machine learning jobs. For example, in more advanced data analytics work, feature engineering and feature extraction may need to be separated from data preparation as a significant, separate step. Also, for people who are more focused on studying instead of applying algorithms, the algorithm development and model evaluation will be emphasized. In more complex industrial AI applications, we may also need to involve more steps for model testing and deployment, for example, deploying the model in a hyper care model before it goes live, during which model evaluation may happen multiple times.

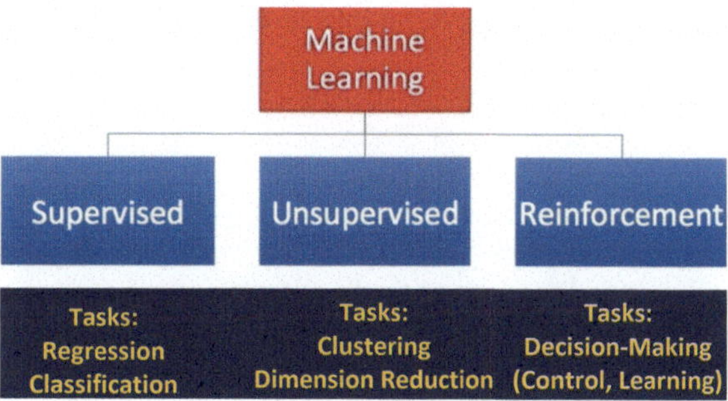

Fig. 1.8 Major categories and tasks of machine learning

1.3.2 Common Algorithms

Overview and Machine Learning Tasks

This subsection presents a very quick overview of common machine learning algorithms in the order of major categories: supervised, unsupervised, semi-supervised, and reinforcement learning. However, prior to the introduction to detailed algorithms, it is necessary to differentiate the types of algorithms from the types of tasks these algorithms can be applied to. This is because, when selecting algorithms, both pros and cons of algorithms and the jobs that they can do are what we need to first consider. Unfortunately, the algorithms and the tasks that they can handle have complicated relationships. For example, some algorithms may handle a task that most other algorithms in the same category cannot. Moreover, such relationships are not fixed, as researchers may extend an algorithm to perform tasks that it could not. This is possibly one reason why such relationships are usually not explained adequately in typical machine learning literature.

Here, we try to lay down the relationships in a simple and general way and leave out the exceptions. The major categories of machine learning are shown in Fig. 1.8. As can be seen, regression tasks and classification tasks are usually discussed in supervised learning. This is because these two tasks both require labeled data: regression requires a continuous number or an array of continuous numbers as the label for each sample, while classification needs a discrete number (or another symbol) or an array of discrete numbers (other symbols) as the label. Unsupervised learning does not contain labels, so it is usually used to process data by analyzing the relationships between data points, such as distances between points. Typical tasks for unsupervised machine include clustering, dimension reduction, anomaly detection, and association rule learning. Reinforcement learning is more for planning and decision-making. It helps the learning agent improve its decision-making ability by letting the agent interact with an environment. Therefore, reinforcement learning is also adopted for "learning": helping models created by other algorithms gain a higher learning ability.

In a more abstract way, it can be interpreted as that machine learning can be used for predictive, descriptive, and prescriptive tasks. The predictive function helps

Week	Topics: Algorithms	Categories
1	AI Basics: Definition, History, Terms, Types & Math	Background Knowledge
2	AI Implementations: Tasks, Metrics & Tools	
3	Linear Models (Basic, Ridge, Lasso, Elastic Net)	
4	Decision Trees (ID3, C4.5, CART)	
5	Support Vector Machines (linear, nonlinear (kernel))	
6	Bayesian Methods* (Naïve, Semi, Bayesian Network, Gaussian Process)	Supervised Learning
7	Artificial Neural Network (Perceptron, Feed Forward NN, RBF)	
8	Deep Learning (DNN, CNN, RNN)	
9	Ensemble Learning (Bagging (Random Forest), Boosting (Adaboost), Stacking)	
10	Clustering (K-Means, Mean-Shift, DBScan, OPTICS, GMM, HAC)	
11	Dimension Reduction (PCA, LDA, ICA, Isomap, MDR)	Unsupervised Learning
12	Anomaly Detection (3sigma, boxplot, KNN, LOF, DBSCAN, iForest, AutoEncoder)	
13	Association Rule Learning* (Apriori, FP Growth, Eclat)	
14	Value-Based RL (Q-learning, Sarsa, Monte Carlo)	
15	Policy-Based RL (Policy Gradient)	Reinforcement Learning
16	Advanced RL* (DQN, DDPG, Actor-Critic, A3C)	

Fig. 1.9 Machine learning topics and popular algorithms

predict what will happen with data. The descriptive function helps explain what has happened with data and what the data conveys. The prescriptive function helps make suggestions about what action to take based on data. These three functions correspond to the above three categories to some extent but not completely. An algorithm from any of the three categories may provide multiple functions depending on how the algorithm is applied.

Figure 1.9 lists the machine learning topics (or types of algorithms) and the popular algorithms in each topic within the three major machine learning categories as well as the background knowledge. The listed topics and algorithms are selected based on a pragmatic criterion: what are the most needed and valuable contents for a typical semester-long course in college? Accordingly, such topics correspond to sixteen weeks, in which some topics are marked with * to indicate that they can be optional if the semester is shorter. In the following, common algorithm types and representative algorithms for each type will be explained using language with as little technical detail as possible to help everybody get a quick understanding of the ideas behind the algorithms.

Supervised Learning

Linear Models refer to a category of algorithms that can be used for regression analysis or curve fitting, which engineers are familiar with. Thus, we can simply understand this as the use of a simple mathematical function to fit the measurement data, in which attributes are the independent variables (xs) and the label (if only one label) is the dependent variables (y). In a narrow sense, such a model adopts a linear function: y is a linear function of xs (or linear combination of xs). In the simplest case, a linear function parameterized by the weights for different dependent variables and the bias, which can be directly obtained via an analytical solution, forms the basic linear model. To overcome overfitting issues, L2, L1, and the combination of L2 and L1 norms of the model parameters can be considered when searching for the best curve fitting function, leading to Ridge, Lasso, and elastic net algorithm, respectively. These linear models are only suitable for linear regression problems. In a broader sense, these linear models can be extended with kernel functions to deal with nonlinear problems. Linear models are initially proposed for regression or curve fitting problems, so they are of a numeric nature—both attribute and label values are numeric. Notwithstanding, we can make further changes like adding a logistic function or a Softmax function to convert the output of linear models to a probability and finally to a discrete label. In this way, linear models can also be applied to classification problems.

Decision Trees are a family of algorithms that use a treelike structure to guide/mimic humans' decision-making process. Starting from a node representing the root of the tree, each node, including the root node and any node on the following layers except the leaf nodes, is associated with an attribute. The possible values of the attribute will lead to the generation of "son" nodes on the next layer (or splitting). Thus, in a decision-making process like classifying a sample, the sample will start from the root node. Then, depending on the attribute values, the sample will move from one node to another one on the next layer based on the value of the attribute associated with the node and eventually to a leaf node. Each leaf node is associated with a class. This is how a classification decision is used for prediction. For training, all labeled samples will start from the root node as well. Then, an attribute is selected at each non-leaf node according to a general criterion of reducing the chaos (uncertainty) in the data the best. The node-by-node selection of attributes and concurrent splitting of the dataset into subsets belonging to nodes on the next level continue until we reach a node where further splitting is not possible, e.g., reaching a leaf node. This process will generate a decision tree according to this dataset. Different criteria for selecting the attributes lead to different decision tree algorithms: ID3, C4.5, and CART. Pre- and post-pruning techniques are very important in controlling the tree size, e.g., depth, to deal with overfitting. These techniques are also frequently included as part of the modern decision tree algorithms.

Support Vector Machines, or more commonly referred to as SVMs, are a machine learning topic that once held a predominant role before deep learning rose. Such machine learning algorithms stem from the idea of performing binary classification by finding the widest margin to separate the samples or the corresponding data points from the two classes. This conceptualization leads to a typical constrained optimization problem: maximize the width of the margin by searching for the margin boundaries controlled by both the margin direction and the locations of samples on the boundaries, which are the "support vectors." Accordingly, no samples can get into the margin or the region belonging to the other class. This strict constraint called "hard margin" can be loosened as "soft margin" to allow for noisy data points. This unconstrained optimization problem, though can be solved using an optimization package, is more commonly converted to a dual problem to facilitate a more convenient computer solution to the corresponding quadratic convex optimization problem. Besides, the basic linear SVM can be extended with kernels to address nonlinear problems. The basic version of SVM was proposed for binary classification tasks. However, it is not difficult to extend it to multiclass classification and regression tasks.

Bayesian Methods are also called Bayesian algorithms, Bayesian machine learning, and probabilistic machine learning in the literature, though the exact meanings of these terms may slightly differ depending on the context and research areas. Bayesian methods were intrinsically constructed for classification tasks. Possibly due to this reason, the use of the "Bayesian classifier" is very common or even predominant in some technical publications. Such machine learning algorithms stem from Bayesianism, which uses probabilities to quantify the level of belief and consequently updates such beliefs in the evidence of new data. Accordingly, such methods correlate the probability of a sample with a certain combination of attribute values and target values with the probabilities that the different classes and attribute values appear conditionally or simultaneously. Depending on how to calculate such probabilities, we have different types of Bayesian algorithms: from naive Bayes, which assumes strong (naive) independence between attributes, to Bayesian network, which can formulate complicated interdependence between different attributes with a network-like probabilistic graphical model. It is worthwhile to mention that such Bayesian classifiers, which belong to parametric machine learning methods, can also be extended to handle regression tasks. Other nonparametric Bayesian methods like Gaussian process are proposed more for regression and may appear much different from traditional Bayesian classifiers.

Artificial Neural Networks (ANNs or simply called NN in machine learning) use mathematical operations like inner products, element-wise products, and activation functions that can work on arrays to mimic the working mechanisms of biological neural networks. In particular, most modern ANNs adopt the M-P neuron model: the input to a neuron is multiplied by the neuron's weights and then the difference between this product and a threshold is fed into an activation function to generate the neuron output. A network that typically consists of multiple layers of such neurons treats and processes the data according to the architecture of the network, e.g., direction and interneuron/interlayer connections, and finally outputs the predicted label. One key piece of knowledge in ANN studies is how to train an ANN. So far, backpropagation has been accepted as the most common method for training ANNs, in both shallow ANNs and deep learning. A typical backpropagation process includes a forward pass for predicting the label and a backward pass for passing the "gradients" of loss to different model parameters, e.g., network weights, to update these parameters so that the ANN can make better predictions in the next forward pass. Multilayer feed-forward neural network, especially 3-layer, is the most widely known ANN architecture, though other types, such as RBF, are also available.

Deep Learning is about the development and use of deep NNs. The "deep" in this definition, in general, denotes the depth of layers in an ANN. A neural network that consists of more than three layers—which would be inclusive of the input and the output layers—can be considered a deep learning model. Thus, in theory, it can be viewed as a subset of or an extension to traditional ANNs. However, the breakthroughs and widespread applications of deep neural network have gained knowledge that makes deep learning much different from traditional (shallow) ANN studies. Some of the typical factors that contribute to the development of ANNs especially the transition from shallow NNs to deep learning, including breakthroughs specific to deep learning and a few advances from the general field of machine learning such as pre-training, transfer learning, solvers, and regularizers are listed as follows:

- Better data: more data, preprocessing, normalization
- Better weights: initialization, pre-training, transfer learning
- Better network structure: activation, batch normalization, CNN, LSTM, NIN, residual network, transformer
- Better solvers
- Better regularizers
- Better computing resources: GPU, parallel computing

Deep learning in the third wave of AI exhibits its success via convolutional neural network (CNN) in computer vision, recurrent neural network (like long short-term memory) in natural language processing, integration with reinforcement learning for learning and control, and more recently in large language models (LLMs).

Ensemble Learning refers to algorithms that employ models generated by other algorithms as constituent models to obtain an ensemble model with performance that is better than what can be obtained with individual constituent models. These constituent models, which are called base models, can be generated by one of those basic machine learning algorithms, like linear model, SVM, decision tree, KNN and NN, or a combination of them. The former is called homogeneous, while the latter is heterogeneous. Currently, homogeneous base learners are more frequently used, and among them, CART decision tree and neural networks are the most common algorithms for generating homogeneous base learners. The idea behind ensemble learning can be described as "union is strength" or "many hands provide great strength." Therefore, ensemble learning is viewed as an optimization method that generates a strong learner from several weak learners. There are three common categories of ensemble learning methods: bagging (e.g., meta-estimator [basic bagging] and random forest), boosting (e.g., AdaBoost and Gradient Boosting), and stacking. Bagging is focused on the utilization of democracy to reduce variance. Boosting, which features elitism, generates better or elite models by focusing on samples associated with wrong predictions and gives elite models more weights in decision to improve bias. Stacking replaces simple combination rules with a machine learning model as a second layer learner to process the results from the first layer learners for better predictions.

Unsupervised Learning

Clustering is a machine learning topic aiming to divide unlabeled data (data points or samples) into different groups. A general goal is to generate groups so that samples within the same group are similar to others while samples from different groups are different from each other. These groups or groupings are referred to as "clusters." Both the ways of generating clusters and evaluating the generated groups (or clustering process) are the content of this unsupervised learning topic. For the former, many different types of clustering algorithms have been proposed based on the patterns that the data points need to be arranged in. Centroid models refer to clustering algorithms wherein the clusters are formed by the proximity of the data points to the cluster center or centroid. Data points are clustered based on multiple centroids in the data. K-Means clustering and mean-shift clustering are the most popular algorithms in this category. Density models are generated by clustering algorithms that group data by areas with a high concentration of data points surrounded by areas with a low concentration of data points. DBScan and OPTICS are two popular density-based algorithms. In distribution models, data points are grouped together based on the probability that they may belong to the same distribution. In a particular distribution, the distance of a data point from a center point is determined to infer the probability of being in that cluster. Gaussian mixture theory (GMM) is the most common option in this category. Hierarchical (connectivity) models involve top-to-bottom or bottom-up hierarchies. Agglomerative hierarchical algorithm is the most popular example in this category.

Dimensionality Reduction, or called dimension reduction, is the transformation of data from a high-dimensional space, e.g., with a large number of attributes, into a relatively low-dimensional space, e.g., with fewer attributes than the original data, while retaining essential properties of the original data. Therefore, dimensionality reduction helps remove redundant or less significant variables. These methods can be classified into two major categories: feature selection and feature projection. Feature selection seeks to find a subset of the input variables (or features, attributes, dimensions). Therefore, we select features directly according to some criteria, such as filter strategy (e.g., information gain) and wrapper strategy (e.g., search guided by accuracy), and drop the less desired features to reduce data dimensionality. By contrast, feature extraction tends to project the data in a high-dimensional space to a space of fewer dimensions, in which the more desired features or their combinations are extracted. Feature selection methods primarily refer to those data reduction techniques studied in statistics for variable selection and now mostly in high-dimensional regression analysis. Classical methods are missing value ratio, low variance filter, high correlation filter, random forest, backward feature extraction, and forward feature selection. Feature projection methods are more popular in the dimensionality reduction literature and are even used to represent dimensionality reduction in a narrow sense. This category includes the most popular dimensionality reduction methods such as principal component(s) analysis (PCA), linear discriminant analysis (LDA), independent component analysis (ICA), Isomap, and MDR.

Dimensionality reduction methods can also be classified based on whether they are for labeled or unlabeled data. A typical example is that PCA was developed for unlabeled data. Thus, PCA is mostly used for clustering in unsupervised learning. On the contrary, LDA was proposed for processing labeled data. Another way of classifying dimensionality reduction methods is based on whether these methods are linear or nonlinear in nature. For example, PCA, ICA, and LDA are linear, while LLE, Isomap, MDR, and kernel PCA are nonlinear.

Anomaly Detection, also called novelty detection, outlier detection, forgery detection, or out-of-distribution detection in different areas, is intended to identify rare items, events, or observations that significantly deviate from the majority of the data and do not conform to a well-defined notion of normal behavior. Anomaly detection has been applied to a variety of areas such as fraud detection, web hack detection, medical (disease) detection, sensor network anomaly detection, IoT bid data anomaly detection, log anomaly detection, and industrial hazard detection. In a broad sense, the available methods for anomaly detection can be roughly grouped into rule-based methods, statistics-based methods, and machine learning-based methods. Among them, the anomaly detection methods based on machine learning algorithms are anomaly detection in a narrow sense and represent the state of the art. The machine learning-based methods can be further categorized into supervised, unsupervised, and semi-supervised methods. In unsupervised learning, common methods can be divided into five groups: statistics-based, distance-based, density-based, clustering-based, and tree-based. In semi-supervised learning, popular methods include one-class SVM, AutoEncoder (or autoencoder), and GMM. In supervised meaning, we usually need to pay attention to data labeling and imbalanced data for possible issues, and such methods are suitable for considering data with new classes. Common methods in this category include linear models, SVM, and ANN.

Association Rule Learning, which is also called association rule analysis and association rule mining in many publications, is a rule-based type of machine learning for discovering interesting relations between variables. This definition is not that straightforward to understand, especially considering the vague meanings of terms like "relations between variables," "rule," and "interesting." Also, association rule learning and algorithms (or methods) for it contain various new parameters/-concepts, such as those embedded in the definition. This fact makes it difficult for people without some expertise in data mining to understand the topic and implement association rule learning algorithms. In addition to these parameters and concepts, association rule learning treats data that are usually formulated in a format slightly different from data dealt with in other machine learning areas due to historical and practical reasons. Therefore, association rule learning may appear much different from other supervised and unsupervised machine learning topics in many aspects, which can further confuse learners. Popular association rule learning algorithms include Apriori, FP growth, and Eclat.

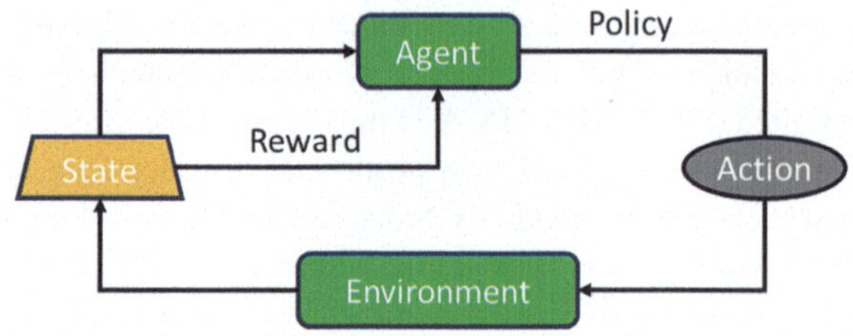

Fig. 1.10 Workflow of reinforcement learning

Reinforcement Learning

Reinforcement Learning is the third category of machine learning, in which no raw data is given as input. Instead, the reinforcement learning algorithm needs to design a way to generate and label data in the training process. Reinforcement learning is frequently used for robotics, gaming, and navigation. With reinforcement learning, the algorithm discovers through trial and error to identify actions yielding the most significant rewards. Thus, actions and rewards are similar to the attributes and labels in supervised machine learning in some way. As illustrated in Fig. 1.10, this type of training has three main components: an agent that can be described as the learner or decision maker, an environment that the agent lives in, and the actions that the agent takes for rewards. The objective is to let the agent take actions that maximize the expected reward over a given measure of time. The agent will reach the goal much quicker by following a good policy. So the purpose of reinforcement learning is to learn the best policy. Reinforcement learning can be roughly categorized into value-based, policy-based, and hybrid (involving both value and policy) algorithms.

Typical examples of value-based reinforcement learning algorithms are Q-learning and Sarsa. In Q-learning, the agent learns by updating its own Q table, which quantifies the values of different actions in specific states. The agent interacts with the environment and generates a sequence of state-action-reward values called a trajectory. The yielded rewards will help the agent update its own Q table and consequently improve the decision-making ability. Q-Learning is an off-policy method because it learns an optimal policy no matter which strategy it carries out. Sarsa follows a very similar procedure for learning, but it adopts the current policy for decision-making during the learning process (or call episode) and thus is an on-policy algorithm. By contrast, policy-based reinforcement learning does not use action and state values to determine the optimal action. But instead, it adopts a probability function for selecting an action. The selected action will be evaluated via the reward function to update the policy parameterized by this distribution. A typical example of policy-based algorithms is policy gradient. More complicated reinforcement learning integrates both the values and policy such as critic-actor and later variations like C3A. Such hybrid algorithms use one type of reinforcement learning to generate an action and use the other type of algorithm to assess the algorithm. In addition, reinforcement learning has been integrated with deep learning. For example, deep neural networks can be used to replace the Q table as the mapping from actions to values, leading to algorithms like deep Q-learning. Many others are being proposed following this direction like deep deterministic policy gradient (DDPG).

Semi-supervised Learning

As the fourth type of machine learning, **semi-supervised learning** uses both labeled data and unlabeled data. Though semi-supervised learning can be viewed as a hybrid of supervised and unsupervised machine learning, it is mostly used for the same purposes as supervised learning, e.g., classification, regression, and prediction. Semi-supervised learning usually employs a small amount of labeled data with a large amount of unlabeled data. Semi-supervised learning brought obvious benefits such as the save in the effort of labeling massive amounts of data and the reduction in the bias caused by labeling. Such algorithms still need to pose strict requirements on the data, such as the accuracy of the labels for labeled data and class balance for the unlabeled data. Semi-supervised learning has sub-categories like simple self-training, co-training, label propagation algorithm, semi-supervised SVM, and, more recently, semi-supervised deep learning. The workflows of different semi-supervised learning algorithms may be considerably different. Figure 1.11 illustrates a typical workflow in self training.

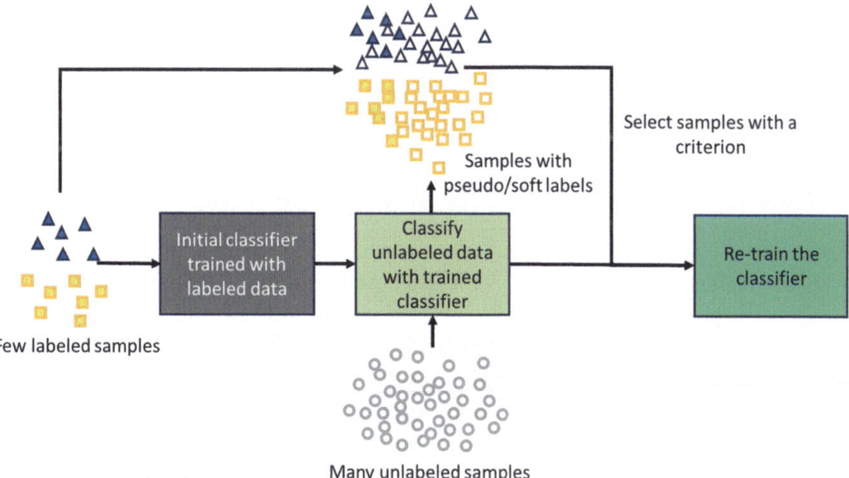

Fig. 1.11 Typical workflow of semi-supervised machine learning

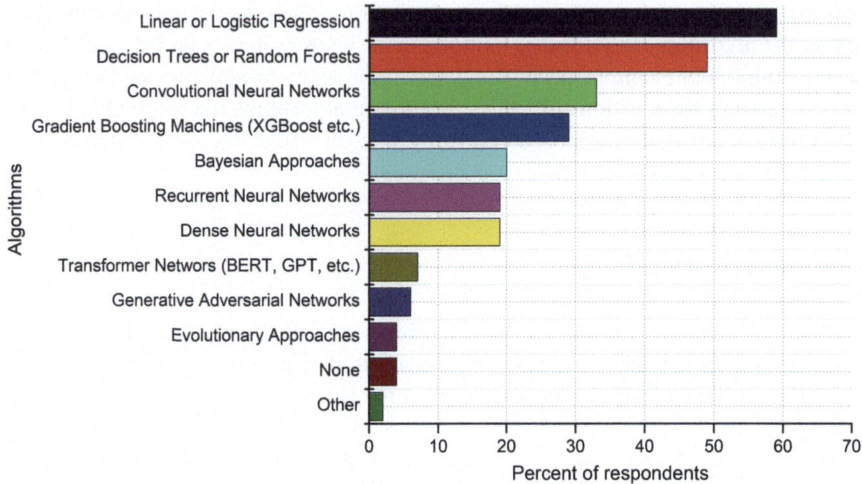

Fig. 1.12 2020 Kaggle survey on most commonly used machine learning algorithms (18,996 respondents)

Summary

AI is a highly dynamic field. In the third wave of AI represented by machine learning algorithms, the popularity of the algorithms is changing every day. Figure 1.12 presents is a list of common algorithms (top-10 ranking) according to a survey conducted by Kaggle with 18,996 respondents. This presents a snapshot of this field for a condition around 2020.

1.3.3 Challenges and Issues in Machine Learning

Data Issues

As the "food" for machine learning, data, especially its health, can easily affect or even determine the development, validation, and application of machine learning models. Unfortunately, real-world problems usually do not provide datasets that have been well assessed, structured, and annotated as those well-documented datasets from machine learning packages, such as the Iris dataset from Scikit-learn. In engineering, we usually need to use data collected in different environments in different ways, which may lead to highly heterogeneous, highly unstructured, incomplete, erroneous, and unknown data. Such data may consume a lot or even most of the project time and threaten the validity of machine learning models built on it. In short, common data issues can be summarized as inadequate data, immature data, incorrect data, noisy data, and biased data.

Inadequate data is a very common issue in traditional machine learning, especially before the advent of big data. In many cases, a major cause of poor performance of machine learning models is an inadequate amount of data. That is why much effort in traditional machine learning was devoted to the utilization of data such as sampling techniques for cross-validation like bagging. Even after we entered the era of big data, inadequate data is still haunting over many machine learning tasks. This is because, as the complexity/capacity of a model increases, the amount of data that is needed for the model to reach the same performance also increases. Besides, the inadequacy of data sometimes does not necessarily mean that the number of samples is not enough, but instead, the number of usable samples is not adequate. For example, it is usually not difficult to obtain images for computer vision; however, labeled images, which may need labor- and expertise-intensive work to generate, can be hard to obtain. Data augmentation, innovative labeling techniques, and semi-supervised machine learning are common solutions to data inadequacy.

Immature data refers to data that will need a significant amount of preprocessing work before it can be used for training or testing a model. This is very common as data may be incomplete, heterogeneous, and structured in ways that are not compatible with a model. For example, many deep learning models will require a specific shape of the input arrays and a certain way of labeling the data (e.g., one-hot labeling). This usually requires us to spend a lot of time converting every sample and label into the required format. The workload can be astonishing when we need to deal with a great amount of such data. Under this condition, it could be even more difficult to tell what data immaturity issues the data may have, because it is hard to check the samples one by one as it may take minutes or hours to load the dataset. It will be helpful to develop code that can automatically assess and preprocess the data, such as format check, trimming, resizing, and removal, though the development of such code may also be time-consuming.

Incorrect data is another type of data issue. Compared with other issues, this type of issue is hard to detect and, if overlooked, can cause serious outcomes such as incorrect models. One typical example is mislabeled data. For example, some samples may be assigned wrong labels due to a variety of reasons. Unfortunately, when such data is used for training, the misinformation will also be learned by the model. In particular, for instance-based algorithms such as KNN, the data will be included as part of the model. Errors will also be integrated into such models, leading to problematic predictions for future data. In regression tasks, wrong label values due to systematic errors such as sensor drift are also this type. Manual data assessment by human experts and algorithms that can detect such issues. For example, anomaly detection algorithms can be used to identify these issues.

Noisy data is very similar to incorrect data but can be different in some ways. It is characterized by the existence of a small amount of data that exhibits trends different from the others. So, it can be caused by random errors like those mislabels due to accidental operations, which affect specific data points, rather than systematic errors, which can lead to offset in all the data. Furthermore, it is also possible that the noise appearance is not caused by an error, but instead, due to the distributions associated with the data. Such issues can be handled by both processing the data to screen out the noisy samples and developing more rigorous algorithms. For example, the soft margin in the SVM can help consider samples that do not meet the basic assumptions of data distributions.

Biased data is produced when certain samples are heavily weighted or need more importance than others. Such data causes a typical issue: the data cannot represent the real problem or cannot be representative of new cases that we need to generalize. For example, a training dataset does not cover all cases that have already occurred and/or are occurring. Biased data may lead to inaccurate predictions, skewed outcomes, and other analytical errors. In other words, the model may learn from data that only represents a part or an aspect of the problem and extend the knowledge to the whole problem. Such issues can be resolved by determining where data is actually biased in the dataset and countermeasures can be proposed to rectify the bias.

Inductive Bias

Strictly speaking, inductive bias is not an issue. However, it can cause issues if we do not understand it and treat it properly. It is among the concepts that are the most difficult to understand in machine learning. Meanwhile, it is an essential element of machine learning, though it may not even be noticed by many machine learning practitioners. However, a better understanding of it can help us search for more suitable models and avoid issues due to inappropriate selection or treatment of the inductive bias.

Inductive bias can be formally defined as the assumption(s) that a machine learning algorithm adopts to generalize a limited set of observations (training samples) into a general model. As introduced in the section for symbolic versus numerical AI, machine learning as a numerical AI boils down to induction: the process of moving from special observations to general rules or models.

Inductive bias is needed because we will need to provide information to describe what is "general" in induction. Take the regression problem in Fig. 1.13 as an example. Two models, i.e., Model 1 (linear, green) and Model 2 (nonlinear, blue), can be obtained based on the same training dataset (black circles). These two models exhibit the same performance if we use typical regression metrics like mean absolute error because both curves pass the centers of all the training data points. In this case, how can we tell which model is better or reflect more general rules?

We can rephrase the above problem using machine learning terms to obtain a strict description. First, supervised learning can be viewed as a process of searching in a set of all possible mappings, or, more broadly, hypotheses. This set is called the hypothesis space. The learning goal is to find a hypothesis that can match or provide the best description of the training data. However, in many cases, there is more than one hypothesis from the hypothesis space that is compatible with training data. These compatible hypotheses constitute the version space. Just like the above example, both Model 1 and Model 2 are the best fitting models. In this

Fig. 1.13 Example of need for inductive bias

case, if no information about future data is provided, we will need a bias to indicate our inclination or preference for selecting a model. "Occam's razor" principle is a common inductive bias. This principle states that we should choose the simpler model when two models exhibit comparable performance.

The word "bias" already indicates that inductive biases present priori and subjective information. So, it does not mean that they are always correct. Getting back to the above example, we can see that if data that will appear in future applications for the model is more like the blue dots, then the nonlinear model is better. On the contrary, if the future data is more like the green squares, the linear model is better. This leads to an extremely important idea in machine learning: ***it makes little sense to talk about models without mentioning the data***. Thus, instead of saying a model is good or not, we may need to say whether it is suitable for an application or the data associated with that application. When the target application changes, or more essentially, the probability distribution of the possible data associated with the application changes, we may also need to adjust the model so that it can maintain its performance.

We can see from the above example that it will be helpful to get a better understanding of possible future data to avoid issues that may be caused by inductive bias. That is also the reason for introducing testing data. However, even if we do that, the use of inductive bias is also inevitable in many cases. An extreme example is illustrated in Fig. 1.14. In this example, the two models have the same level of complexity and performance; thus, we will need another hypothesis on top of Occam's razor rule. Though this example is too simple and special, it shows the idea that we may need to use multiple hypotheses on different levels to help determine which model is better without knowing anything about the data that we will encounter.

Inductive bias is hard to understand also because it may appear in different formats in different algorithms. In the above regression problem, Occam's razor is widely accepted as an inductive bias in regression and has been incorporated

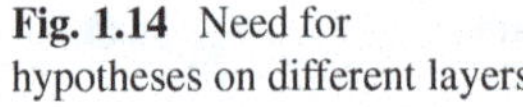

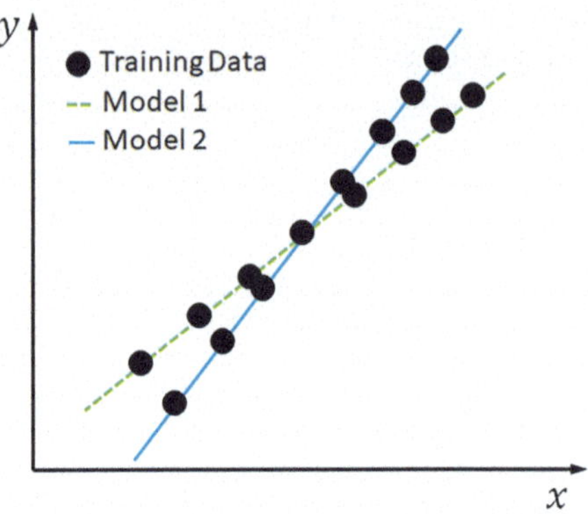

Fig. 1.14 Need for hypotheses on different layers

in many regression algorithms via regularization terms. However, it is neither the only option nor a must-have inductive bias. For example, a unique inductive bias—selecting the SVM with the widest margin—is explicitly specified in most SVM algorithms. Extra examples of inductive biases include the maximum conditional independence in naive Bayes classier, minimum number of features in feature selection, and nearest neighbor in KNN.

It is worthwhile to mention that inductive bias not only determines which solution will be selected, but also affects whether we can efficiently find a solution. From this perspective, we can also understand it as constraints that we place there, which may affect both the solution result and solution process. Let us take a look at deep learning as an example. The inductive bias of CNN can be locality (elements in the space show higher correlation as they get closer) and spatial variance (the kernel weights are shared). The inductive bias of RNN is sequentiality (points that are close to each other in time are related) and time invariance (RNN share weight cross time steps). Thus, these two types of deep NNs can be viewed as special cases of fully connected deep NNs, which assume all elements can be related. The extra inductive biases help CNN and RNN search for solutions along the biased directions for computer vision and natural language processing problems, respectively. They generate faster and more accurate results than those general deep NNs without such biases, because computer vision and natural language processing problems exhibit locality and sequentiality, respectively.

Thus, the selection of inductive bias does not only affect the usefulness of the model, but also determines how a model can be constructed and identified. An opinion in recent computer vision research is that traditional CNNs involve too much inductive bias. Thus, deep learning algorithms like self-attention in the ViT (vision transformer) can provide better functions by loosening the constraints placed by the inductive bias. More recently, multilayer perception, which has less inductive bias, can be used to achieve the accuracy of SOTA model in ImageNet. This leads to a controversial conclusion: is inductive bias in CNN not needed? In fact, a better way to understand this is that inductive bias helps reach a balance between fast and accurate solution and flexibility. When we do not have strong power to obtain a solution, e.g., better data, higher computing power, and more efficient algorithm, it is better to use inductive bias to help us stay more focused so that we can find an acceptable solution or reach it more quickly. But when our power for solution is satisfactory, we can remove some inductive bias or loosen the constraints so that we can find better solutions or better ways of reaching the solutions.

Underfitting and Overfitting

Generalization
As explained, machine learning represents a process to learn general rules from specific observations. From this perspective, the goal of machine learning is to generalize from the training data to any data from the problem domain. A good model will allow us to make predictions for data that will appear in future applications, which the model cannot see in the training stage. Thus, we use the concept of "generalization" to tell how well a model trained with specific observations can perform on the data that it will be applied to.

Overfitting and Underfitting
Two issues, or outcomes of poor generalization, of machine learning models are overfitting and underfitting. In fact, overfitting and underfitting are also the two major causes of poor performance of machine learning algorithms. As we showed in the previous section, without any knowledge about the future data, we can only rely on inductive bias to assess/select models. To improve the model selection, we usually split the available data in supervised learning into a training dataset and a testing dataset. In this way, we can have the testing data as a representation of future data. The trained model can then be assessed using the testing data. Let us assume both the training and testing data can perfectly represent all the possible data. Then, a model that is trained with the training dataset and can achieve comparable performance on the testing dataset is believed to have good generalization.

Figure 1.15 gives an illustration of underfitting, good fitting, and overfitting as well as training and testing data. If a model's performance on training data is poorer than on testing data, e.g., lower accuracy and higher loss, we can infer the low generalization may be caused by overfitting. If poor model performance is observed on both datasets, then underfitting may be the reason. High bias (overall offset) and low variance (high scatter), which will be introduced in detail in the chapter for ensemble learning, are two common indicators of underfitting. Good fitting is associated with good results, e.g., high accuracy and low loss, on both the training and testing datasets.

It is worthwhile to mention that, usually, training data and testing data cannot perfectly represent all the possible data. However, to ensure the above process for assessing and improving generalization is valid, we will need to make sure the training data and testing data are independent and identically distributed (written

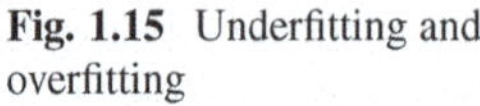

Fig. 1.15 Underfitting and overfitting

as i.i.d. or IID). The independence requirement ensures the two datasets are not the same thing. Otherwise, it does not make sense to separate data into these two sets. The requirement of identical distribution helps us enforce that the training data and testing data have the same nature or come from the same data domain. If we consider the random variables corresponding to different attributes follow certain distributions for the problem(s) that we want to address, then we want to make sure such distributions in the training data, testing data, and all the possible data are the same.

Underfitting is easier to address compared with overfitting. A straightforward way is to increase the model complexity or switch to a new model with higher capacity (complexity). The detailed changes will be different in different algorithms. Taking NN for example, we can add in more layers and neurons. Sometimes, the capacity of the model may be enough. What is truly needed is to perform more thorough training. For example, some algorithms use iterative optimization to search for the best model. In such a case, we will need to wait until enough iterations or epochs are finished so that the loss can gradually decrease to an acceptable value. If any technique for addressing overfitting is used, we may also need to reduce the effects of such techniques to alleviate underfitting, e.g., reducing the weight of the regularization term in the loss function.

Techniques to Prevent Overfitting

Overfitting can be addressed or controlled from three different angles: (1) controlling/reducing model complexity, (2) better monitoring and controlling the training process to avoid over-training, and (3) making data better represent the whole data domain (or sample space).

For the model complexity control, different types of machine learning algorithms may use different techniques. But the use of regularization and model trimming techniques is very common in most supervised learning algorithms.

Regularization constrains the model complexity by including functions of model parameters as a penalty term in the loss function. In this way, the complexity of the model needs to be considered when searching for the optimal model during training.

$$\ell(\vec{w}, b, \vec{\alpha}) = \alpha\|\vec{w}\| + f(err)] \tag{1.3}$$

where $f(err)$ represents the original loss as a function of error, $\|\vec{w}\|$ is the norm of the vector (or higher order tensor) containing model parameters, and α is a parameter that we use to adjust the significance of the complexity term in the optimization. The higher the α value, the more likely that simpler models will be sought. A norm is a measure of the magnitude of a vector—in this case, it tells the size of the array composed of all the model parameters. Usually, we use L1 or L2 norms, leading to L1 and L2 regularization, respectively.

Model Trimming can be implemented in different ways depending on the algorithms. For example, in decision trees, this can be done by removing part of the tree and limiting the maximum depth (or maximum layers) of the tree in the so-called pre-pruning or post-pruning process, respectively. In deep learning, we can add in dropout and pooling layers to intentionally leave out and combine information, respectively.

Overfitting can also be addressed by better understanding and controlling the training process. The following are some widely used techniques.

Holdout is very useful for controlling overfitting by better informing the extent of overfitting. This is actually what we did for splitting data into training and testing sets, e.g., 60%/40% and 80%/20%. As introduced above, the difference between the performance indicators of the two datasets serves as a measure of overfitting. If overfitting occurs, we then should stop training and take some other actions. This overfitting prevention technique has been widely accepted as an essential step of training; thus, sometimes, it is not even considered a special technique for dealing with overfitting.

Cross-Validation is very similar to testing, but it is counted as part of the training process. Similarly, we split our dataset into k groups, i.e., k-fold cross-validation. Then we use one group for testing and the others for training during a training time unit, e.g., a certain number of iterations or epochs (one epoch means looping over all the samples once). Next, another group will be used for testing, while the remaining groups will be used for training during the next training time unit. This process is continued until the training is finished. We can see that this can allow us to check the extent of overfitting even before the testing. Compared with holdout, cross-validation allows all data to be eventually used for training while being more computationally expensive.

Early Stop is an action we usually take when we find overfitting occurs. In the typical loss versus iterations/epochs plot during cross-validation, once the validation loss stops decreasing but rather begins increasing and far exceeds that of the training loss, we stop the training and save the current model. An early stopping trigger can be set to stop the training automatically.

We can also help prevent or eliminate overfitting by better preparing the data.

One goal of **data improvement** is to make the data better represent the sample space. This can be done directly by increasing the amount of data. Data sampling techniques that can generate data to better reflect the distributions of the real data would also help. Data augmentation is another popular technique. Data augmentation adds noise to the data. In deep learning, this can be performed by flipping, rotating, trimming, rescaling, or shifting the image data.

Feature Selection is a good option when we have only a limited number of training samples while each sample contains lots of features. In this case, we can select the most important features for training, so that the model does not need to learn that many features, which can more easily lead to overfitting. For this purpose, different combinations of features can be selected for training models with the best generalization. Or alternatively, we can resort to a feature selection method and use the feature selected by this model for training.

1.4 Practice: Gain First Experience with AI via a Machine Learning Task

>>> More and Up-to-Date Course Materials including Practices @ AI-engineer.org <<<

1. Get familiar with the Anaconda environment (Spyder, Python, Scikit-learn, Matplotlib). Write code to plot $Y = X^2 * \sin(X)$, in which $X = [0, 0.1, \dots, 4]$.
   Hints:
   In the process, you will import packages, try basic algebraic operations with NumPy arrays, and plot using Matplotlib.
2. Problem-solving with Scikit-learn: classification (using the Iris dataset) and regression (using the Diabetes dataset) tasks.
   Hints:

2a. Load datasets (check https://scikit-learn.org/stable/datasets/toy_dataset.html, link to an external site to see how to load Diabetes and Linnerud separately). Check the structure of the data in Spyder and think about why they can be used for the intended classification/regression task.
   For example,

```
from sklearn.datasets import load_diabetes
X, y = load_diabetes(return_X_y=True)
```

2b. Select a machine learning algorithm (corresponding to a command/function in Scikit-learning) to perform the classification/regression task.
   Hints:

Check https://scikit-learn.org/stable/supervised_learning.html; link to an external site for a list of algorithms/commands/functions. You can start with support vector machines (SVCs or SVR in sklearn.svm).

2c. Find a way to check the results (e.g., plot, evaluate using a metric [e.g., score method coming with SVC when using SVC for classification], and *try predictions with new data).

Chapter 2
Tools for Artificial Intelligence

2.1 Overview of Tools for AI

This chapter provides information about common tools for implementing AI. The predominant AI implementations require extensive coding for implementing or extending existing machine learning algorithms. Because AI is a fast-evolving field, the tools for such implementations usually cannot be done using software packages with a user-friendly graphical user interface (GUI). Instead, we will need to resort to a programming language and/or packages built with it. There may be a pragmatic reason for this fact: at an AI high tide, AI develops so fast that commercial software with GUI can hardly catch up, while during a low tide, AI becomes inadequately attractive to software developers.

Typical computer language only provides very basic functions. It is rare that we only rely on these basic functions for implementing AI. It is more common that we use functions built by others for AI implementations to save effort at "reinventing the wheel." Some of such functions have been wrapped up as packages shared across the community. As a result, when we are talking about AI tools, we mostly refer to a development environment or a hierarchical ecosystem consisting of a basic programming language and the packages built on it. Here, hierarchy connotes that some tools may rely on others.

The above fact exhibits the concept of "dependencies," which is very common in Linux environments but obscure in other operating systems. Therefore, such concepts may be new to you if you are only familiar with GUI-based environments such as those in Windows. In addition, because such AI development environments may be contributed more by open-source communities, you can still easily run into dependency or compatibility issues even if you use such tools in Windows. For example, this could happen when you use a software package that manages all the tools including the programming language, such as Anaconda. Anaconda

Z. "L." Liu, *Artificial Intelligence for Engineers*,
https://doi.org/10.1007/978-3-031-75953-6_2

includes the basic Python and popular packages built with Python, which can help us easily construct an AI coding environment running in various operating systems including Windows. Therefore, one of the first jobs to do in AI implementation is to be prepared for dealing with such dependency and compatibility issues.

Regarding the basic programming language, we need a language with versatility to handle the complexity inherent in AI projects. According to statistics, most AI developers prefer Python, while a much less common choice is JAVA with close competition by R, Prolog, and Lisp. Some other languages like Scala, Julia, and C++ have also been used for AI development.

Python is a general-purpose, object-oriented language with much emphasis on code readability. The selection of this language for AI by major AI developers helped this language become more popular and powerful. Python secured the first place in Tiobe's language popularity ranking [29] as of August 2024, which exhibits its predominance since 2022. In fact, Python is likely to be at the top of most other language popularity rankings as the third wave of AI continues.

There are many reasons for Python's popularity in AI. First, Python is very easy to learn. Because it uses an English-like syntax, it can be written much faster than other major languages like C/C++ and Java. Second, as an interpreted language, you can run the code on any platform with a Python interpreter. Third, Python's vast library support and huge community make Python an excellent choice for beginners. In particular, in areas involving data, Python offers many powerful libraries such as NumPy, Pandas, and Matplotlib, which provide what you need for the acquisition, manipulation, analysis, and visualization of data. What is more, Python is great for use in large-scale machine learning attributed to multiple out-of-box deep learning and machine learning libraries such as Scikit-learn, Keras, TensorFlow, and PyTorch. In summary, the versatility, easiness, and availability of powerful libraries, as well as the excellent integration and the vast and active community, helped Python secure a predominant role in AI and relevant areas.

Figure 2.1 presents an illustration of the structures of Python-based AI tools. We can divide the AI development environments into multiple layers: coding language, data manipulation and visualization tools, machine learning and data analysis tools, and deep learning tools, among which the latter/upper layers likely rely on the former/lower layers.

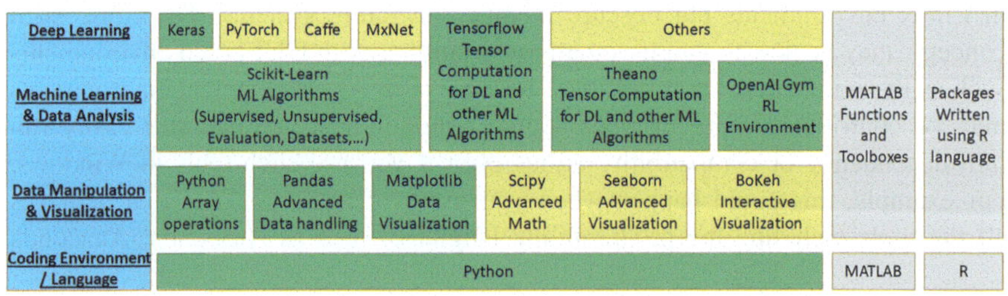

Fig. 2.1 Structure of Python-based AI tools

Many packages are available for data manipulation and visualization. NumPy is preferred for array operations [30, 31], while Scipy can provide extra complicated math functions and algorithms [32]. Pandas provides powerful tools for data treatment like the organization, rearrangement, storage, and input/output of data [33]. Visualization for viewing the structures and trends of the data can be performed with Matplotlib [34]. Extra visualization work can be finished using packages like Seaborn [35] and Bokeh.

Mature packages are also available for machine learning and deep learning. Scikit-learn is one of the most popular packages for machine learning [36]. For statistics-intensive work like probability distributions and advanced statistical models, we can resort to packages like Scipy.stats and Statmodels, which can achieve most goals similar to the R framework. Many packages are available for deep learning. On a "lower" level, we have popular options like TensorFlow [37], Theono, and PyTorch [38]. Deep learning can be practiced with relatively high flexibility on this level. On a "higher" level, packages like Keras further wrap up functions from lower levels like TensorFlow to conveniently implement deep learning at the cost of sacrificing some flexibility [39].

Software for reinforcement learning has not been that well developed compared with deep learning. But packages like OpenAI Gym can still be good tools for beginners to perform reinforcement learning [40].

It is worthwhile to mention that the software pool for AI is a highly dynamic area considering the fast technology iteration in the world of AI, especially when AI is at a high tide. For example, the popularity shift from Caffe [41] to TensorFlow to PyTorch in deep learning occurs only within several years. The development is not necessarily a one-way process and could be very complex. However, even if newer and fancier players come to the playground, the above packages can still provide an adequate environment for beginners to learn and practice AI.

In the following sections, some essential functions of Python and popular Python-based AI tools including NumPy, Pandas, Matplotlib, Scikit-learn, TensorFlow, Keras, and OpenAI Gym, which will be used in later chapters of this book for AI implementations, will be briefly introduced one by one. It is noted that only the most essential functionalities of these tools will be mentioned. Hence, the materials to be presented can be similar to many online tutorials titled "Learn Python in *** minutes." In fact, it is not difficult to find out that engineers' AI practices may only need to frequently use a small set of functions from these packages, while other functions that may be used can be easily grasped by checking the official manuals or other learning materials for these well-documented packages. The purpose of the following sections is to provide engineers with a quick entry to the world of AI tools in Python, so that beginners would not be discouraged or overwhelmed by the explosive information that is available online.

2.2 Python

2.2.1 Introduction to Python Coding Environment

The way(s) of using Python and Python-based AI tools can appear confusing to beginners. There are at least two widely accepted ways, both of which can be adopted for common computer operating systems like Windows, Linux, and Mac.

In the first way, one can install Python in some GUI-equipped operating systems by downloading and double-clicking an installation executable from the Python official website or other sources. It is also common to install Python and its package management tools like Pip using command lines with the system's software management tool in systems that heavily rely on command lines, e.g., Yum and Apt in Linux. After that, one can start a command line terminal, e.g., Command Prompt in Windows or a terminal in Linux, and use Pip to install the other packages, e.g., "pip install numpy". Python can be used in two different modes in both ways: interactive and script modes. To use the interactive mode in the first way, one will need to type in "python" in the terminal and run until the sign ">>>" appears. Then, any Python command can be typed in and run line by line (or block by block), for which the execution of the command will return results in the terminal or new windows (e.g., for new figures) immediately. This is because Python is an interpreted language, which can be executed right away; this is distinct from compiled languages like C, which need to be compiled before execution. By contrast, the second mode allows us to put lots of commands together as a script file and run this file like those in compiled languages. We will just need to write the script file and then run "python script_file_name" in the terminal.

The second way is to adopt Python distributions such as Anaconda instead of directly working with Python together with its packages in the operating system. Such a distribution, which also appears as a software package, can work as a platform that can provide everything that we need. This setup will save us trouble in dealing with operating systems, finding packages, and addressing compatibility issues between different versions of the packages. In addition, such packages usually have good GUIs and can be run similarly across different operating systems. After a package of this type like Anaconda is installed, you can install other AI packages using Anaconda's package management tool in either a command line terminal (using commands "conda install" and others) or in the Anaconda GUI. To improve efficiency, we can also install an integrated development environment (IDE) package such as Spyder, PyCharm, and Jupiter. Among them, tools like Spyder can provide an IDE that is very similar to MATLAB [42], which can be a good tool to start with for engineers who are familiar with MATLAB. Thus, the second way is recommended for beginners.

Another thing that could easily puzzle beginners is the versions of Python and its packages. In general, newer versions provide better functions and fewer bugs and thus are installed by default. But stories may be much more complicated in some cases. For Python, Versions 2.*.* and Versions 3.*.* are much different, while minor versions in Python 2 (or 3) are much more similar and mutually compatible. Many people have to stick to old versions because some old script files were written using old versions of Python or Python packages, for example, the Python 2 syntax. Compatibility issues may appear in some scripts that use syntax or commands that are abandoned in newer versions. It is recommended to use new versions unless old versions are needed due to other reasons. The versions of Python and its packages all have interdependencies, which usually can be handled by software/package management tools, e.g., via virtual environments, to different extents.

Readers are suggested to go through the above process and explore more functions like checking the installed packages, versions, and interdependencies to obtain better first-hand experience, which will not take too much time with platforms like Anaconda. After Python and the needed packages are installed, you can try a simple command like "print("AI for Engineers")" in both the interactive and script modes (in Spyder or a terminal) to see if the environment is working. In the script file mode, we will always need to pay attention to the directory for saving the file and make sure we run the file from that directory to avoid "File Not Found" issues.

Python only provides a limited number of basic commands (or viewed as functions). More functions can be obtained from packages written using Python. This can be done using a command like "from package_name import function_name" or "from package_name import *". The latter is not recommended as it will import everything from the packages. This may include functions or constants with names that you plan to use for your own variables or functions, which can cause conflicts and issues that are hard to detect. Another thing that needs to be mentioned is the hierarchical concepts of "a package," "a module," and "a function," which can be understood as a file folder, a script file, and a function in the script file, respectively. Because of this, it is also very common to use "from matplotlib import pyplot as plt" or "import matplotlib.pyplot as plt", in which Matplotlib is a package and Pyplot is a module in the package. With the execution of the above command, we can use "plt" to replace the module "matplotlib.pyplot". The period "." is a common way in object-oriented programming languages like Python to indicate the concept of "belonging to." So, we can easily call a function "plot" in the module "pyplot" of the package "matplotlib" using "plt.plot". We can use the functions or constants from a package only after we execute a command for importing the function or the module containing that function.

2.2.2 Basics

First, like any other language, Python reserves a small set of keywords that designate special functionality, which can be viewed as the commands of basic Python. Therefore, we cannot use any of these words to define our own variables or functions: "False," "None," "True," "and," "as," "assert," "break," "class," "continue," "def," "del," "elif," "else," "except,' "finally," "for," "from," "global," "if," "import," "in," "is," "lambda," "nonlocal," "not,' "or,' "pass," "raise," "return," "try," "while," "with," and "yield." For example, one common error in engineering applications with Python is the use of "lambda" to define a new variable. This is because λ is a very common Greek symbol used in modern publications, so it is intuitive to define a variable using its English spelling, without noticing this word is reserved by Python for a specific function.

Second, some rules need to be followed when defining your own identifiers, i.e., names given to variables and functions. These rules include the following: (1) a variable name must start with a letter or the underscore character, (2) a variable name cannot start with a number, (3) a variable name can only contain alpha-numeric characters and underscores (A–z, 0–9, and _), and (4) variable names are **case-sensitive** (age, Age, and AGE are three different variables). Besides, by convention (not a strict rule), variables should be lowercase, with words separated by underscores as necessary to improve readability, while identifiers for classes, which will be introduced later, usually start with an uppercase letter.

The ways of organizing the code are another thing that we need to understand first. Spaces do not matter within or after a statement (i.e., a line of commands for fulfilling a task) as long as they are added between or after independent words, e.g., "a=2" is the same as "a = 2". However, the number of spaces at the beginning of any line, i.e., indent, is significant because indentation is used to define the level/layer of the statement, e.g., for loops and conditional stations. Indenting a line by four spaces will move the statement to the next (inner) level. This needs special attention if you use a general system terminal to run Python. However, if you use an IDE like Spyder, you can use the Tab key on the keyboard, which will add four spaces by default each time you click the key. In that way, you do not need to worry about the number of spaces.

It is recommended that one line contains one statement. However, multiple statements can be put into the same line using semicolons (";"). By contrast, a multiline statement may be necessary when a statement is really long, i.e., a long equation. In that case, we can use the back sign slash ("\") at the end of the line where the statement needs to pause and then start from the next line to continue. Alternatively, we can use braces { }, parentheses (), or square [] to extend a one-line statement to multiple lines.

Another basic skill needed for understanding and writing Python code is for making comments. You can easily comment on a line or part of a line so that it will take no effect by putting a hash (or pound) sign ("#") in front of the line or the part of the line. In Spyder, this can be more easily done using a shortcut key "Ctrl + 1" after selecting the line(s) that you plan to comment. For multiline comments, we can two " """ " on two sides of the lines (above and below) to comment the lines between them.

2.2.3 *Variables and Data Types*

Next, let us take a look at variables, which are like containers where you can store different types of Python objects. Such an object can be a string of characters, an integer, a float number, a list of objects, a dictionary, and a more complicated user-defined data structure. These different types are called data types.

Unlike some languages such as FORTRAN, a variable in Python does not need to be declared, i.e., defining the variable with its data type, before its first use. We can use "a = 5" to both declare, i.e., defining a new variable named "a" to store an integer (using "a = 5.0" will define a float instead), and initialize the variable with a numeric value 5, which can be updated later if needed.

Data types may have methods associated with them. When we create a variable, e.g., using a command "action = 'warn'" to define a string variable named "action" as a Python string data type called "str", we actually assign an object of a certain type to it. Then, the variable, which instantiates the data type, also inherits the data type's methods. For example, action.replace("a","o") or 'warn'.replace("a","o") will replace the letter "a" with the letter "o". The following are examples of defining common types of variables in Python:

- Integers: a=5.
- Floats: b=5.0.
- Lists: c=[5,2.1,"engineer", 3]. We can later use any element from the list using the corresponding index. For example, c[0] will give us the first element of c, i.e., 5. It is worthwhile to mention that the indices in Python, e.g., for lists, strings, tuples, and even data structures defined in Python packages (e.g., arrays in NumPy), start from 0 instead 1. This is much different from MATLAB and may cause difficulties for proficient MATLAB users when transitioning to the Python environment.
- Strings: d="Monday". A string can also be viewed as a special list consisting of characters (between a pair of symbols like ' ' or " "). The element (or character) in it can be cited using the same way as lists, e.g., d[1] gives out "o".
- Tuples: e=(1,2.1,"engineer", 3). Tuples can be viewed as a list whose values will not change after the definition. Their use is less common than lists. It is not difficult to notice that tuples are defined using "()", making them different from the "[]" for lists.
- Dictionaries: f={"Name":"Jason Bourne","Height": 1.78,"age":54}. A dictionary is more like a list of key and key value pairs/sets. Each key and key value set, which is an element in the dictionary, are defined as "key: value". We cannot use indices to reference a dictionary, but we can use dictionary_name[key] directly, e.g., f["Height"], to use the key value of 1.78.

2.2.4 Operators

Operators are used to perform operations on variables and values. The most common operators in Python are arithmetic operators, assignment operators, comparison operators, and logical operators.

Table 2.1 lists Python's arithmetic operators, which are used with numeric values to perform common mathematical operations. One operator that MATLAB users can easily get in trouble with is the exponentiation operator, which is "**" in Python but "∧" in MATLAB.

Assignment operators are used to assign values to variables. The ones that we will mostly see in coding for AI are given out in Table 2.2.

Comparison operators in Table 2.3 are used to compare two values for making a conditional statement.

When there are more than two conditions, we can consider logical operations shown in Table 2.4. More complicated conditions can be handled using conditional statements to be introduced.

In addition to the above types of operators, there are still identity operators, membership operators, and bitwise operators, which are less common.

Table 2.1 Arithmetic operators

Operator	Name	Example
+	Addition	x + y
−	Subtraction	x − y
*	Multiplication	x * y
/	Division	x / y
%	Modulus	x % y
**	Exponentiation	x ** y
//	Floor division	x // y

Table 2.2 Assignment operators

Operator	Example	Meaning
=	x = 5	x = 5
+=	x += 3	x = x + 3
−=	x −= 3	x = x − 3
*=	x *= 3	x = x * 3
/=	x /= 3	x = x / 3
%=	x %= 3	x = x % 3
//=	x //= 3	x = x // 3
**=	x **= 3	x = x ** 3
^=	x ^= 3	x = x ^3

Table 2.3 Comparison operators

Operator	Name	Example
==	Equal	x == y
!=	Not equal	x != y
>	Greater than	x >y
<	Less than	x <y
>=	Greater than or equal to	x >= y
<=	Less than or equal to	x <= y
**=	x **= 3	x = x ** 3
^=	x ^= 3	x = x ^3

Table 2.4 Logical operators

Operator	Description	Example
and	Returns True if both statements are true	x <5 and x <4
or	Returns True if one of the statements is true	x <5 or x <4
not	Reverse the result, returns False if the result is true	not(x <5 and x <10)

2.2.5 Conditional Control Statements

Conditional (control) statements are essential in computer languages, and Python is no exception. Three types of if-statements can be used in Python to control the conditional execution of a statement or a group of statements based on the value of an expression.

The simplest type of if-statement is as follows:

if expression:

    statement(s)

where “expression” is an expression evaluated in a Boolean context, i.e., True or False. If the expression is judged to be True, then the statement is executed. If the expression is False, then the statement is skipped over and not executed. The above “statement(s)” can be multiple statements, e.g., statements in multiple lines, and must be indented. It is noted that the use of a colon after the expression is required.

It is very common to have two conditions, each of which is associated with the execution of some statements. In that case, we just need to add an else statement. A colon is also needed after “else”.

if expression:

    statement(s)

else:

    statement(s)

If more than two conditions exist, we can add in one or more elif clauses.

if expression:

```
    statement(s)
elif expression:
    statement(s)
elif expression:
    statement(s)
    ...
else:
    statement(s)
```

2.2.6 Sequential Control Statements

Python has two primitive loop commands for sequential control: **for** loops and **while** loops.

A for loop is used to iterate over a sequence, e.g., a list, a tuple, a dictionary, a set, and a string. With the for loop, we can execute a set of statements, once for each item in a sequence.

```
for iteration_item in sequence:
    statement(s)
```

'iteration_item' can be any valid variable name. In the ith iteration, iteration_item will take the value of the ith item in the sequence. For example, if the sequence is ['a','b','c'], then iteration_item will be "b" in the second iteration. It is very common to use range() to generate a sequence. For example, if we use range(3), which will generate a list [0,1,2], then iteration_item in the second iteration of the loop will be equal to the second item, i.e., 1.

In the while loop, a set of statements will be executed as long as the condition specified by the expression is true.

```
while expression:
    statement(s)
```

In a few cases, two other statements "break" and "continue" can be used with the above two types of loops. The break statement can stop the **for** loop before it has looped through all the items. Likewise, the break command can stop the **while** loop even if the while condition is true. The continue statement can stop the current iteration of the **for** or **while** loop, i.e., skipping the statements after the stop statement in the current iteration, and continue with the next iteration.

2.2.7 Functions

A function is a block of code that can serve a certain purpose. A function usually takes some data known as arguments or parameters as input and may or may not return some results as the output. There are three types of functions: built-in functions, user-defined functions, and functions from packages. The Python interpreter provides a number of built-in functions as listed in Table 2.5, which do not need to be imported before use. User-defined functions and functions from packages can be used after being imported as described earlier.

Table 2.5 Built-in functions

Function	Description
abs()	Returns the absolute value of a number
bool()	Converts a value to a Boolean
chr()	Returns a character (a string) from an integer
complex()	Creates a complex number
dict()	Creates a dictionary
dir()	Tries to return attributes of an object
eval()	Runs Python code within a program
float()	Returns a floating point number from a number or a string
format()	Returns a formatted representation of a value
help()	Invokes the built-in help system
input()	Reads and returns a line of string
int()	Returns an integer from a number or string
len()	Returns a length of an object
list()	Creates a list in Python
max()	Returns the largest item
min()	Returns the smallest value
open()	Returns a file object
pow()	Returns the power of a number
print()	Prints the given object
range()	Returns a sequence of integers
reversed()	Returns the reversed iterator of a sequence
round()	Rounds a number to specified decimals
set()	Constructs and returns a set
str()	Returns the string version of the object
sum()	Adds items of an iterable
tuple()	Returns a tuple
type()	Returns the type of the object
zip()	Returns an iterator of tuples
__import__()	Function called by the import statement

It is not difficult to notice that functions usually carry a pair of parentheses, i.e., "()", which may contain no or a few arguments, depending on how the functions are defined.

User-defined functions can be constructed by any user and later reused by this and other users. The following code shows how to create a function for unit conversion.

```
def unit_converter(new_old_ratio,unit_to_be_converted):
    new_unit=unit_to_be_converted*new_old_ratio
    return new_unit
```

The code shows how to define a function. We can see that the lines below the def line are indented with (four) blank spaces. In this case, all the indented lines belong to the function we are creating. Once a function is created, we can generate output from it. The following are two examples for executing an instance of the function:

```
print(unit_converter(2.54,10))
b = unit_converter(2.54,10)
```

In these two examples, unit_converter(2.54,10) converts 10 inches to 25.4 centimeters: the first statement prints the new value in cm while the latter assigns the new value to a variable named b.

Functions from packages are essentially also user-defined functions, both of which are called custom functions. However, functions from a package do not need to be defined as above because they have been defined by the developer of the package containing these functions.

In addition to the above three types of functions, we also have methods belonging to certain types of variables (i.e., data types or more broadly as classes), which can work in a way similar to functions. The major difference is that methods are functions defined for classes, which are used to define objects in object-oriented programming languages such as Python. For example, data types like lists are defined as classes. Class is a virtual concept that defines a type of data structure and its behavior. Each class has attributes, i.e., data values within them, and methods, i.e., the behavior of an object, which tell how to access the data. When we instantiate a class, e.g., a = ["u","s","t"], an object belonging to the class of Python list is created, and this list object named "a" has the attributes and method belonging to the list class. In this case, lists are a class defined by Python. We can also define classes ourselves, which will not be discussed here. After an object is instantiated, the methods defined for the corresponding class then become something that can be used like a function for the object. Such functions are used as a.sort(). Again, the period in front of the function (or method) sort() here means "belonging to."

For example, common methods for strings include upper(), lower(), replace(), and find(). Lists have methods like remove(), append(),insert(), and sort(). Dictionaries can be operated using methods like get(), pop(), update, and values(). For example, an element in a dictionary object can be removed using the pop method, e.g., dictionary_name.pop["key_name"].

2.2.8 *Input and Output*

The input functionality in Python can be easily realized using input(" "). This function displays the text message between " " on the monitor to prompt the user to enter something using a keyboard, which can be saved to a variable using an assignment statement ('='). By contrast, output is usually handled using the print() function. This function can output a string or a combination of string and other variable(s) with designated format(s). In addition to such input and output via a keyboard and a monitor, input and output can also be processed with files. This can be easily done via various functions provided by Python packages such as NumPy and Pandas, which will be introduced later.

```
name = input("Please enter the your age!\n")
print("Welcome"+name+"!")
x = 50; y= 100
print("The value of x is {} and y is {}." .format(x,y))
```

2.2.9 *Advanced Python Functionality*

Python has some advanced functionalities that can extend the flexibility and reusability of the code written using Python such as classes/objects for object-oriented coding. These advanced functionalities will not be discussed here.

2.3 Data Manipulation and Visualization

In this section, NumPy, Pandas, and Matplotlib will be introduced as basic Python packages for data manipulation and visualization.

2.3.1 *NumPy*

NumPy Array

The core of NumPy is the NumPy array. The NumPy array is a data structure (data type, object, or Python class) that is similar to but much more powerful than the Python list in terms of array operations. Unlike Python lists, NumPy arrays can perform the calculation on entire arrays. In fact, NumPy arrays also serve as an essential base for other packages such as Scikit-learn.

In addition to algebraic operations on multidimensional arrays, NumPy is preferred also because of its powerful functions for logical and shape manipulation,

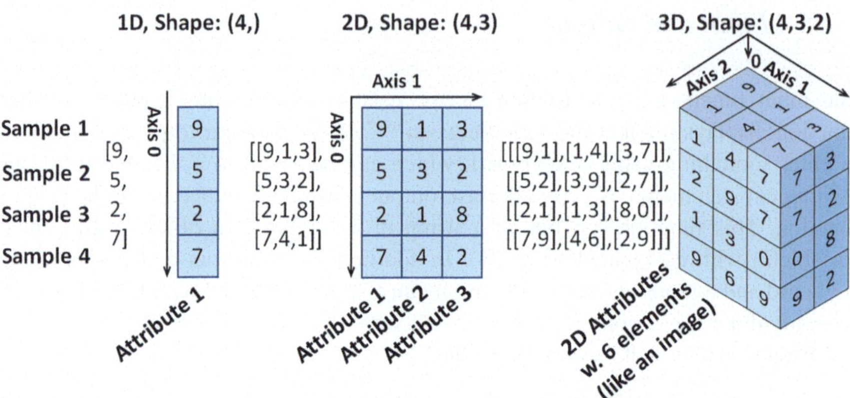

Fig. 2.2 Visualization of Python arrays

sorting, selecting, I/O, discrete Fourier transforms, and tools for integrating C/C++ and Fortran code, random number capabilities, and abilities to integrate with other Python packages. Many functionalities of NumPy arrays are designed for more convenient usage and higher computational efficiency.

Some basic knowledge needs to be covered before we can easily understand how n-dimensional ($n \geqslant 2$) arrays are represented and manipulated. Each NumPy array is a table of elements (usually numbers), which all have the same type and are indexed by a tuple of nonnegative integers. In NumPy, dimensions of an array are also called axes. Figure 2.2 shows an example of 1D, 2D, and 3D NumPy arrays. The written format and schematic of the arrays, the meanings of axes and shape, and the related concepts in machine learning (samples and attributes) are illustrated. This illustration is slightly adjusted to facilitate the use of NumPy for machine learning. It is worth mentioning that the above understanding of the data structure in NumPy is different from that in MATLAB, in which we usually start with a 1D array as a row and extend it to 2D by adding more rows of data. Also, the shape for 1D, which is represented using a tuple, i.e., (4,), is different from a 2D array with the same number of elements, e.g., (4,1).

Array Constructions

In the following, we will start the introduction to NumPy with topics like importing the package, creating NumPy arrays, and performing basic operations. Then, more operations like indexing, slicing, shape manipulation, stacking, splitting, and copying as well as some common linear algebraic operations will be introduced via a table. Finally, more information will be provided about a few items that can easily confuse beginners like broadcasting.

We can easily import NumPy by executing a statement "import numpy as np" (interactive mode) and adding this statement in the script file (script mode), in which "np" is just a short nickname that most people choose. A common way of generating a NumPy array is to create one from a Python list:

```
my_list = [1,2,3,4]
my_array1 = np.array(my_list)
my_array2 = np.array([1,2,3,4])
my_array = np.array([my_array1, my_array2])
```

There are a few other functions in NumPy that we can use to create special arrays such as np.ones() (arrays in which all the elements are 1s), np.zeros() (array of zeros), np.eyes() (identity matrix), and np.arange() (same as np.array(range())). The following are some examples.

```
np.zeros(5)
```

Output:

```
array([0., 0., 0., 0., 0.])
```

```
np.ones([5, 5])
```

Output:

```
array([[1., 1., 1., 1., 1.],
       [1., 1., 1., 1., 1.],
       [1., 1., 1., 1., 1.],
       [1., 1., 1., 1., 1.],
       [1., 1., 1., 1., 1.]])
```

```
np.eye(5)
```

Output:

```
array([[1., 0., 0., 0., 0.],
       [0., 1., 0., 0., 0.],
       [0., 0., 1., 0., 0.],
       [0., 0., 0., 1., 0.],
       [0., 0., 0., 0., 1.]])
```

The type (data type of the array elements), shape (a tuple of numbers representing the numbers of elements along different dimensions/axes), size (total number of elements), and number of dimensions (axes) are the things we will need to check frequently. Table 2.6 shows some common methods to obtain information about arrays. It is noted that a.transpose() and x.flatten() only help us check the transpose and flattened version of the array instead of making changes to the original array. To make the change, we need to use a=a.transpose() and x=x.flatten().

Table 2.7 shows methods that are handy for constructing new arrays. In particular, those functions for slicing, i.e., extracting a portion of an array, are very useful in the use of arrays.

Array Operations

Basic operations between or on arrays are listed in Table 2.8. "a @ b" or "np.dot(a,b)" both refer to the dot product between two tensors: $a@b = \sum a_i \cdot b_i$.

Table 2.6 Lookup and reference

Operation	NumPy	MATLAB
Number of dimensions of array a	np.ndim(a) or a.ndim	ndims(a)
Number of elements of array a	np.size(a) or a.size	numel(a)
"Size" of array a	np.shape(a) or a.shape	size(a)
Number of elements of nth dimension of a	a.shape[n-1]	size(a,n)
Max element of a	a.max() or np.nanmax(a)	max(max(a))
Max element of each column of array a	a.max(0)	max(a)
Max element of each row of array a	a.max(1)	max(a,[],2)
Max value from a and b element-wise	np.maximum(a, b)	max(a,b)
Matrix whose i,jth element is (a_ij >0.5).	(a >0.5)	(a >0.5)
Find the indices where (a >0.5)	np.nonzero(a >0.5)	find(a >0.5)
A vector of unique values in array a	np.unique(a)	unique(a)
Transpose of a	a.transpose() or a.T	a.'
Turn array into vector	y = x.flatten()	y=x(:)

Thus, this operation generates a sum (a scalar) of the products of the corresponding elements from the two arrays. In a more general case, when a and b are tensors (arrays with any number of dimensions), this operation is an outer or tensor product with a single contraction involving the last axis (or dimension) of the first tensor and the first axis of the second tensor. It is equivalent to a more general NumPy function "np.tensordot(a,b,axes=1)", in which "axes=1" means a single contraction. "np.tensordot(a,b,axes=0)" means a tensor product (no contraction), which is written using symbol "$\otimes$" in math. In fact, "np.tensordot()" can be used to do double or higher-order contraction using axes=2 or greater numbers. Furthermore, this operator can be used to specify the contraction of any axes from two tensors such as "np.tensordot(a,b,((1,2)(3,2)))", in which the second and third axes from a will be contracted with the fourth and third axes of b. More information about tensor analysis and tensor notation is available in the book appendices.

It is also worthwhile to mention that putting an arithmetic operator between two arrays, e.g., "a * b", means an element-wise operation, which is different from the MATLAB convention, in which an extra period is used to indicate element-wise operations, e.g., "a.*b".

Common linear algebraic operations for NumPy arrays can be found in Table 2.9. These functions are needed for solving algebraic equations and eigenvalue problems.

Table 2.7 Construction of arrays

Operation	NumPy	MATLAB
Define a 2 × 3 2D array	np.array([[1. ,2. ,3.], [4. ,5. ,6.]])	[1 2 3; 4 5 6]
Construct a matrix from blocks a, b, c, and d	np.block([[a, b], [c, d]])	[a b; c d]
Last element of a 1D array a	a[−1]	a(end)
2nd row, 5th column in a (2D)	a[1, 4]	a(2,5)
Entire 2nd row of 2D array a	a[1] or a[1, :]	a(2,:)
First 5 rows of 2D array a	a[0:5] or a[:5] or a[0:5, :]	a(1:5,:)
Last 5 rows of 2D array a	a[−5:]	a(end-4:end,:)
1st–3rd rows and 3rd–9th columns of a (2D)	a[0:3, 4:9]	a(1:3,5:9)
rows 2,4, and 5 and columns 1 and 3	a[np.ix_([1, 3, 4], [0, 2])]	a([2,4,5],[1,3])
Every other row of a, 3rd-21st	a[2:21:2,:]	a(3:2:21,:)
Every other row of a, starting with the first	a[::2,:]	a(1:2:end,:)
a with rows in reverse order	a[::−1,:]	a(end:−1:1,:) or flipud(a)
Copy the values of array x to make y	y = x.copy()	y=x
Copy part of x to make y	y = x[1, :].copy()	y=x(2,:)
4 equally spaced samples between 1 and 3	np.linspace(1,3,4)	linspace(1,3,4)
Return a vector of the diagonal of a (2D)	np.diag(a)	diag(a)
Returns a square diagonal matrix with vector v	np.diag(v, 0)	diag(v,0)
Generate a random array with seed = 42	rng = default_rng(42),rng.random(3, 4)	rng(42,'twister'), random.rand((3, 4))(old)
Create m by n copies of a	np.tile(a, (m, n))	repmat(a, m, n)

2.3.2 Pandas

From NumPy to Pandas

Built on top of NumPy, Pandas is one of the fastest and most effective tools for manipulating real-world messy data in data science and machine learning. In general, NumPy works well for a small amount of structured data. But when there are more and complicated unstructured data, Pandas may be a better option. This tool can easily handle missing and heterogeneous data, check and clean up the data, and extract, reformat, and create new datasets. This tool supports multiple file formats like CSV, Excel, SQL, etc.

Table 2.8 Operations between and on arrays

Operation	NumPy	MATLAB
Matrix multiply	a @ b	a * b
Element-wise multiply	a * b	a .* b
Element-wise divide	a/b	a./b
Element-wise exponentiation	a**3	a.^3
a with elements less than 0.5 zeroed out	a[a <0.5]=0	a(a<0.5)=0
a with elements less than 0.5 zeroed out	a * (a >0.5)	a .* (a>0.5)
Sort each column of a 2D array a	np.sort(a) or a.sort(axis=0)	sort(a)
Sort the each row of 2D array a	np.sort(a, axis = 1) or a.sort(axis = 1)	sort(a, 2)
Set all values to the same scalar value	a[:] = 3	a(:) = 3
Concatenate columns of a and b	np.concatenate((a,b),1), np.hstack((a,b)), np.column_stack((a,b)) or np.c_[a,b]	[a b]
Concatenate rows of a and b	np.concatenate((a,b)), np.vstack((a,b)) or np.r_[a,b]	[a; b]

Table 2.9 Linear algebraic operations of NumPy arrays

Operation	NumPy	MATLAB
Inverse of square 2D array a	linalg.inv(a)	inv(a)
Pseudo-inverse of 2D array a	linalg.pinv(a)	pinv(a)
Matrix rank of a 2D array a	linalg.matrix_rank(a)	rank(a)
Solution of $ax = b$ for x	linalg.solve(a, b) if a is square; linalg.lstsq(a, b) otherwise	a\b
Solution of $xa = b$ for x	Solve a.T x.T = b.T instead	b/a
Eigenvalues λ and eigenvectors v of a, $\lambda v = av$	D,V = linalg.eig(a)	[V,D]=eig(a)

Like the NumPy array in NumPy, the core of Pandas is two data structures called Series and DataFrame. The Pandas dataframe can be viewed as a table similar to an Excel spreadsheet. As illustrated in Fig. 2.3, one major difference between this table and that represented by a 2D NumPy array is that the table comes with two extra elements: indices and column labels. Each index is unique and marks a row, while each column label assigns a name to a column. Besides indices and column labels, the Pandas dataframe can have different data types for different columns. The Pandas series is the combination of a 1D array of data with its corresponding indices. Therefore, the series can be simply viewed as a column of data in the dataframe (without label) together with the corresponding indices. The word "axes" in Pandas has the same meaning as that in the NumPy array.

From the machine learning perspective, we can view each row in the dataframe as a sample and each column as an attribute. Pandas offers convenient ways to change any data point or any part of a dataframe.

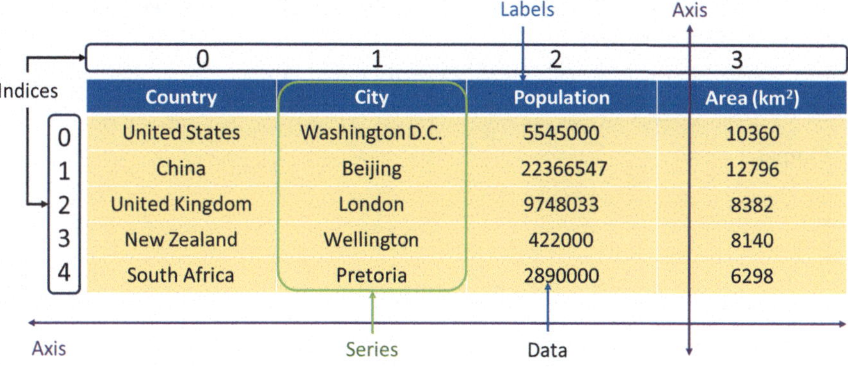

Fig. 2.3 Structure of data in Pandas

Series

As mentioned, the series object (or data structure) in Pandas, or written as Series, is similar to the 1D array in NumPy. But Series is different from 1D NumPy array in that each element in the Series has a unique index. A Series can be created from a 1D NumPy array. If not otherwise specified, the indices will be integers increasing from 0 by an increment 1 until the last element. As shown below, if we print a Series, we will see a column of indices on the left and a column of values on the right.

```
import pandas as pd
from pandas import Series,DataFrame
obj = Series([9,5,2,7])
print(obj)
```

Output:

```
0 9
1 5
2 2
3 7
dtype: int64
```

The indices and real data can be accessed or cited using the index or values attributes of a Series object.

```
print(obj.index)
```

Output:

```
RangeIndex(start=0, stop=4, step=1)
```

```
print(obj.values)
```

Output:

```
[9 5 2 7]
```

If we want to use indices different from the default integers, we can create a Series in the following way.

```
obj2 = Series([9,5,2,7],index=['a','b','c','d'])
print(obj2)
```

Output:

```
a 9
b 5
c 2
d 7
dtype: int64
```

We can easily use any data points or a group of points in Series using the corresponding index.

```
print(obj2['c'])
obj2['a'] = -9
print(obj2[['a','b']])
```

Output:

```
2
a -9
b 5
dtype: int64
```

We can apply various arithmetic, Boolean, comparative, and other math applications to a Series, in which the relationship between the values and indices will be kept.

```
print(obj2*2)
```

Output:

```
a -18
b 10
c 4
d 14
dtype: int64
```

```
import numpy as np
print(np.exp(obj2))
```

Output:

```
a 0.000123
b 148.413159
c 7.389056
d 1096.633158
dtype: float64
```

```
print(obj2>0)
```

Output:

```
a False
b True
c True
d True
dtype: bool
```

Also, we can view Series as an ordered dictionary. Accordingly, we can create a Series from a Python dictionary.

```
states = {'Michigan':44, 'Ohio': 53, 'Virginia': 60, 'Florida'
    :75}
obj3 =Series(states)
print(obj3)
```

Output:

```
Michigan 44
Ohio 53
Virginia 60
Florida 75
dtype: int64
```

Dataframe

The Pandas dataframe data structure, or written as DataFrame (as a class in Pandas), can be viewed as being evolved from Series. Like Series, DataFrame carries the default indices (or implicit indices) if not otherwise specified. To get a better idea of this data type, we can take a look at the command for creating a DataFrame object:

```
import pandas as pd
pd.DataFrame(data, index, columns, dtype, copy)
```

The parameters in the above command have the following meanings.

- Data: the real data points, which can be ndarray, series, list, dict, scalar, or DataFrame.
- Index: the identity tag for each row. It will default to RangeIndex (np.arange(n), in which n is the number of data points/elements) if no indexing information as part of the input data or no index is provided.
- Columns: the identity tag for each row. It will default to RangeIndex (np.arange(n), in which n is the number of columns or data attributes) if no indexing information as part of the input data or no index is provided.
- dtype: the Python data type of each column.
- Copy: bool or None, default None. It copies data from inputs. If data is a dict containing one or more Series (possibly of different dtypes), "copy=False" will ensure that these inputs are not copied. For dict data, the default of None behaves like "copy=True". For DataFrame or 2d ndarray input, the default of "None" behaves like "copy=False".

A DataFrame can be created in multiple ways. In the simplest way, an empty DataFrame object can be created after we import the package, "import pandas as pd", and then use the command "df = pd.DataFrame()", in which df is the name of the DataFrame object. More commonly, it can be created using Python list(s), Python dictionary(ies), NumPy array(s), and Pandas Series in different ways.

First, using a Python list (including nested list as the following), we can generate a DataFrame:

```
data = [['Alex',10],['Bob',12],['Clarke',13]]
df = pd.DataFrame(data,columns=['Name','Age'])
print(df)
```

Output:

```
  Name Age
0 Alex 10
1 Bob 12
2 Clarke 13
```

In the above example, if we specify the data types

```
df = pd.DataFrame(data,columns=['Name','Age'],dtype=float)
print(df)
```

Output:

```
  Name Age
0 Alex 10.0
1 Bob 12.0
2 Clarke 13.0
```

Second, if we use dictionaries (including nested dictionaries as the following), we can generate a DataFrame:

```
data = [{'a': 1, 'b': 2},{'a': 5, 'b': 10, 'c': 20}]
df = pd.DataFrame(data)
print(df)
```

Output:

```
  a b c
0 1 2 NaN
1 5 10 20.0
```

Next, we can also create a DataFrame object based on a Series object.

```
d = {'one' : pd.Series([1, 2, 3], index=['a', 'b', 'c']),
'two' : pd.Series([1, 2, 3, 4], index=['a', 'b', 'c', 'd'])}
df = pd.DataFrame(d)
print(df)
```

Output:

```
  one two
a 1.0 1
b 2.0 2
c 3.0 3
d NaN 4
```

We can visit the data in any specific row, i.e., corresponding to a sample, using the corresponding index in a way similar to Python lists, tuples, and dictionaries:

```
print(df ['one'])
```

Output:

```
a 1.0
b 2.0
c 3.0
d NaN
```

After the creation of a DataFrame object, we can have a quick overview of useful functions for typical operations: data view, selection, Boolean indexing, missing data treatment, basic operations, data change (remove, add, and merge), function application, and data reorganization. Common functions for these operations are introduced in the following.

View Data

Let us assume "df" is the name of a DataFrame object that we are handling. df.head() and df.tail() can be used when there are lots of samples (long columns). A number can be used, e.g., using df.tail(3), to only show the last three lines. The indices, column labels, and data can be checked using df.index, df.columns, and df.values. In addition, df.describe() can be used to show more statistics of the data in the DataFrame object, e.g., number of rows, average, and mean.

Data Selection

In the simplest way, we can select data from certain rows using df[0:3] (first three samples) and from certain columns using df['column1','column3'] (columns or attribute values of all the samples with labels "column1" and "column3"). For a single column, e.g., a column with a label "A", we can use df['A'] or df.A. If the index is date, e.g., created using pd.date_range('20221003',period=31), we can also obtain the data values corresponding to certain dates (rows) such as df['20221003':'20221005'].

In addition, two other methods "loc" and "iloc" are also very popular. "loc" is used when we specify the name of the index and/or column labels. For example, we can use df.lob[:,['column1','column3']]. "iloc" is preferred when we want to use numeric values to specify the location(s). The following are a few examples: df.iloc[3], df.iloc[3:5,0:2], and df[[1,2,4], [0,2]]. A few more methods such as "at" and "iat" are also available. Part or all the data from a DataFrame object can be used for other purposes, e.g., df2=df.copy() for creating a new DataFrame object.

Boolean Indexing

We can use df.A>0 directly to generate a DataFrame with only Boolean values that are generated by the comparison of the corresponding data with 0. We can get samples (rows) whose values in Column "A" is great than 0 using df[df.A>0].

Missing Data

Missing data is common and can be easily processed in DataFrame objects. We can remove the missing data (marked with "NaN" in Pandas) using df.dropna(how='any'). Another common task is to replace all the missing data with a certain number like 0: df.fillna(0). In some cases, we will need to find out the locations of missing data in a DataFrame object: pf.isnull(df), which will give out a DataFrame object with Boolean values.

Basic Operations
Some common basic operations include the following:

Sort by an axis: df.sort_index(axis=0, ascending=True).
Sort by a column: df.sort_value(by='A').
Obtain the mean along an axis: df.mean(); using 0 (default, axis=0) apply to every column.
Shift all the values downward and add zero to the emptied locations: df.shift(3).

Data Change (Remove, Add, and Merge)
The drop() method removes the specified row or column. We can directly enter labels or indexes as the argument to drop the corresponding columns and rows, respectively. In more advanced use, we can use axis=0 (or "index") or axis=1 (or "columns") and a string or list for the index and columns to remove. pop() is another function that can be adopted to quickly remove a column, for which the label of the column will be used as the argument.

Two DataFrame objects or their parts can be merged using functions such as pd.concat(). Besides, we also employ DataFrame methods such as merge() for merging columns (default) or merging columns and rows in more complicated ways and use append() for appending rows of other DataFrame objects. Typical examples include pd.concat([df[:3],df[3:7]],axis=0), df1.merge(df2,how='inner'), and df1.append(df2,ignore_index=True).

2.3.3 *Matplotlib*

Matplotlib provides two distinct ways (or interfaces) of drawing scientific figures. This first way is functional or procedural via the Pyplot interface, which can be used to imitate MATLAB's functionality for drawing. The second way is essentially an object-oriented approach. By comparison, the former is easier to work with, while the latter is suitable for more complex tasks.

Pyplot: Procedural Plotting Interface

Similar to MATLAB, Matplotlib's Pyplot enables easy creation of simple and interactive plots. We can start with an example for plotting multiple 2D curves. After importing the Pyplot module from Matplotlib by executing "import matplotlib.pyplot as plt", we can generate a curve in the simplest way as follows:

```
x = np.linspace(0,10,100) # Create a NumPy array with 100 points
    from 0 to 10
y1 = np.sin(x)
```

```
plt.plot(x,y1) # plotting the line 1 points
plt.show() # Display a figure. This is needed in some cases.
```

In a more general case, we can plot two curves in the same figure with more information such as axis labels, figure title, and a legend.

```
x = np.linspace(0,10,100) # Create a NumPy array with 100 points
    from 0 to 10
y1 = np.sin(x)
y2 = np.cos(x)
plt.plot(x,y1,'r*',label = "sin") # plotting the line 1 points
plt.plot(x,y2,'g.',label = "cos") # plotting the line 2 points
plt.xlabel('x - axis')
plt.ylabel('y - axis') # Set the y axis label of the current axis
    .
plt.title('Two or more lines on the same plot with suitable
    legends ') # Set a title of the current axes.
plt.legend(loc='upper left') # Show a legend on the plot
plt.show() # Display a figure. This is needed in some cases.
```

Besides the basic line plot in the above code, we can also plot histograms, vertical and horizontal bar charts, paths, 3D graphs, filled curves, and so on. For example, the above continuous line plot can be changed to scatter, bar, and pie plots using plt.scatter(), plt.bar(), and plt.pie(), respectively. There is a long list of extra functionality for Pyplot. This allows us to make different changes such as creating legends, creating a title, creating axis labels, changing the appearance of ticks and grid-lines, setting axis limits, and much more. Common examples include plt.xlim(), plt.xticks, and plt.annotate().

Object-Oriented Plotting Interface

We may need to resort to object-oriented plotting when we get more complicated plotting assignments like subplots and 3D plots. To understand object-oriented plotting in Matplotlib, let us first take a closer look at the objects: the hierarchical structure of a typical figure.

As shown in Fig. 2.4, a figure object is a container for all of the elements that make up the image. This object controls foundational settings such as the image size and the inclusion of legends. Each element rendered on the figure, i.e., each labeled part, is known as an artist. Inside the figure, we can see one or more axes. An axes in Matplotlib is essentially a (sub)plot, which holds the information for what we are graphing. To use Matplotlib in an object-oriented fashion, we have to generate our own figure and axes.

First, let us see how to plot a simple plot, e.g., 2D curves as what we did using Pyplot, using object-oriented plotting.

```
fig, ax1 = plt.subplots() # Initializing a figure and axes object
ax1.plot(x, y1, 'r*', label = 'sin') # Plotting the data
ax1.plot(x, y2, 'g.', label = 'cos')
ax1.set_title("My Plot")
```

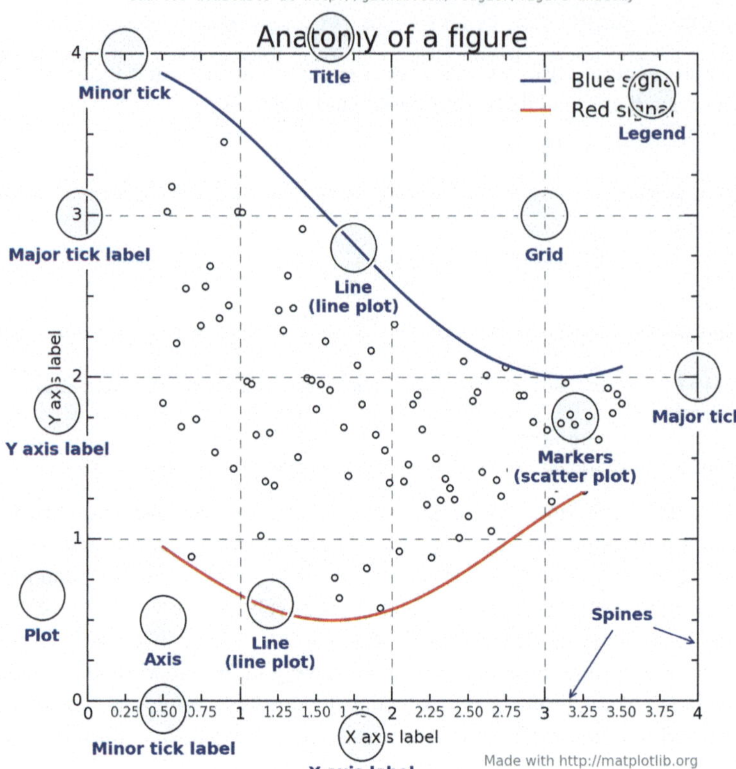

Fig. 2.4 Hierarchical structure of a typical figure

```
ax1.set_xlabel("x - axis")
ax1.set_ylabel("y - axis")
ax1.set_title('Two or more lines on same plot with suitable
    legends ')
ax1.legend()
```

Next, let us take a look at how subplots can be easily handled with object-oriented plotting. One of the key advantages of the object-oriented approach is the simplicity of creating additional axes. Whenever we call plt.subplots(), it returns a figure and one axes or subplot. However, there are multiple ways of creating more subplots, which may involve objects to different extents. Two slightly different ways are given in the following.

The first way is exemplified by the following code:

```
import matplotlib.pyplot as plt
import numpy as np
x = np.linspace(0,10,100) # Create a NumPy array with 100 points
    from 0 to 10
y1 = np.sin(x)
```

```
y2 = np.cos(x)

fig, (ax1,ax2) = plt.subplots(2,1) # Create 2 subplots in 2 rows
    and 1 column; number them from left to right, up to down. #1
    is the first
plt.sca(ax1) # Set the current axes, i.e., ax1, to the current (
    active) subplot.
ax1.plot(x, y1, 'r*', label = 'sin')
plt.title("My Plot") # Or ax1.set_title("My Plot")
ax2.plot(x, y2, 'g.', label = 'cos')
```

The second way is slightly different:

```
fig, ax = plt.subplots(2,1)
ax[0].plot(x, y1, 'r*', label = 'sin')
ax[1].plot(x, y2, 'g.', label = 'cos')
```

Another way that can be easily confused with the above two is as follows. Please be aware that here we use "subplot" instead of "subplots."

```
ax1 = plt.subplot(211) # Equivalent to more general subplot
    (2,1,1)
ax1.plot(x, y1, 'r*', label = 'sin')
ax2 = plt.subplot(212)
ax2.plot(x, y2, 'g.', label = 'cos')
```

Last, let us work on 3D plots. This is usually needed for visualizing data with three axes or dimensions, in which two axes are for the attributes and the remaining axis for the label, e.g., in a multivariate regression problem with two variables and one target/label. If we know the coordinates of all the points are in a 3D space, i.e., x, y, and z as three 1D NumPy arrays, then we can make 3D plots directly using methods like plot() for 3D curves, scatter() for scatter plots, plot_wireframe() for 3D wireframes, and plot_surface() for surface plots, plot_trisur for triangulation plots, etc. This can be directly done using the following code:

```
fig = plt.figure()
ax = fig.gca(projection='3d')
theta = np.linspace(-4 * np.pi, 4 * np.pi, 100)
z = np.linspace(-2, 2, 100)
r = z**2 + 1
x = r * np.sin(theta)
y = r * np.cos(theta)
ax.plot(x, y, z, label='parametric curve')
```

The above 3D plot is for curves in a 3D space. In many cases, we need to plot a surface in a 3D space, which is much different and could be confusing to beginners. For such 3D plots, we will need to first create a meshgrid, which defines the coordinates of all of the grid points of the surface. For example, we can use X = np.arange(−5, 5, 0.25) and Y = np.arange(−5, 5, 0.25) to create two 1D NumPy arrays first, which are more like the "ticks" on the axes. Next, a command like "X, Y = np.meshgrid(X, Y)" can turn them into 2D NumPy arrays marking the x and y coordinates of the grid points. This process is similar to the generation of a grid (or mesh) by sweeping from control points on the boundaries of the grid area. With the meshgrid, a surface can be plotted using the following code.

```
fig = plt.figure()
ax = fig.gca(projection='3d')
X = np.arange(-5, 5, 0.25)
Y = np.arange(-5, 5, 0.25)
X, Y = np.meshgrid(X, Y)
Z = np.sin(np.sqrt(X**2 + Y**2))
surf = ax.plot_surface(X, Y, Z, cmap=cm.coolwarm, linewidth=0,
    antialiased=False)
ax.set_zlim(-1.01, 1.01) # Customize the z axis.
```

2.4 General Machine Learning

2.4.1 Scikit-Learn

Scikit-learn (Sklearn) is one of the most useful and robust libraries for machine learning in Python. It provides a collection of common algorithms for classification, regression, clustering, and dimensionality reduction via a consistent interface. As a library, Scikit-learn is written in Python and relies on other packages like NumPy, SciPy, and Matplotlib. Scikit-learn is preferred due to many reasons, such as its clean, uniform, and streamlined APIs and useful and complete online documentation. In particular, the APIs enable users to easily pick up new models or algorithms once they understand the basic use and syntax of Scikit-learn for one model.

Scikit-learn library is intended more for data analysis instead of the acquisition and manipulation data, which can be done with packages introduced in the previous section. This library groups functions for implementing common machine learning algorithms (excluding reinforcement learning) into four categories according to the application types: regression and classification in supervised learning and clustering and dimensionality reduction in unsupervised learning. Functions in other categories such as feature extraction, data processing, and model assessment are also available. In addition, Scikit-learn also provides many datasets to facilitate the development, comparison, and assessment of algorithms.

Let us get familiar with Scikit-learn by following a typical machine learning procedure that consists of five steps: data import, data preprocessing, model definition, training, and testing (or called predictions). We will do the introduction using a few common functions (algorithms). Advanced use of such functions or more machine learning algorithms can be easily learned with the user guide from Scikit-learn's official website.

Data Import

Data is needed for training, testing, comparing, and validating algorithms. In some studies, the preparation of data may account for a major portion of the work. In Scikit-learn, we can utilize data from the Scikit-learn datasets or create our own datasets.

The following is one example of importing Sklearn datasets:

```
from sklearn.datasets import load_iris
iris = load_iris() # Import the famous Iris dataset
X = iris.data # Obtain the features/attributes of the Iris
    samples as X
y = iris.target # Obtain the labels of the samples as y
```

Alternatively, we can generate some data for quick testing:

```
from sklearn.datasets.samples_generator import
    make_classification
X, y = make_classification(n_samples=6, n_features=5,
    n_informative=2,
n_redundant=2, n_classes=2, n_clusters_per_class=2, scale=1.0,
random_state=20)

# n_samples: number of samples
# n_features: number of features/attributes
# n_classes: number of classes
# random_state: seed that we use to generate data with random
    numbers; using the same seed to ensure we will get the same
    data despite the fact that the data is generated using random
     numbers.
```

The generated data can be easily checked using

```
for x_,y_ in zip(X,y):
    print(y_,end=': ')
    print(x_)
```

In engineering practice, it is more common to import data from an external data file that we prepared earlier. The following code shows a simple way of importing such data:

```
X = np.loadtxt('./engineInputs.csv', delimiter=',',skiprows=1)
y = np.loadtxt('./engineTargets.csv', delimiter=',',skiprows=1)
```

where "engineInputs.csv" and "engineTargets.csv" are two data files in the same folder (as specified by "./") as the script file.

Data Preprocessing

Data preprocessing may be needed or desired for many machine learning tasks. Common preprocessing operations include normalization (scaling), standardization (mean and standard error), regularization, One-Hot coding, dataset splitting, and data augmentation.

Normalization, which scales the data including the input (X) and/or output (y), is desired in most supervised learning tasks because it helps ensure the data values are distributed in a certain range and facilitates the comparison of the model performance across different datasets. Due to the same reason, it enables the trained models to work for data with different value ranges. In fact, normalization is necessary for machine learning tasks in which a fixed range of input or output is adopted. For example, some models use a surrogate function to compress the final model output to a range between -1 and 1. Normalization tools are available in many data analysis and machine learning packages. Scikit-learn's normalization functions can be seen in the following code:

```
from sklearn.preprocessing import MinMaxScaler
scaler = MinMaxScaler(feature_range=(-1, 1))
scaler_X = scaler.fit(X)
X = scaler_X.transform(X)
scaler_y = scaler.fit(y)
y = scaler_y.transform(y)
```

Standardization based on the mean and standard error can sometimes help us improve the learning results. This operation is illustrated using the following code:

```
from sklearn.preprocessing import StandardScaler
scaler = StandardScaler().fit(train_data)
scaler.transform(train_data)
scaler.transform(test_data)
```

Regularization can be implemented using "normalize":

```
from sklearn.preprocessing import normalize
X = [[ 1., -1., 2.],
[ 2., 0., 0.],
[ 0., 1., -1.]]
X_normalized = normalize(X, norm='l2')
```

One-Hot is a preprocessing step that is frequently needed for handling data for multi-class classification in deep learning. The following code presents an example in Scikit-learn:

```
from sklearn.preprocessing import OneHotEncoder
data = [[0, 0, 3], [1, 1, 0], [0, 2, 1], [1, 0, 2]]
encoder = OneHotEncoder().fit(data)
encoder.transform(data).toarray()
```

As explained in the previous chapter, a typical supervised machine learning task involves selecting an algorithm, training a model that functions according to the algorithm with training data, testing whether the trained model works satisfactorily using testing data, and finally applying the model to predict new data. In this process, both the training and testing data come from the data that we prepare. Thus, a step that we usually need before starting the training and testing process is to split the original datasets into a training set and a testing set. This can be performed using the function "train_test_split(arrays, options)":

```
from sklearn.model_selection import train_test_split
X_train, X_test, y_train, y_test = train_test_split(X, y,
    test_size=0.3, random_state=42)
```

In the above example, "arrays" are the arrays of samples including features and labels, test_size is the ratio of the testing samples in the total dataset specified either with a floating point number as the ratio or an integer as the number of testing samples, random_state is specified using an integer so that the data splitting performed with random numbers can be reproducible, and shuffle (True by default, not specified in the code) tells whether the samples will be shuffled before splitting.

Using Models

This subsection illustrates the use of machine learning models in Scikit-learn, including algorithm selection, training, testing, and prediction. First, we need to select an algorithm for creating the model. The following code shows a typical example from the "linear_model" module in Scikit-learn:

```
from sklearn import linear_model as LM
regr = LM.Lasso(alpha = 0.5)
regr.fit(X,y)
regr.predict(np.array([[3,5]]))
```

In this example, the model created using the LM.Lasso algorithm is named "regr." Scikit-learn provides multiple categories of algorithms as different modules: linear models, logistical regression, Bayesian methods, decision trees, support vector machines, K-nearest neighbors, and multilayer perceptions (ANNs). The use of such algorithms via the functions provided in Scikit-learn is very straightforward and can be learned quickly by checking the user guide when needed. Thus, we will not go into details.

The remaining steps of the procedure can be very easily implemented using the following functions.

```
# Training
model.fit(X_train, y_train)
# Testing
model.fit(X_test, y_test)
# Prediction
model.predict(X_test)
# Obtain parameters about the model
model.get_params()
# Evaluate the model
model.score(data_X, data_y)
```

"model" in the above code is the name of the model, which is essentially an instance of the algorithm. We can replace "model" with "regr" if we choose to use a linear regression model and name it as "regr."

Saving Models

Again, a machine learning model is the instance of an algorithm, whose parameters are determined by the training data. In this way, the trained model can be used to predict new data. Thus, after a model is trained, we usually need to find a way to save the trained model, or, more specifically, save the values of the parameters. For example, such parameters in a machine learning task with a typical deep neural network are the hyperparameters controlling the training process and the network weights that control the behavior of the model, though only the latter will be considered as the model in a narrow sense. In Scikit-learn, model parameters can be saved using the joblib module as

```
from sklearn.externals import joblib
# Save a model
joblib.dump(model, 'model.pickle')
# Load a saved model
model = joblib.load('model.pickle')
```

Alternatively, we can use other tools such as pickle:

```
import pickle
# Save a model
with open('model.pickle', 'wb') as f:
    pickle.dump(model, f)
# Load a saved model
with open('model.pickle', 'rb') as f:
    model = pickle.load(f)
model.predict(X_test)
```

2.5 Deep Learning

2.5.1 Deep Learning Frameworks

The world of deep learning software (e.g., frameworks and libraries) is evolving fast. Right after the start of the third AI wave, tools from academia like Caffe and Theano were the major players. But a few years later, after industrial giants stepped in, tools developed by these industry players started taking predominant roles. Most of the big players in the IT world have their own frameworks: Google launched TensorFlow (TF), Amazon supported MXNet, Microsoft created CNTK, Amazon and Microsoft jointly contributed to Gluon, Facebook created PyTorch, and so on.

TensorFlow attracted lots of users since Version 1.*, which was proposed to facilitate the computation with arrays. The computation in deep learning can be viewed as the flow of such arrays, which can be called tensors from a tensor analysis perspective, through the artificial neural network. This concept of "tensor flow" is actually much broader than merely deep learning, though one of its major purposes is to facilitate the computation in deep learning.

However, Versions 1.* (TF1) suffers from criticisms for its shortage of usability. Especially, concepts like (computing) graphs, placeholders, and sessions could both steepen the learning curve and reduce the efficiency of implementing deep learning. Possibly due to this reason, some of its competitors like PyTorch rose quickly. Then in TensorFlow Version 2.* (TF2), Keras was introduced as a more convenient higher-level library for the use of the TensorFlow framework for performing deep learning, and eager execution was strengthened to better integrate with the Python environment, ease the learning process, and enable easier debugging. In particular, though it may not be as computationally efficient as Graph, eager execution inherits many advantages of graph execution and significantly improves the usability and learning curves. This helped TensorFlow Version 2.* regain a position in the competition against others especially PyTorch, which gained more popularity in academia. Despite the trend, TensorFlow and its high-level deep learning API, Keras, still provide a classic and easy way for us to learn and practice deep learning.

In the following sections, we will introduce TensorFlow and Keras. These two tools can cover major code and tasks that deep learning practitioners and learners would encounter in the current state of the practice. Even if they are replaced in some way by their successors or competitors, they still can be decent tools and examples for learning and implementing classic deep learning topics.

2.5.2 TensorFlow

Overview of APIs

The architecture of the TensorFlow framework will be introduced first to offer an overview of the tool. Next, considering the above state of the practice, we will introduce TensorFlow including difficult but efficient concepts like computing graphs. New concepts in Version 2.* will also be explained. Finally, an example of using TensorFlow for conducting a deep learning task will be offered to show how to use this tool in a production environment. A similar example for Keras will be provided so that we also see the difference between TensorFlow and Keras.

TensorFlow's Application Programming Interfaces (APIs) are arranged hierarchically, with the high-level APIs built on the low-level APIs. In this section, we will use a high-level API named tf.keras to define and train machine learning models and to make predictions. tf.keras is the TensorFlow variant of the open-source Keras API. Figure 2.5 shows the hierarchy of the TensorFlow toolkit:

TensorFlow's APIs can be classified into two major categories. The low-level APIs:

- Can complete programming control
- Are recommended for machine learning researchers
- Provide fine levels of control over the models
- Form the TensorFlow Core

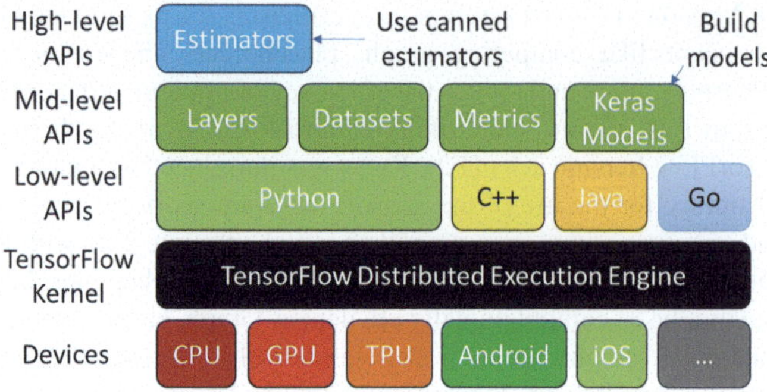

Fig. 2.5 Architecture of TensorFlow

The low-level or core APIs provide access to low-level functionality within the TensorFlow ecosystem. Compared to high-level APIs, such APIs provide more flexibility and control for building machine learning models, applications, and tools. The core APIs can be used as an alternative to high-level machine learning APIs like Keras. These high-level APIs are best suited for general machine learning needs.

The common low-level APIs include:

- Data structures: tf.Tensor, tf.Variable, tf.TensorArray
- Primitive APIs: tf.shape, slicing, tf.concat, tf.bitwise
- Numerical: tf.math, tf.linalg, tf.random
- Functional components: tf.function, tf.GradientTape
- Distribution: DTensor
- Export: tf.saved_model

By contrast, the high-level APIs are:

- Built on top of TensorFlow Core
- Easier to learn and use than TensorFlow Core
- Advantageous in saving repetitive tasks
- More consistent between different users

The following are some high-level APIs:

- Keras: deep learning for humans
- TensorFlow Model Optimization Toolkit: a suite of tools to optimize machine learning models
- TensorFlow Graphics: a library that makes useful graphics functions widely accessible

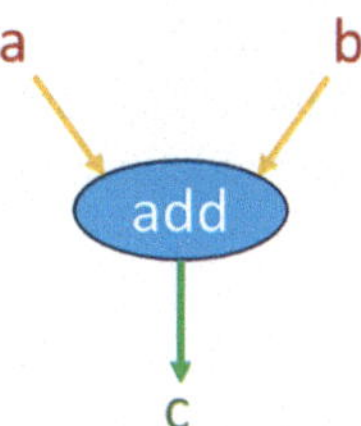

Fig. 2.6 Typical unit of computational graph

Computational Graph

Basic Concepts

Let us first take a look at a few essential, basic concepts in TensorFlow: graph, variable, and placeholder. TensorFlow is basically a software library for numerical computation using data flow graphs. Therefore, it has a much wider use than deep learning, though it is mostly known for deep learning especially the implementation of deep learning algorithms with low-level APIs.

A graph is the main path for the flow of tensors. A typical graph consists of:

(1) Nodes in the graph that represent mathematical operations
(2) Edges in the graph that represent the multidimensional data arrays (called tensors) communicated between nodes

The concept and use of graph can be illustrated using Fig. 2.6, which is a widely cited figure from the TensorFlow documentation.

Here, "add" is a node which represents the addition operation, "a" and "b" are input tensors, and "c" is the resultant tensor. Such a graph outlines the way the tensors flow before the actual computation. One another advantage of such graphs is that they allow us to deploy computation to one or more CPUs or GPUs in different devices using a single API.

Procedure for Using Graph

TensorFlow Core programs are organized around graph, which can be divided into two steps:

1. Building the computational graph. A computational graph is a series of TensorFlow operations that are arranged into a graph of nodes.
2. Running the computational graph. We must run the computational graph within a session to evaluate the nodes. A session entails the control and state of the TensorFlow runtime.

This is distinct from regular computer programs, in which the computation is performed line by line. In regular programs, each line represents one or more operations on the given data. Consequently, the data can be operated without a structured process. Such traditional programs are much less structured, can be changed anytime, and are executed based on the order of command line instead of a predefined graph.

Example

To deepen the understanding of graphs, let us check a very simple TensorFlow program as follows. This program shows the original (classic) way of defining and executing a computation graph.

```
# Import TensorFlow
import tensorflow as tf

# Create nodes in computation graph
node1 = tf.constant(3, dtype=tf.int32)
node2 = tf.constant(5, dtype=tf.int32)
node3 = tf.add(node1, node2)

# Create TensorFlow session object (tested in TF 2, eager
    execution disabled; use sess = tf.session() to replace the
    following two lines in TF 1)
tf.compat.v1.disable_eager_execution()
sess = tf.compat.v1.Session() # sess = tf.Session() in TF1

# Evaluate node3 and print the result
print("sum of node1 and node2 is :",sess.run(node3))
# Close the session
sess.close()
```

Output:

```
Sum of node1 and node2 is: 8
```

Let us try to understand the above code step by step.

Step 1

Create a computational graph by defining the nodes. TensorFlow provides different types of nodes for a variety of tasks. Each node takes zero or more tensors as input and produces a tensor as output.

It is noted that node1 and node2 in the above code are of "tf.constant" type. A constant node takes no inputs and only outputs a constant. The data type of the output tensor can be specified using a dtype argument.

```
node1 = tf.constant(3, dtype=tf.int32)
node2 = tf.constant(5, dtype=tf.int32)
node3 = tf.add(node1, node2)
```

"node3" is of the "tf.add" type. This node takes two tensors as the input and computes their sum as the output tensor.

Step 2

Run the computational graph via a session.

```
sess =  tf.compat.v1.Session() # sess = tf.Session() in TF1
print("Sum of node1 and node2 is:",sess.run(node3))
sess.close()
```

Here, a session is created or instantiated first. Then, the run method of the session object is invoked to perform computations on node3. In the last line, the session is closed.

Alternatively, we can adopt another way of working with sessions as follows:

```
with  tf.compat.v1.Session() as sess: # with tf.Session() as sess
    :
    print("Sum of node1 and node2 is:",sess.run(node3))
```

In this way, we do not need to close the session explicitly because it is automatically closed once the control extends out of the scope of the block.

Eager Execution

The above example shows the classic procedure of handling a computation graph. Despite the mentioned advantages, it brings more difficulty understanding and implementing computation. Due to the reason, starting in TensorFlow 2.0, eager execution is introduced to simplify the above procedure to make it closer to the way that we typically handle computation. This simplified approach is similar to PyTorch and helped TensorFlow catch up to some extent in its competition with PyTorch.

```
# Import TensorFlow
import tensorflow as tf

# Launch the graph in a session.
with tf.compat.v1.Session() as sess: # with tf.Session() as sess:

    # Create nodes in computation graph
    node1 = tf.constant(3, dtype=tf.int32)
    node2 = tf.constant(5, dtype=tf.int32)
    node3 = tf.add(node1, node2)

    # Evaluate node3 and print the result
    print("sum of node1 and node2 is :",sess.run(node3))
```

Variables

TensorFlow uses variable nodes to hold and update the parameters of a training model. Variables need to be explicitly initialized. Their values can be saved to disk during and after training and later restored from disk. It is noted that a constant's value is stored in the graph and replicated wherever the graph is loaded. By contrast, a variable is stored separately and may live on a parameter server.

The following code presents an example of using variable.

```
# Import TensorFlow
import tensorflow as tf

# Create a node of type Variable and assign it an initial value.
node = tf.Variable(tf.zeros([2,2]))

# Run computation graph
with tf.compat.v1.Session() as sess: # with tf.Session() as sess:
    in TF1

    # Initialize the (global) variable node in the current
    session's scope
    sess.run(tf.compat.v1.global_variables_initializer()) # sess.
    run(tf.global_variables_initializer()) in TF1

    # Evaluate the node
    print("Tensor value before addition:\n",sess.run(node))

    # Assign a new value to a variable node by applying element-
    wise addition to tensor
    node = node.assign(node + tf.ones([2,2]))

    # Evaluate node again
    print("Tensor value after addition:\n", sess.run(node))
```

The following outputs are generated before and after addition:

```
[[ 0. 0.]
[ 0. 0.]]

[[ 1. 1.]
[ 1. 1.]]
```

Placeholders and Comprehensive Example

Placeholder is an essential unit in a graph. However, it is removed in TensorFlow2 due to the introduction of Eager execution. A placeholder is graph unit reserved for accepting external inputs. Usually, as a graph with placeholder nodes is evaluated, a "feed_dict" parameter is passed to the session's run method to specify tensors that provide values for these placeholders. The following is example code for defining and using placeholders.

```
import tensorflow.compat.v1 as tf; tf.disable_v2_behavior() # in
    TF1, use import tensorflow as tf
import numpy as np

x1 = tf.placeholder(tf.float32, shape=(2, 3))
x2 = tf.placeholder(tf.float32, shape=(3, 2))

y = tf.matmul(x1, x2)
```

```
with tf.Session() as sess:
  x1_value = np.random.rand(2, 3)
  x2_value = np.random.rand(3, 2)
  print('The value of x1:\n', x1_value)
  print('The value of x2:\n', x2_value)
  print('y: ', sess.run(y, feed_dict={x1: x1_value, x2: x2_value
    }))
```

Running the above code yields the following results:

```
The value of x1:
 [[0.60484767 0.09801068 0.33437585]
 [0.1758365  0.52327325 0.04377763]]
The value of x2:
 [[0.49392097 0.1164585 ]
 [0.4980184  0.73880872]
 [0.78039387 0.98722004]]
y:   [[0.608503   0.47295338]
 [0.38161284 0.45029467]]
```

In TensorFlow2, the use of placeholders can be replaced by other ways such as functions. For example, the following code with a placeholder,

```
x = tf.placeholder(...)
y = x * x
```

can be replaced with a function:

```
@tf.function
def my_function(x):
  y = x * x
  return y
```

Comprehensive Example

The following example code shows the use of TensorFlow for constructing a linear model, in which the aforementioned graph concepts were used.

Example Code

```
# Import the dependencies
import tensorflow.compat.v1 as tf # in TF1, use import tensorflow
     as tf
import numpy as np
import matplotlib.pyplot as plt

# Model parameters
learning_rate = 0.01
training_epochs = 2000
display_step = 200

# Training data
train_X = np.asarray
    ([3.3,4.4,5.5,6.71,6.93,4.168,9.779,6.182,7.59,2.167,7.042,
```

```
                        10.791,5.313,7.997,5.654,9.27,3.1])
train_y = np.asarray
    ([1.7,2.76,2.09,3.19,1.694,1.573,3.366,2.596,2.53,1.221,2.827,

                        3.465,1.65,2.904,2.42,2.94,1.3])
n_samples = train_X.shape[0]

# Testing data
test_X = np.asarray([6.83, 4.668, 8.9, 7.91, 5.7, 8.7, 3.1, 2.1])
test_y = np.asarray([1.84, 2.273, 3.2, 2.831, 2.92, 3.24, 1.35,
    1.03])

# Set placeholders for the feature and target vectors
X = tf.placeholder(tf.float32)
y = tf.placeholder(tf.float32)

# Set model weights and bias
W = tf.Variable(np.random.randn(), name="weight")
b = tf.Variable(np.random.randn(), name="bias")

# Construct a linear model
linear_model = W*X + b

# Mean squared error
cost = tf.reduce_sum(tf.square(linear_model - y)) / (2*n_samples)

# Gradient descent
optimizer = tf.train.GradientDescentOptimizer(learning_rate).
    minimize(cost)

# Initialize the variables
init = tf.global_variables_initializer()

# Launch the graph
with tf.compat.v1.Session() as sess: # with tf.Session() as sess:
     in TF1
    # Load initialized variables in the current session
    sess.run(init)

    # Fit all training data
    for epoch in range(training_epochs):

        # perform gradient descent step
        sess.run(optimizer, feed_dict={X: train_X, y: train_y})

        # Display logs per epoch step
        if (epoch+1) % display_step == 0:
            c = sess.run(cost, feed_dict={X: train_X, y: train_y
    })
            print("Epoch:{0:6} \t Cost:{1:10.4} \t W:{2:6.4} \t b
    :{3:6.4}".format(epoch+1, c, sess.run(W), sess.run(b)))

    # Print final parameter values
    print("Optimization Finished!")
```

```
training_cost = sess.run(cost, feed_dict={X: train_X, y:
train_y})
print("Final training cost:", training_cost, "W:", sess.run(W
), "b:", sess.run(b), '\n')

# Graphic display
plt.plot(train_X, train_y, 'ro', label='Original data')
plt.plot(train_X, sess.run(W) * train_X + sess.run(b), label=
'Fitted line')
plt.legend()
plt.show()

# Display fitted line on training data
testing_cost = sess.run(tf.reduce_sum(tf.square(linear_model
- y)) / (2 * test_X.shape[0]),feed_dict={X: test_X, y: test_y
})
print("Final testing cost:", testing_cost)
print("Absolute mean square loss difference:", abs(
training_cost - testing_cost))

# Display fitted line on testing data
plt.plot(test_X, test_y, 'bo', label='Testing data')
plt.plot(train_X, sess.run(W) * train_X + sess.run(b), label=
'Fitted line')
plt.legend()
plt.show()
```

The following is a screenshot of the results:

```
Epoch: 200 Cost: 0.1715 W: 0.426 b:-0.4371
Epoch: 400 Cost: 0.1351 W:0.3884 b:-0.1706
Epoch: 600 Cost: 0.1127 W:0.3589 b:0.03849
Epoch: 800 Cost: 0.09894 W:0.3358 b:0.2025
Epoch: 1000 Cost: 0.09047 W:0.3176 b:0.3311
Epoch: 1200 Cost: 0.08526 W:0.3034 b:0.4319
Epoch: 1400 Cost: 0.08205 W:0.2922 b:0.5111
Epoch: 1600 Cost: 0.08008 W:0.2835 b:0.5731
Epoch: 1800 Cost: 0.07887 W:0.2766 b:0.6218
Epoch: 2000 Cost: 0.07812 W:0.2712 b: 0.66
Optimization Finished!
Final training cost: 0.0781221 W: 0.271219 b: 0.65996
Final testing cost: 0.0756337
Absolute mean square loss difference: 0.00248838
```

Code Explanation

Let us try to understand the above code. First, we define some hyperparameters.

```
learning_rate = 0.01
training_epochs = 2000
display_step = 200
```

Then, we define placeholder nodes for the feature and target vectors.

```
X = tf.placeholder(tf.float32)
y = tf.placeholder(tf.float32)
```

Next, we define variable nodes for weight and bias.

```
W = tf.Variable(np.random.randn(), name="weight")
b = tf.Variable(np.random.randn(), name="bias")
linear_model is an operational node that calculates the
    hypothesis for the linear regression model.
linear_model = W*X + b
```

The loss (or cost) per gradient descent is calculated as the mean squared error.

```
cost = tf.reduce_sum(tf.square(linear_model - y)) / (2*n_samples)
```

The optimizer node is defined with the gradient descent algorithm.

```
optimizer = tf.train.GradientDescentOptimizer(learning_rate).
    minimize(cost)
```

The linear model is fit to the training data by performing the optimization. The optimization is repeated training_epochs times.

```
sess.run(optimizer, feed_dict={X: train_X, y: train_y})
```

After every display_step number of epochs, we print the value of the current loss.

```
c = sess.run(cost, feed_dict={X: train_X, y: train_y})
```

The model is evaluated on test data with testing_cost.

```
testing_cost = sess.run(tf.reduce_sum(tf.square(linear_model - y)
    ) / (2 * test_X.shape[0]),feed_dict={X: test_X, y: test_y})
```

2.5.3 *Keras*

Keras is a high-level, user-friendly, modular, and extensible neural network API running on top of TensorFlow. Keras can also be run on both CPU and GPU. In this following, we will go through the basics of Keras, including the two widely used types of Keras models, i.e., sequential and functional, the core layers, and some preprocessing functionalities.

Installation and Data Preparation

Installing Keras

Keras is a central part of TensorFlow 2, though it can also be installed as a separate package. After TensorFlow is installed, we can import Keras easily as.

```
from tensorflow import keras
```

Loading Dataset

Keras provides several well-documented datasets. In the following, we will use the MNIST dataset from Keras, which contains 70,000 28 × 28 grayscale images with ten different classes. Keras splits it into a training set with 60,000 samples and a testing set with 10,000 samples.

```
from tensorflow.keras.datasets import mnist
(x_train, y_train), (x_test, y_test) = mnist.load_data()
```

The image data first needs to be transformed into 4D arrays so that it can be fed to a convolutional neural network. NumPy's reshape method is applied after we transform the data into floats and normalize it.

```
X_train = x_train.astype('float32')
X_test = x_test.astype('float32')
X_train /= 255
X_test /= 255
X_train = X_train.reshape(X_train.shape[0], 28, 28, 1)
X_test = X_test.reshape(X_test.shape[0], 28, 28, 1)
```

The labels need to be transformed via one-hot encoding using the to_categorical method from Keras so that they can be compatible with the selected neural network for classification.

```
from tensorflow.keras.utils import to_categorical

y_train = to_categorical(y_train, 10)
y_test = to_categorical(y_test, 10)
```

Model Establishment with the Sequential API

The sequential API in Keras, which lets us stack one layer on the other, provides a relatively easy way for creating a model. A few constraints of the sequential API are that it does not allow multiple inputs or outputs and limits the possibility of creating complicated structures between different layers. However, it could be a good option for most simple problems. As shown in the following code, we can use the "add" function to add layers for building a deep neural network.

```
from tensorflow.keras.models import Sequential
from tensorflow.keras.layers import Conv2D, MaxPool2D, Dense,
    Flatten, Dropout
model = Sequential()
model.add(Conv2D(filters=32, kernel_size=(5,5), activation='relu'
    , input_shape=X_train.shape[1:]))
model.add(Conv2D(filters=32, kernel_size=(5,5), activation='relu'
    ))
model.add(MaxPool2D(pool_size=(2, 2)))
model.add(Dropout(rate=0.25))
model.add(Conv2D(filters=64, kernel_size=(3, 3), activation='relu
    '))
```

```
model.add(Conv2D(filters=64, kernel_size=(3, 3), activation='relu
    '))
model.add(MaxPool2D(pool_size=(2, 2)))
model.add(Dropout(rate=0.25))
model.add(Flatten())
model.add(Dense(256, activation='relu'))
model.add(Dropout(rate=0.5))
model.add(Dense(10, activation='softmax'))
```

In the above code, we created a sequential object and added convolutional, max-pooling, and dropout layers. The generated data was then flattened and outputted to a dense and dropout layer before it enters the output layer.

The sequential API also supports another syntax where the layers are passed to the constructor directly.

```
from tensorflow.keras.models import Sequential
from tensorflow.keras.layers import Conv2D, MaxPool2D, Dense,
    Flatten, Dropout
model = Sequential([
    Conv2D(filters=32, kernel_size=(5,5), activation='relu',
    input_shape=X_train.shape[1:]),
    Conv2D(filters=32, kernel_size=(5,5), activation='relu'),
    MaxPool2D(pool_size=(2, 2)),
    Dropout(rate=0.25),
    Conv2D(filters=64, kernel_size=(3,3), activation='relu'),
    Conv2D(filters=64, kernel_size=(3,3), activation='relu'),
    MaxPool2D(pool_size=(2, 2)),
    Dropout(rate=0.25),
    Flatten(),
    Dense(256, activation='relu'),
    Dropout(rate=0.5),
    Dense(10, activation='softmax')
])
```

Model Establishment with the Functional API

Alternatively, the functional API allows us to create the same model while offering more flexibility at the cost of simplicity and readability. This API can be used with multiple input and output layers and shared layers, enabling us to build complex network structures.

Model Construction

In the functional API, we need to pass the output of the previous layer to serve as the input of the current layer. This connection mechanism enables the stacking of multiple layers into a deep neural network. Each stacking process starts with the use of an input layer and ends at the output layer.

```
from tensorflow.keras.models import Model
from tensorflow.keras.layers import Conv2D, MaxPool2D, Dense,
    Flatten, Dropout, Input
```

```
inputs = Input(shape=X_train.shape[1:])
x = Conv2D(filters=32, kernel_size=(5,5), activation='relu')(
    inputs)
x = Conv2D(filters=32, kernel_size=(5,5), activation='relu')(x)
x = MaxPool2D(pool_size=(2, 2))(x)
x = Dropout(rate=0.25)(x)
x = Conv2D(filters=64, kernel_size=(3,3), activation='relu')(x)
x = Conv2D(filters=64, kernel_size=(3,3), activation='relu')(x)
x = MaxPool2D(pool_size=(2, 2))(x)
x = Dropout(rate=0.25)(x)
x = Flatten()(x)
x = Dense(256, activation='relu')(x)
x = Dropout(rate=0.5)(x)
predictions = Dense(10, activation='softmax')(x)
model = Model(inputs=inputs, outputs=predictions)
```

Model Compilation

Before we can start training our model, we need to configure the learning process. For this purpose, we need to specify an optimizer, a loss function, and optionally some metrics like accuracy. The loss function is a measure of how well our model achieves the given objective.

An optimizer is used to minimize the loss (objective) function by updating the weights using the gradients.

```
model.compile(
    loss='categorical_crossentropy',
    optimizer='adam',
    metrics=['accuracy']
)
```

Augmentation of Image Data

Augmentation for images can involve minor transformations like rotating, scaling, adding noise, and so on. The augmented data helps address the problem of insufficient data and improve the robustness and generalization of the model. Keras provides a method called ImageDataGenerator for augmenting images.

```
from tensorflow.keras.preprocessing.image import
    ImageDataGenerator
datagen = ImageDataGenerator(
    rotation_range=10,
    zoom_range=0.1,
    width_shift_range=0.1,
    height_shift_range=0.1
)
```

The above code shows the use of rotation, zooming, and shifts from ImageDataGenerator.

Training and Result Visualization

Training

Now, let us train the model that has been defined and compiled. We can easily use the "fit" method to train a model. However, as we use a data generator, we will need to use "fit_generator" and pass parameters to our generator: X data, y data, the number of epochs, and the batch size. The input also includes a validation set and "steps_per_epoch", which is set to the length of the training set divided by "batch_size".

```
epochs = 5
batch_size = 32
history = model.fit_generator(datagen.flow(X_train, y_train,
    batch_size=batch_size), epochs=epochs,
validation_data=(X_test, y_test), steps_per_epoch=X_train.shape
    [0]/batch_size)
```

The execution of the above code generates the following outputs:

```
Epoch 1/5
1875/1875 [==============================] - 31s 16ms/step -
    batch: 937.0000 - size: 32.0000 - loss: 0.3450 - accuracy:
    0.8901 - val_loss: 0.0324 - val_accuracy: 0.9889
Epoch 2/5
1875/1875 [==============================] - 25s 13ms/step -
    batch: 937.0000 - size: 32.0000 - loss: 0.1198 - accuracy:
    0.9643 - val_loss: 0.0302 - val_accuracy: 0.9906
Epoch 3/5
1875/1875 [==============================] - 24s 13ms/step -
    batch: 937.0000 - size: 32.0000 - loss: 0.0933 - accuracy:
    0.9737 - val_loss: 0.0243 - val_accuracy: 0.9924
Epoch 4/5
1875/1875 [==============================] - 31s 16ms/step -
    batch: 937.0000 - size: 32.0000 - loss: 0.0791 - accuracy:
    0.9774 - val_loss: 0.0272 - val_accuracy: 0.9919
Epoch 5/5
1875/1875 [==============================] - 31s 16ms/step -
    batch: 937.0000 - size: 32.0000 - loss: 0.0752 - accuracy:
    0.9779 - val_loss: 0.0173 - val_accuracy: 0.9943
```

Visualizing the Training Process

The performance of the model can be checked by visualizing the accuracy and loss in training and testing for each epoch. The accuracy and loss can be saved in the history variable for later visualization with Matplotlib. The following code plots the loss and accuracy for training and testing in the cross-validation as shown in Fig. 2.7.

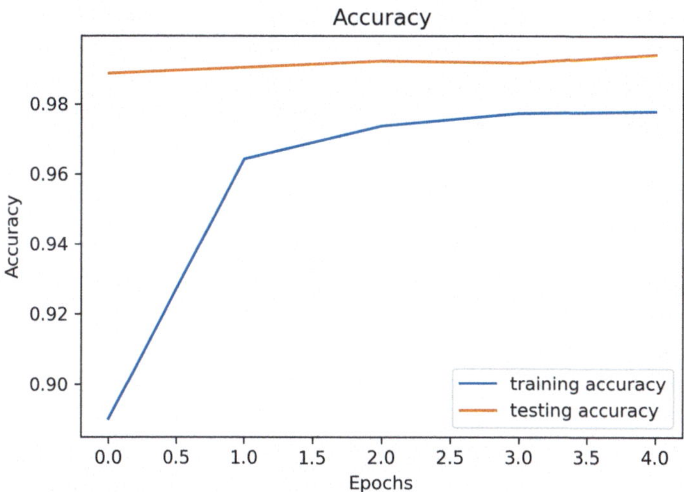

Fig. 2.7 Variation of training and testing accuracy with epochs

```
import matplotlib.pyplot as plt
plt.plot(history.history['accuracy'], label='training accuracy')
plt.plot(history.history['val_accuracy'], label='testing accuracy
    ')
plt.title('Accuracy')
plt.xlabel('Epochs')
plt.ylabel('Accuracy')
plt.legend()
```

2.6 Reinforcement Learning

2.6.1 Overview of RL Tools

Reinforcement learning (RL) frameworks provide higher-level abstractions of the core components of an RL algorithm. Such tools can significantly simplify the development, readability, and efficiency of RL code. Like other machine learning tools, the modularity, ease of use, flexibility, and maturity should be considered. However, simplicity is usually mutually exclusive with modularity and flexibility.

Unfortunately, the subdomain of RL frameworks has been less developed than and less attached to other machine learning topics, possibly because modularity and abstraction of reinforcement learning are much more difficult. So far, there has been no consensus on the use of RL frameworks, while the top-selected list of such tools can evolve dramatically. Besides, sometimes the tools for providing the RL environments and functions for realizing RL algorithms are delivered separately. These facts made the selection of RL tools much different from other machine learning packages.

To provide a rough idea about RL tools, some major frameworks and environment libraries were compiled into the following list according to GitHub and other online trending topics as of 2024:

(1) OpenAI Gym
(2) Google Dopamine
(3) Stable-Baselines (SB3, as fork of and improvement up OpenAI Baselines)
(4) RLLib (Ray)
5) TensorForce
(6) TensorFlow Agents
(7) Garage
(8) Coach by Intel AI Lab
(9) Facebook Horizon
(10) CleanRL
(11) skrl (on top of PyTorch and JAX)

This chapter introduces a most widely accepted RL tool, OpenAI Gym, which provides environments instead of RL algorithms. As for algorithms, tools such as Google Dopamine, RLLib, and Stable-Baselines can be considered.

2.6.2 OpenAI Gym

According to the OpenAI Gym GitHub repository, "OpenAI Gym is a toolkit for developing and comparing reinforcement learning algorithms..., which gives you access to a standardized set of environments."

Open AI Gym has an environment-agent arrangement. It simply means Gym gives you access to an "agent" that can perform specific actions in an "environment." In return, the agent gets observations and rewards as a consequence of performing a particular action in the environment.

There are four values returned by the environment for every "step" taken by the agent.

(1) Observation (object): an environment-specific object representing your observation of the environment, for example, board state in a board game.
(2) Reward (float): the amount of reward/score achieved by the previous action. The scale varies between environments, but the goal is always to increase your total reward/score.
(3) Done (boolean): whether it is time to reset the environment, for example, when the player dies in the game.
(4) Info (dict): diagnostic information useful for debugging.

The following are the available types of environments in Gym:

- Classic control and toy text
- Algorithmic

- Atari
- 2D and 3D robots

Let us see what the agent-environment loop looks like in Gym. The following example runs an instance of LunarLander-v2 environment for 1000 time steps. Since we pass render_mode="human", you should see that a window pops up rendering the environment.

```
import gym
env = gym.make("LunarLander-v2", render_mode="human")
env.action_space.seed(42)

observation, info = env.reset(seed=42)

for _ in range(1000):
    observation, reward, terminated, truncated, info = env.step(
    env.action_space.sample())

    if terminated or truncated:
        observation, info = env.reset()

env.close()
```

Every environment specifies the format of valid actions by providing an "env.action_space" attribute. Similarly, the format of valid observations is specified by "env.observation_space". In the above example, we sampled random actions via env.action_space.sample(). Note that we need to seed the action space separately from the environment to ensure reproducible samples.

2.7 Practice: Use, Compare, and Understand TensorFlow and Keras for Problem-Solving

>>> More and Up-to-Date Course Materials including Practices @ AI-engineer.org <<<

Try to use TensorFlow and Keras to solve one problem. You can try to find and copy code from Internet sources such as https://stackoverflow.com/questions/57273888/keras-vs-tensorflow-code-comparison-sources. Please use Chap. 2 in the textbook to help you understand the code. It is okay to skip some details, as long as you understand what each line of code does for you. You will need to run the code, try to get the expected results, compare the two software packages, and draw some conclusions based on your testing and observations.

Chapter 3
Linear Models

3.1 Overview

This chapter introduces one of the simplest yet very useful types of machine learning algorithms: linear models. First, a theory for the basic linear model algorithm will be laid down. Then, the employment of regularization to generate other common types of linear model algorithms, e.g., Ridge and Lasso, will be explained. Linear models are, by nature, proposed for regression. The extension of such algorithms to classification, including both binary classification and multiclass classification, will be introduced next. Finally, we will show the use of kernel functions to turn linear models into nonlinear methods for regression and classification.

It is worthwhile to clarify some common confusing points before we start the journey.

First, "linear model" or "linear models" here refer to the algorithm(s) that helps us obtain the actual models. As explained in previous chapters, algorithms are methods to obtain models, while models are algorithms that have been instantiated or fixed with model parameters, which are determined by specific datasets during training. In a simple way, an algorithm can be analogous to the use of a polynomial function to fit a dataset, while a model is the polynomial with parameters fixed in this curve fitting process. But it is kind of a convention to call such algorithms "linear model" or "linear models" in machine learning. This convention is still kept here to stay aligned with most machine learning literature.

Second, in a narrow sense, linear model as a machine learning topic contains only those linear model algorithms in which the independent variables (representing the data attributes), e.g., $x_i, x_2, \cdots, x_i, \cdots$ in $\vec{x}$, maintain a linear relationship. That is, the dependent variable (representing the label) y is a linear combination of these independent variables x_i. However, it is quite common to adopt kernel functions to give linear models nonlinear capacities. Thus, the linear models with the use of kernel functions are still introduced as part of the linear model topic in a broad sense in this book.

Z. "L." Liu, *Artificial Intelligence for Engineers*,
https://doi.org/10.1007/978-3-031-75953-6_3

Third, there are big overlaps between linear model, linear regression, and curve fitting with linear functions, and they may even refer to the same thing in some cases. This confusion is caused by the fact that these terms are proposed and used by people from different disciplines, e.g., computer science, math, and engineering. In this book, the use of linear models is mostly inherited from machine learning and data analysis; thus, it focuses on the use of linear models for addressing regression and classification problems. In this context, an analytical solution can be obtained directly due to the intentional confinement to simple linear models. By contrast, relevant topics like curve fitting (especially nonlinear curve learning), which may require the use of complicated math functions and optimization techniques to iteratively obtain the optimal solution, will not be discussed here.

3.2 Basics of Linear Models

3.2.1 Simple Explanation of Linear Models

The idea of linear models is to use a linear mathematical function of one or more variables to approach some given function value(s) of these variable(s). In machine learning, such a function value correspond to the label of a sample, while such variables correspond to different attributes of the sample. Frequently, in the context of math functions, independent variables and dependent variables are used to refer to the sample attributes and labels, respectively. In some cases, especially when the context of statistics and probability is adopted, it is also common to see the use of predictors and response variables.

Let us first outline a typical problem for linear models using the math language. In the simple case, let us only consider one variable x. Then, the function $f(x)$ can be formulated as

$$f(x) = w \cdot x + b \tag{3.1}$$

where $[x, y]$ represents one data point in a space consisting of 2 axes (or dimensions) and $\cdot$ is the (arithmetic) product between two scalars.

In a regression problem, there are usually multiple data points in this space, e.g., I points $[(x_1, y_1), (x_2, y_2), \cdots, (x_I, y_I)]$, which are the "measurements" that need to be approximated with Eq. 3.1. The goal of using a linear model for regression is to find the function $f(x) = w \cdot x + b$ closest to all the measurements. The closest curve can be associated with the smallest total distance between the measured values y_i and predicted values $f(x_i)$. Usually, the Euclidean distance is selected as the measure of distance in regression tasks. Under this condition, a regression task using a linear model can be mathematically described as

$$(w^*, b^*) = \arg\min_{(w,b)} \sum_{i=1}^{I} (f(x_i) - y_i)^2 = \arg\min_{(w,b)} \sum_{i=1}^{I} (w \cdot x_i + b - y_i)^2 \tag{3.2}$$

where w^* and b^* are the optimal model parameters w and b, and $\arg\min_{(w,b)}$ means to minimize a function parameterized by w and b. This equation presents a formulation of linear regression with the ordinary least squares.

3.2.2 General Formulation of Basic Linear Model

The data in general regression applications can be much more complicated than that in the above simple case. In particular, each data sample may contain more than one attribute, which can be viewed as multiple random variables from a statistics perspective. We can understand the typical data that we will deal with in linear model applications using a data structure that is compatible with those adopted in tools like NumPy and Pandas.

As shown in Fig. 3.1, the unlabeled data $\bar{X}$ has I samples (rows), and each sample $\vec{x}_i^T$ consists of J attributes (columns). In fact, any sample in its original shape, $\vec{x}_i$, is essentially a column vector as follows:

$$\vec{x}_i = \begin{bmatrix} x_{i1} \\ x_{i2} \\ \vdots \\ x_{iJ} \end{bmatrix} \tag{3.3}$$

and all the unlabeled data form $\bar{X} = [\vec{x}_1{}^T, \vec{x}_2{}^T, \cdots, \vec{x}_I{}^T]^T$ (or written as $\bar{X} = [\vec{x}_1{}^T; \vec{x}_2{}^T; \cdots; \vec{x}_I{}^T]$). Note that each sample i in Fig. 3.1 is $\vec{x}_i{}^T$ (a row) instead of $\vec{x}_i$ (a column). This data structure and tensor notation (a hybrid notation, explained in the book Appendices) are adopted throughout this book.

For any sample, $\vec{x}_i$, the linear function of the variables, $\vec{x}_{ij}$ ($j \in \{1, \cdots, J\}$), can be formulated as the following linear combination:

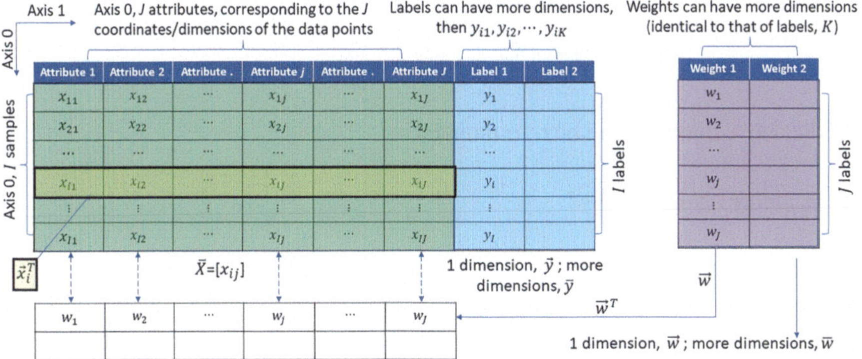

Fig. 3.1 Illustration of general data structure

$$f(\vec{x}_i) = \vec{w}^T \cdot \vec{x}_i + b = \vec{x}_i^T \cdot \vec{w} + b = w_1 \cdot x_{i1} + w_2 \cdot x_{i2} + \cdots + w_J \cdot x_{iJ} + b$$
$$= \begin{bmatrix} w_1 \; w_2 \cdots w_J \end{bmatrix} \begin{bmatrix} x_{i1} \\ x_{i2} \\ \vdots \\ x_{iJ} \end{bmatrix} + b \tag{3.4}$$

where $[x_{i1}, x_{i2}, \cdots, x_{iJ}, y_i]^T$ represents one labeled data point in a space consisting of $J + 1$ axes, in which J axes represent the J variables, while the remaining axis represents the label. In some places, people consider the label separately and thus state the data has J dimensions. $f(\vec{x}_i)$ is still used to map the input $\bar{X}$ to the output $\vec{y}$. The meaning of the "·" sign changes from the multiplication in arithmetic to dot product in tensor algebra as the variables change from scalars x_i's to vectors $\vec{x}_i$'s.

In a regression problem, usually, multiple points, e.g., I points in this case, are available as "measurements." The goal of using a linear model for regression is to find a function (e.g., a curve in 2D), $f(\vec{x}) = \vec{w}^T \cdot \vec{x} + b$, that is the closest to all the measurements. Again, the closest curve is usually associated with the smallest total distance between the measurement points and the curve, which is usually formulated using the Euclidean distance. Under this condition, the regression task using a linear model can be mathematically described as

$$(\vec{w}^*, b^*) = \arg \min_{(\vec{w},b)} \sum_{i=1}^{I} (f(\vec{x}_i) - y_i)^2 = \arg \min_{(\vec{w}^T,b)} \sum_{i=1}^{I} (\vec{w}^T \cdot \vec{x}_i + b - \vec{y}_i)^2 \tag{3.5}$$

The linear model can be written using arrays as

$$f(\bar{X}) = X \cdot \vec{w} + b$$
$$= \begin{bmatrix} x_{11} \; x_{12} \cdots x_{1J} \\ \vdots \quad \ddots \quad \vdots \\ x_{Ii} \; x_{I2} \cdots x_{IJ} \end{bmatrix}_{I \times J} \begin{bmatrix} w_1 \\ w_2 \\ \vdots \\ w_J \end{bmatrix}_{J \times 1} + b \tag{3.6}$$

The above equation can be reformulated as

$$f(\hat{X}) = \hat{X}\hat{w}$$
$$= \begin{bmatrix} x_{11} \; x_{12} \cdots x_{1J} & 1 \\ \vdots \quad \ddots \quad \vdots & 1 \\ x_{I1} \; x_{I2} \cdots x_{IJ} & 1 \end{bmatrix}_{I \times (J+1)} \begin{bmatrix} w_1 \\ w_2 \\ \vdots \\ w_J \\ b \end{bmatrix}_{(J+1) \times 1} \tag{3.7}$$

where

$$\hat{X}_{(J+1)\times I} = \left[\hat{X}, [1; 1 \ldots; 1]^T\right]$$

$$\hat{w}_{(J+1)\times 1} = \left[\vec{w}; b\right]$$

The loss function (cost function or objective function) can be written as

$$\ell(\hat{w}) = \|f(\hat{X}) - \vec{y}\|^2 = (\hat{X} \cdot \hat{w} - \vec{y})^T \cdot (\hat{X} \cdot \hat{w} - \vec{y}) \tag{3.8}$$

The use of the sign "·" can be confusing: this sign can refer to either scalar product (regular multiplication between scalars) or dot product. As introduced in the previous chapters, concepts like vectors and tensors are widely used in engineering, and sometimes, the tensor notation can be more convenient than the index notation, which is widely used in traditional machine learning. For example, algorithms formulated using the tensor notation can be more easily expressed using the tensor notation without worrying about the operations between tensors if the operators are clearly defined. Also, tensors can be more straightforwardly related to the arrays used in coding and computing. Thus, the tensor notation is adopted with minor modifications in most places of this book, including the majority of this chapter. More information about notation and operations can be found in Chap. 1 and the appendices at the end of the book.

To facilitate the use of tensors and the tensor notation, it is clarified that scalars are viewed as 0th order tensors, vectors is 1st-order tensors, while 2nd- and higher-order tensors, which are called tensors in a narrow sense, are termed tensors or high-order tensors. Also, as for operators or operation signs, in this book, when "·" is used between a scalar and any other tensors (scalars, vectors or high-order tensors), it is used for scalar product: the product between the scalar and the other scalar or elements in the tensors. When it is used between a vector and another 1st- or higher-order tensors, it refers to a dot (vector) product or equivalently a tensor product with a single contraction. In Python coding, "*" is used for the former type of product (scalar product), while "@" is used for the latter (dot product). Following this rule will make the implementation of the theories in this book much simpler than what can be seen in other literature.

Then, the mathematical description of the regression problem using a linear model is

$$\hat{w}^* = \arg\min_{(w,b)} (\hat{X} \cdot \hat{w} - \vec{y})^T \cdot (\hat{X} \cdot \hat{w} - \vec{y}) \tag{3.9}$$

The following closed-form solution to the above equation can be obtained:

$$\hat{w}^*_{(J+1)\times 1} = \left(\hat{X}^T_{(J+1)\times I} \cdot \hat{X}_{I\times(J+1)}\right)^{-1} \cdot \hat{X}^T_{(J+1)\times I} \cdot \vec{y}_{I\times 1} \tag{3.10}$$

Python code is provided in the following to exhibit the construction of a linear model based on the above theory. This model was created with a dataset consisting of eight data points, each of which has two attributes and one label. These points are linearly related: distributed on a plane in a 3D space. Random noise was added to make the data more like measurements with minor errors.

```
import numpy as np
import matplotlib.pyplot as plt
from mpl_toolkits import mplot3d

# Create data
X_list = [[1.,1.],[1.,2.],[2.,2.],[2.,3.],[1.5,2.5],
[2.,4.],[1.,3.],[3.,1.5]] # Create unlabeled data point/instance
using a Python list
X = np.array(X_list) # Turn instance into a NumPy array
Proj = np.array([1,2]) # Create perfect labels
y = np.dot(X,Proj)+3  # Add intercept
y = y + np.random.rand(y.size) # Add noise to mimic experimental
    measurements

# Define a function to perform analytical solution to linear
    model problems
def LinearRegression(X,y):
    X_bar = np.hstack((X,np.ones([X.shape[0],1]))) # I = 8 by J+1
      = 3
    W_hat = np.linalg.inv(X_bar.T @ X_bar) @ X_bar.T @ y # J+1 =
    3 by 1
    return W_hat, X_bar

# Symbols are the same as those in the textbook
W_hat, X_bar= LinearRegression(X,y)
W = W_hat[:-2]
b = W_hat[-2:]

# Solution
y_fit = np.dot(X_bar,W_hat)

# Plot and compare results
fig = plt.figure()
ax = plt.axes(projection='3d')
ax.scatter(X[:,0],X[:,1],y,'b*')
ax.scatter(X[:,0],X[:,1],y_fit,'rd')
```

The original data points and predictions made by the trained model are plotted in Fig. 3.2.

3.3 Other Linear Regression Algorithms

When the number of samples (I) is relatively small and the number of features (J) is relatively large, basic linear models such as the above one with the ordinary least squares are susceptible to overfitting issues. To avoid or prevent overfitting, we can resort to several methods. One typical method is to reduce the number of features.

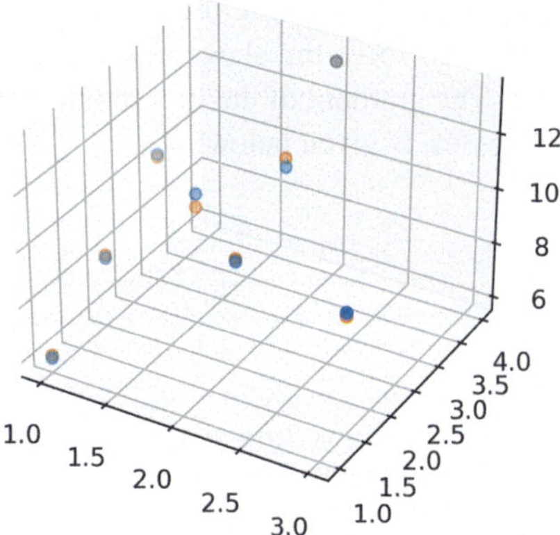

Fig. 3.2 Training result of a linear model

For this purpose, significant features will be retained while some insignificant ones will be abandoned. Another popular method for avoiding overfitting in linear models is regularization. The goal of regularization is to obtain models with both high accuracy and low complexity. The addition of a regularization term to the above basic linear model can generate some other popular linear models such as Ridge, Lasso, and elastic net regression models. In the following, information for Ridge and Lasso linear models will be presented.

3.3.1 Ridge

The Ridge regression is a linear model with L_2 norm regularization [43]. With a regularization term, the loss function of Ridge is usually constructed as follows:

$$\ell(\hat{w}) = (\hat{X} \cdot \hat{w} - \vec{y})^T \cdot (\hat{X} \cdot \hat{w} - \vec{y}) + \lambda \hat{w}^T \cdot \hat{w} \tag{3.11}$$

Then, the Ridge regression problem can be formulated as

$$\hat{w}^* = \arg \min_{(\hat{w})} \left[(\hat{X} \cdot \hat{w} - \vec{y})^T \cdot (\hat{X} \cdot \hat{w} - \vec{y}) + \lambda \sum_{i=1}^{J} \hat{w}_i^2 \right] \tag{3.12}$$

where $\lambda \hat{w}^T \cdot \hat{w} = \lambda \sum_{i=1}^{J} \hat{w}_i^2$.

The closed-form solution to the above Ridge regression problem can be derived as

$$\hat{w}^* = (\hat{X}^T \cdot \hat{X} + \lambda \bar{I})^{-1} \cdot \hat{X}^T \cdot \vec{y} \tag{3.13}$$

where $\bar{I}$ is a identify matrix with dimensions of $(J + 1)$ by $(J + 1)$.

In addition to the closed-form solution, we can also use the gradient descent method to solve the above optimization problem associated with the Rigid regression. The gradient of the loss function, which is a key in such optimization solution processes, is given below:

$$\frac{\partial \ell(\hat{w})}{\partial w_j} = \frac{1}{I}\sum_{i=1}^{I}(\hat{w}^T x_i - y_i)x_{ij} + 2\lambda\hat{w}_j \tag{3.14}$$

where j is the jth element in the weight array and i is the ith sample.

With the above loss gradient, the equation for updating the coefficient $\hat{w}$ can be formulated as follows:

$$\begin{aligned} \hat{w}_j &= \hat{w}_j - \frac{\alpha}{I}\sum_{i=1}^{I}(\hat{w}^T x_i - y_i)x_{ij} + 2\lambda\hat{w}_j \\ &= (1-2\lambda)\hat{w}_j - \frac{\alpha}{I}\sum_{i=1}^{I}(\hat{w}^T x_i - y_i)x_{ij} \end{aligned} \tag{3.15}$$

where α is the learning rate.

3.3.2 *Lasso*

The Lasso regression is the linear model with L_1 norm regularization [44]. Thus, it uses the absolute value, leading to a loss function that is neither continuous nor differentiable. That is, the gradient descent and Newton's method do not work well. Meanwhile, a closed-form solution is not available. To use Lasso regression, we can apply solution techniques like the proximal gradient descent (PGD) and Taylor's expansion to help address the problem.

The loss function (cost function or objective function) of Lasso can be written as

$$\ell(\hat{w}) = (\hat{X}\cdot\hat{w} - \vec{y})^T \cdot (\hat{X}\cdot\hat{w} - \vec{y}) + \lambda\sum_{j=1}^{J}\|w_j\| \tag{3.16}$$

Accordingly, Lasso can be mathematically formulated as

$$\hat{w}^* = \underset{(\hat{w})}{\arg\min}\left[(\hat{X}\cdot\hat{w} - \vec{y})^T \cdot (\hat{X}\cdot\hat{w} - \vec{y}) + \lambda\sum_{j=1}^{J}\|w_j\|\right] \tag{3.17}$$

To illustrate the implementation of a more complicate linear model, a Lasso model was trained using an optimizer (or solver) from the SciPy package. We can still use the problem given at the end of Sect. 3.2.2. The following code shows the key part for building and training this Lasso model.

```
def LinearRegression(X,y,alpha):
    X_bar = np.hstack((X,np.ones([X.shape[0],1])))
    W_hat = Optimize(X_bar,y,alpha)
    return W_hat, X_bar

def Optimize(X_bar,y,alpha):
    from scipy.optimize import minimize
    e = 1e-10 # Tolerance
    fun = lambda W_hat : np.sum( (np.dot(X_bar,W_hat) - y)**2 ) +
     alpha*np.sum(np.absolute(W_hat[:-2]))  # L1 regularization
    is added
    W_hat0 = np.ones(X_bar.shape[1]) # Set initial values for
    optimization
    res = minimize(fun, W_hat0, method='SLSQP')
    W_hat = res.x
    return W_hat

# Training
W_hat, X_bar= LinearRegression(X,y,alpha)
W = W_hat[:-2]
b = W_hat[-2:]

y_fit = np.dot(X_bar,W_hat) # Prediction
```

3.4 Logistic Regression for Classification

3.4.1 Binary Classification

The goal of classification is to map the input $\bar{X}$ to a set of discrete numbers or symbols for labeling different categories instead of a range of continuous numbers. In order to extend linear models to classification tasks, surrogate functions are usually introduced to map continuous function values to discrete labels for different categories [45]. Such surrogate functions, which are usually continuous and differentiable rather than unit-step functions, are selected to facilitate mathematical operations. A typical example of such surrogate functions is the logistic function:

$$z = \frac{1}{1 + e^{-y}} \tag{3.18}$$

where $y = \vec{w}^T \cdot \vec{x} + b$.

As plotted in Fig. 3.1, the logistic function converts or squeezes the input (abscissa) from $[\infty, -\infty]$ to $[0, 1]$. Interestingly, the range of $[0, 1]$ is exactly what we use for describing probabilities (Fig. 3.3).

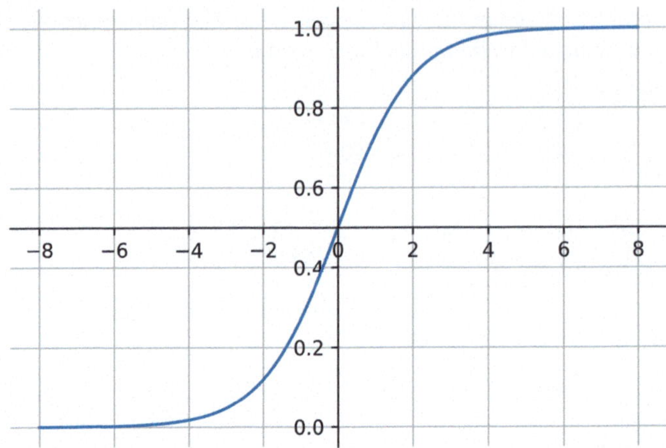

Fig. 3.3 Plot of logistic function

Equation 3.18 can be reformulated to reach a very interesting finding:

$$\frac{z}{1-z} = e^{y} \tag{3.19}$$

In this equation, z can be understood as the probability that a sample belongs to one category. In a binary classification problem with two categories, i.e., Category 1 and Category 0, the probability of a sample belonging to Category 1 is $P_1 = z$, while the probability that this sample belongs to the other category is $P_0 = 1 - z$. Therefore, the exponential of the function value, e^{y}, as the output of the regression problem, represents the ratio between the probabilities of belonging to the two categories in a binary classification task. We typically label the two categories in a binary classification problem as 1 and 0. Then, the above probabilities are formulated as follows:

$$P_1 = P(\tilde{y} = 1|\hat{X}, \hat{w}) = z \tag{3.20a}$$

$$P_2 = P(\tilde{y} = 0|\hat{X}, \hat{w}) = 1 - z \tag{3.20b}$$

where $\tilde{y}$ is the predicted label for the sample. When a linear model is obtained, i.e., with fixed $\vec{w}$ and b, we can calculate z and $1 - z$ for any sample i as its probabilities of belonging to the two categories. Then, the category with the greater probability value can be determined to be the label. This hypothesis can be mathematically formulated as follows:

$$\tilde{y}_i = 1, \text{ if } P_1 \geqslant 0.5 \tag{3.21a}$$

$$\tilde{y}_i = 0, \text{ if } P_1 < 0.5 \tag{3.21b}$$

Thus, the above hypothesis is equivalent to the selection of a threshold value of 0.5 for the z value when determining the category (label). In a more general case, any threshold values can be selected, though 0.5 is widely used without being explained in many publications. $\tilde{y}_i$ is the predicted label determined with Eq. 3.21 in the prediction stage, and y_i is the actual label coming from the labeled data in the training stage.

A loss function needs to be constructed to quantify the accuracy of the binary classification task to replace the loss in the regression task. The cross entropy is usually recalled because it measures the difference between two different distributions. Being adopted here, it can measure the distance or difference between the actual label, e.g., y_i for the ith sample, and the prediction, which can be represented using the z_i value (corresponding to P_1 and P_2) for this sample. In this way, the cross entropy formulates the overall likelihood that the samples can be classified correctly in a binary classification problem. The loss function constructed with the cross entropy is as follows:

$$\ell(\hat{X}, \hat{w}) = -\sum_{i=1}^{I} \left[y_i \cdot \ln P_1(\tilde{y}_i) + (1 - y_i) \ln P_0(\tilde{y}_i) \right] \tag{3.22}$$

The above equation has values from 0 to infinity, and a low value indicates a higher likelihood. As can be seen, the cross entropy function considers the probabilities of one sample belonging to both categories when assessing the distance/difference.

Therefore, the binary classification can be mathematically described as

$$\hat{w}^* = \arg\min_{(\hat{w})} \left[-\sum_{i=1}^{I} \left[y_i \cdot \ln P_1(\tilde{y}_i) + (1 - y_i) \ln P_0(\tilde{y}_i) \right] \right] \tag{3.23}$$

The following code demonstrates the implementation of the above theory for logistic regression. Here, the Iris dataset was used. One hot labeling was performed first to change the label format. The loss function as the key component was embedded in the solver in this implementation.

```
from sklearn import datasets
import numpy as np
np.seterr(divide='ignore', invalid='ignore')
from sklearn.datasets import load_iris
X, y_label = load_iris(return_X_y = True)

# Convert label array to the onehot format
y_onehot = np.zeros((y_label.size, y_label.max()+1))
```

```
y_onehot[np.arange(y_label.size),y_label] = 1
y=y_onehot

def LogisitcRegression(X,y):
    X_bar = np.hstack((X,np.ones([X.shape[0],1]))) # I = 8 by J =
     3
    W_hat = Optimize(X_bar,y)
    return W_hat, X_bar

def Optimize(X_bar,y):
    from scipy.optimize import minimize
    e = 1e-10 # Tolerance
    # Logistic regression cost function
    fun = lambda W_hat : - np.sum( y * np.log( np.exp(
    X_bar@W_hat.reshape(X_bar.shape[1],y.shape[1]) ) / np.sum( np
    .exp(  X_bar@W_hat.reshape(X_bar.shape[1],y.shape[1]) ) ) ) )
    W_hat0 = np.ones(X_bar.shape[1]*y.shape[1]) # Set initial
    values
    res = minimize(fun, W_hat0, method='SLSQP', tol=e)
    W_hat = res.x
    return W_hat

W_hat, X_bar= LogisitcRegression(X,y)
W = W_hat[:-2]
b = W_hat[-2:]

y_fit = np.dot(X_bar,W_hat.reshape(X_bar.shape[1],y.shape[1]))
```

3.4.2 *Multiclass Classification*

There are more than two categories in multiclass classifications. An intuitive way to apply the logistic regression for multiclass classification tasks is to extend the method introduced in the previous subsection for binary classification by dividing a multiclass classification problem into multiple binary classification problems. However, a more widely accepted way is to replace the logistic function with a Softmax function. In a multiclass classification problem, any sample, e.g., sample i, can have K output values generated by the Softmax function $y_{i,k}$, in which $k \in [1, K]$). $y_{i,k}$ determines the likelihood that sample i belongs to a specific category k as follows:

$$P_{i,k} = P(\tilde{y}_{i,k}|\tilde{X}, \hat{w}) = \frac{e^{\tilde{y}_{i,k}}}{\sum_{k=1}^{K} e^{\tilde{y}_{i,k}}} \tag{3.24}$$

Similar to the binary classification, we can compare the probabilities of one sample belonging to different categories and believe the category with the greatest probability is the category this sample belongs to (predicted label). This prompts the wide use of one-hot labels for multiclass classification tasks: for each sample, the predicted y value is 1 only for the category (k) that the sample belongs to in the prediction, i.e., $\tilde{y}_{i,k} = 1$, while all the other $\tilde{y}_{i,*}$ values are 0, i.e., $\tilde{y}_{i,!k} = 0$.

$$\tilde{y} = [0, 0, \cdots, \underbrace{1}_{k\text{th}}, 0, \cdots] \text{ when } P_{i,k} \text{ is the greatest one among } [P_{i,k} \cdots P_{i,K}] \tag{3.25}$$

Accordingly, the loss function can be constructed as

$$\ell(\hat{X}, \hat{w}) = -\sum_{i=1}^{I} \left(\tilde{y}_{i,k} \cdot P_{i,k}\right) = -\sum_{i=1}^{I} \left[\tilde{y}_{i,k} \cdot \frac{e^{\tilde{y}_{i,k}}}{\sum_{k=1}^{K} e^{\tilde{y}_{i,k}}}\right] \tag{3.26}$$

Therefore, for each sample, K function values need to be generated for the K categories. But in binary classification and regression tasks, there is only one function value, z, for each sample (if $1 - z$ does not count as an independent one). To solve the problem, the weight array, $\hat{w}$, needs to be expanded from a $(J + 1) \times 1$ array to a $(J + 1) \times K$ array.

3.5 Making Linear Models Nonlinear via Kernel Functions

A linear model can be turned into a nonlinear one by means of the kernel trick: replacing its features (predictors) with a kernel function [46, 47]. This kernel trick is described as the kernel method or (the use of) kernel functions in many places. In fact, the kernel kick can also be applied to other machine learning methods, such as the Support Vector Machine (SVM) and dimension reduction, to generate nonlinearity capability.

In this section, we will first explain the concept of mapping data from a low-dimensional space to a high-dimensional space. In this way, data that appears non-linear in the lower-dimensional space can become linear in the higher-dimensional space if an appropriate mapping is selected. Then, the methods for linear problems, such as the linear models introduced above, can be applied to the data in the high-dimensional space. In the following, we will first demonstrate how to map data to a higher-dimensional space via stretching (or stretch) functions, which are easy to understand but computationally expensive. Based on that, kernel functions, which are more difficult to understand but computationally efficient, will be introduced. The application of linear models with kernels for the solution of nonlinear problems will be presented last. The use of kernel functions will also be mentioned in later chapters such as the chapter for SVM.

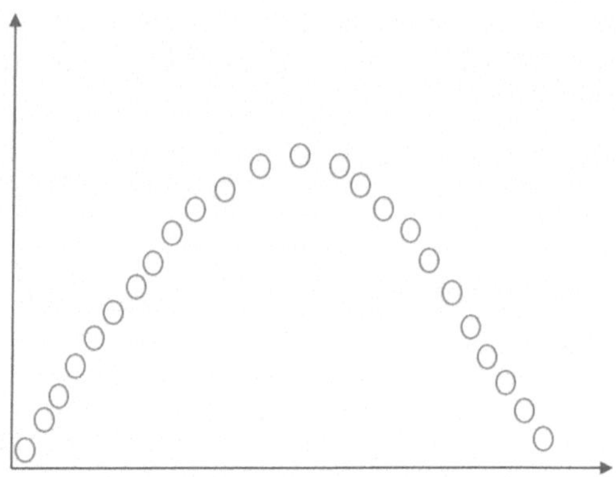

Fig. 3.4 Example of nonlinear data

3.5.1 Mapping Data to Higher-Dimensional Space with Stretching Functions

Let us start with a problem that cannot be solved with the linear models shown in Fig. 3.4. It is not difficult to find out that we cannot get very satisfactory results if we want to fit the data with a linear model, i.e., a straight line. The reason is that the data points are not linearly distributed: the data does not form a straight line in the 2D space. Instead, the data appears to follow a polynomial function.

Let us try a second-order polynomial function:

$$f(x) = w_1 \cdot x + w_2 \cdot x^2 + b \tag{3.27}$$

A key in the mapping of the data to a higher-dimensional space is to reformulate Eq. 3.27 into a form of the linear model:

$$f(x) = \vec{x}^T w + b = \begin{bmatrix} x \; x^2 \end{bmatrix} \begin{bmatrix} w_1 \\ w_2 \end{bmatrix} + b \tag{3.28}$$

In this way, the second-order polynomial function is connected to a linear (first-order) function, and a linear regression form is obtained. However, this is a linear model of variables $[x, x^2]^T$, which are in a space of higher dimensions than that of the original variable x. Let us use $\vec{t}$ to represent this variable in the higher dimension. Then we have

$$\vec{t} = \begin{bmatrix} t_1 \\ t_2 \end{bmatrix} = \begin{bmatrix} x \\ x^2 \end{bmatrix} \tag{3.29}$$

In this example, x is 1D and $[x, x^2]^T$ is 2D. Thus, what we did in the above process is to “stretch” the data from a lower-dimension space to a higher-dimension one. We use a function ϕ to represent the above stretching operation: $\phi(\vec{x}) = \vec{t}$, which has J' axes/dimensions ($J' > J$). Let us consider a general case in which every data point is a vector or a 1D array with J numbers (attributes) corresponding to the J axes in the space. In this case, the above stretching function will stretch the original data vector as follows:

$$\vec{x}_i = \begin{bmatrix} x_{i1} \\ x_{i2} \\ \vdots \\ x_{iJ} \end{bmatrix} \overset{\phi(\vec{x})}{\longrightarrow} \vec{t}_i = \begin{bmatrix} x_{i1} \\ x_{i2} \\ \vdots \\ x_{iJ} \\ x_{i1}^2 \\ x_{i2}^2 \\ \vdots \\ x_{iJ}^2 \end{bmatrix} \tag{3.30}$$

An explicit effect of using such stretching functions is to increase the dimensionality of a vector. However, the use of such functions does more than just mapping data from a lower-dimensional space to a higher-dimensional space. To understand this statement, we can apply a linear combination of the original data $\vec{x} = [x_1, x_2]^T$ as

$$\phi(\vec{x}) = \begin{bmatrix} 1 & 1 \\ 2 & -1 \\ 3 & 2 \end{bmatrix} \vec{x} = \begin{bmatrix} x_1 + x_2 \\ 2x_1 - x_2 \\ 3x_1 - 2x_2 \end{bmatrix} \tag{3.31}$$

As can be seen, stretching the data in this way does increase the dimension from 2 to 3. However, it does not change the nature of the problem (or data), i.e., from linear to nonlinear. The stretched data in the higher-dimensional space is just a linear combination of the data in the lower-dimensional space. Thus, if the original data is linear, the data obtained via this mapping—as a combination of linear data—is thus still linear. Also, if the original data is nonlinear, the data obtained via this mapping is very likely still nonlinear. Therefore, stretching functions valid for our purpose need to change the nonlinearity status in addition to altering the dimensionality of the data.

To sum up, a nonlinear problem of data in a low-dimensional space can possibly be converted to a linear problem by appropriately mapping the data to a high-dimensional space. For simplicity, we can reformulate the above equations (e.g., Eq. 3.28) as what we did in the linear model section by incorporating b into $\vec{w}$, so we will have $\hat{X}$ (consisting of $\hat{x}'s$) and $\hat{w}$. Then, the linear problem in the higher-dimensional space can be formulated using the linear model as follows:

$$f(\hat{x}) = \phi(\hat{x})^T \hat{w} \tag{3.32}$$

where the dimensions of $\phi(\hat{x})$ and $\hat{w}$ correspond to the dimensions of the data in the higher-dimensional space, i.e., $J' + 1$, rather than those in the lower-dimensional problem, i.e., $J + 1$.

Next, we can recall the following solution to the linear model:

$$\hat{w}^* = (\hat{X}\hat{X}^T + \lambda\bar{I})^{-1}\hat{X}\vec{y} \tag{3.33}$$

There is still one more thing to do before we can reformulate this equation for the linear problem in the higher-dimensional space. That is, we will need to relate $\hat{x}$ and $\phi(x)$ for one data point to $\hat{X}$ and $\hat{\Phi}$ for all the data points.

$$\begin{aligned} \hat{X}^T &= \begin{bmatrix} x_{11} & x_{21} & \cdots & x_{I1} \\ \vdots & \ddots & \vdots \\ x_{1J} & x_{2J} & \cdots & x_{IJ} \\ 1 & 1 & \cdots & 1 \end{bmatrix}_{(J+1)\times I} \\ &= \begin{bmatrix} \hat{x}_1 \; \hat{x}_2 \cdots \hat{x}_m \end{bmatrix}_{(J+1)\times I} \end{aligned} \tag{3.34}$$

All the data in the high-dimensional space is

$$\bar{\Phi}^T = \begin{bmatrix} \phi(\hat{x}_1) \; \phi(\hat{x}_2) \cdots \phi(\hat{x}_I) \end{bmatrix}_{(J'+1)\times I} \tag{3.35}$$

where J' in the above equation depends on how the stretching equation will stretch every data point: from J variables (axes) to J' variables (axes). Then, we can replace $\hat{X}$ with $\bar{\Phi}$

$$\hat{w}^* = (\bar{\Phi}^T \cdot \bar{\Phi} + \lambda\bar{I})^{-1} \cdot \bar{\Phi}^T \cdot \vec{y} \tag{3.36}$$

The solution can be employed to solve a nonlinear model with a mapping of the original data in a lower-dimensional space to data in a higher-dimensional space via a stretching function. The following is the typical procedure:

(1) Data preparation: construct $\hat{X}$ with $\hat{x}_i$, each $\hat{x}$ has J rows for J features and 1 row for "1," and then apply the stretching function to $\hat{X}$ to obtain $\bar{\Phi}$.
(2) Training: calculate $\hat{w}^*$ as the model coefficients: $\hat{w}^* = (\bar{\Phi}^T \cdot \bar{\Phi} + \lambda\bar{I})^{-1} \cdot \bar{\Phi}^T \cdot \vec{y}$.
(3) Testing: to predict $\hat{x}_t$'s output y_t, we first apply the stretching function to a testing date point $\hat{x}_t$: $\phi(\hat{x}_t)$; then apply the weights obtained in training to predict y_t: $y_t = \phi(\hat{x}_t)^T \hat{w}$.

3.5.2 *Kernel Functions*

Kernel functions are a very useful trick to skip the use of stretching functions [48]. One problem with the use of stretching functions in the previous subsection is that we need to use the stretching function twice: once in training and once in testing. Unfortunately, the use of stretching functions is very computationally expensive. What is more, it is not easy to identify a good stretching function. The selection of kernel functions is easier than that of stretching functions. In addition, kernel functions are easier to implement. In this following, the concept of kernel functions will be explained.

A kernel function is a function that can convert two vectors as the input to a scalar as the output. It can be related to the stretching function in the following way:

$$k(\vec{a}, \vec{b}) = \phi(\vec{a})^T \cdot \phi(\vec{b}) \tag{3.37}$$

We intentionally define the kernel function in this way so that the stretching function can be easily replaced with it, which can be seen in the following deduction.

To show how to use this function with the linear model and extend the above mapping to a higher-dimensional space, let us first recall the linear model in the higher-dimensional space:

$$y = f(\hat{x}) = \phi(\hat{x})^T \cdot \hat{w} \tag{3.38}$$

where the scalar output y corresponds to input $\hat{x}$. For example, if $\hat{x}$ is $\hat{x}_i$, then y is y_i. The above equation is for one data point. For all the data, we have the following equation:

$$\hat{y}_{I\times 1} = f(\hat{X})_{I\times 1} = \bar{\Phi}^T_{I\times(J'+1)} \cdot \hat{w}_{(J'+1)\times 1} \tag{3.39}$$

In order to construct a kernel function in it, we can reformulate $\hat{w}$: $\hat{w}_{(J'+1)\times 1} = \bar{\Phi}_{(J'+1)\times I}\hat{z}_{I\times 1}$. At this moment, we do not know the meaning of $\hat{z}$, but just know it is a vector of size of I by 1. Accordingly, $\hat{w}$ can be reformulated as

$$\hat{w}_{(J'+1)\times 1} = \bar{\Phi}_{(J'+1)\times m}\hat{z}_{I\times 1} = \sum_{i=1}^{I} z_i \phi(x_i) \tag{3.40}$$

Substituting the above equation into the linear model in the higher-dimensional space (Eq. 3.38), we obtain

$$f(\hat{x}) = \phi(\hat{x})^T \cdot \hat{w} = \phi(\hat{x})^T \left[\sum_{i=1}^{I} z_i \phi(x_i)\right] = \sum_{i=1}^{I} z_i \phi(\hat{x})^T \cdot \phi(\hat{x}_i) \tag{3.41}$$

It can be seen that a kernel function appears on the right-hand side of the equation: $k(\hat{x}, \hat{x}_i) = \phi(\hat{x})^T \cdot \phi(\hat{x}_i)$, in which $\hat{x}_i$ is the ith point in the original data, and $\hat{x}$ could be any given data point such as a data point for testing. With the kernel function, the above equation becomes

$$f(\hat{x}) = \sum_{i=1}^{I} z_i k(\hat{x}, \hat{x}_i) \tag{3.42}$$

For all the data, we can obtain

$$f(\hat{X})_{I\times 1} = \bar{\Phi}^T_{I\times(J'+1)} \bar{\Phi}_{(J'+1)\times I} \hat{z}_{I\times 1} \tag{3.43}$$

Next, we can construct the loss function with L2 regularization (Rigid) as

$$\ell(\hat{w}) = (\bar{\Phi}^T \cdot \bar{\Phi} \cdot \hat{z} - \vec{y})^T (\bar{\Phi}^T \cdot \bar{\Phi} \cdot \hat{z} - \vec{y}) + \lambda(\bar{\Phi} \cdot \hat{z})^T \cdot (\bar{\Phi} \cdot \hat{z}) \tag{3.44}$$

Then, we define the kernel matrix as $\bar{K} = \bar{\Phi}^T \bar{\Phi}$, that is,

$$\begin{aligned} \bar{K}_{I\times I} = \bar{\Phi}^T \bar{\Phi} &= \begin{bmatrix} \phi(\hat{x}_1)^T \phi(\hat{x}_1) \; \phi(\hat{x}_1)^T \phi(\hat{x}_2) \cdots \phi(\hat{x}_1)^T \phi(\hat{x}_I) \\ \vdots \\ \phi(\hat{x}_I)^T \phi(\hat{x}_1) \; \phi(\hat{x}_I)^T \phi(\hat{x}_2) \cdots \phi(\hat{x}_I)^T \phi(\hat{x}_I) \end{bmatrix} \\ &= \begin{bmatrix} k(\hat{x}_1, \hat{x}_1) \; k(\hat{x}_1, \hat{x}_2) \cdots k(\hat{x}_1, \hat{x}_I) \\ \vdots \\ k(\hat{x}_I, \hat{x}_1) \; k(\hat{x}_I, \hat{x}_2) \cdots k(\hat{x}_I, \hat{x}_I) \end{bmatrix} \end{aligned} \tag{3.45}$$

Substituting the kernel matrix into the loss function (Eq. 3.44), we get

$$\ell(\hat{w}) = (\bar{K}\hat{z} - \vec{y})^T \cdot (\bar{K}\hat{z} - \vec{y}) + \lambda(\hat{z}^T \cdot \bar{K} \cdot \hat{z}) \tag{3.46}$$

The optimal loss function will be obtained when the derivative of the above loss function with respect to $\vec{z}$ equals 0, leading to

$$\frac{\ell(\hat{w})}{\partial \hat{z}} = 2(\bar{K}^2 + \lambda \bar{K})\hat{z} - 2\bar{K}\vec{y} = 0 \tag{3.47}$$

Solve the above equation, and we obtain

$$\hat{z} = (\bar{K} + \lambda \bar{I})^{-1} \cdot \vec{y} \tag{3.48}$$

All the stretching functions have been replaced with the kernel function. Now, we can first obtain $\hat{z}$ with Eq. 3.48. After $\hat{z}$ is obtained, we can get the prediction for any given $\hat{x}_t$ as

$$f(\hat{x}_t) = \sum_{i=1}^{m} z_i k(\hat{x}_t, \hat{x}_i) \tag{3.49}$$

An implementation of a linear model with the above kernel trick can be seen in the following Python code. This code can be used to train a model using the simple regression dataset presented above.

```
def KernelMatrix(X_bar,d):
    # Use Polynomial kernal with order d
    global K
    K = ((X_bar@X_bar.T)**d)
    return K

def LinearRegression(X,y):
    X_bar = np.hstack((X,np.ones([X.shape[0],1])))
    K = KernelMatrix(X_bar,d)
    z = np.linalg.inv(K)@y
    return z

z = LinearRegression(X,y)
y_fit = K@z
```

In summary, when using machine learning for regression or classification problems, if the data is not easy to fit or classify, e.g., exhibits nonlinear distributions, we can consider mapping the data to a higher dimension in a specific way to turn the problem into a linear one. The above-introduced kernel trick can help avoid the trouble of searching for appropriate stretching functions and their implementations. Common kernel functions are listed in Table 3.1.

Despite the benefits, it is worthwhile to mention that the use of the kernel trick also brings artifacts. Both the linear model and its use with the stretching functions are parametric methods. Parametric methods, such as linear models and neural networks, assume the form of the function for the mapping from the input to the output. Thus, the number of parameters of such methods will not change with the increase in the data amount. By contrast, in nonparametric methods, such as SVM and Gaussian process, the input is also part of the model. As a result, the

Table 3.1 Common kernel functions

Kernel name	Expression	Parameters
Linear kernel	$\kappa(\vec{x}_i, \vec{x}_j) = \vec{x}_i^T \vec{x}_j$	
Polynomial kernel	$\kappa(\vec{x}_i, \vec{x}_j) = (\vec{x}_i^T \vec{x}_j)^d$	$d \geqslant 1$
Gaussian kernel	$\kappa(\vec{x}_i, \vec{x}_j) = \exp\left(-\frac{\|\vec{x}_i^T - \vec{x}_j)\|^2}{2\sigma^2}\right)$	$\sigma > 0$
Laplacian kernel	$\kappa(\vec{x}_i, \vec{x}_j) = \exp\left(-\frac{\|\vec{x}_i^T - \vec{x}_j)\|}{\sigma}\right)$	$\sigma > 0$
Sigmoid kernel	$\kappa(\vec{x}_i, \vec{x}_j) = \tanh(\beta \vec{x}_i^T \vec{x}_j + \theta)$	$\beta > 0$ and $\theta < 0$

Note: σ in the above kernel functions are the bandwidth of the corresponding kernel

number of parameters is determined by the size of the input. More information about nonparametric methods will be provided in the chapter on Bayesian algorithms.

One major consequence of using the kernel trick with linear models is that it turns the linear model into a nonparametric model. As can be seen in Eq. 3.49, the input data needs to be used to predict new data. Therefore, every prediction will require the use of all the input data. This can place a high demand for storing such data. Also, this demand will increase as the amount of data increases.

3.6 Practice: Develop Code to Implement the Basic Linear Model

>>> More and Up-to-Date Course Materials including Practices @ AI-engineer.org <<<

Develop code for implementing the basic linear model algorithm. Use it to conduct linear regression for the following two problems.

1. Data points: (1.0, 0.9), (2.0, 2.1), (3.0, 2.8), (4.0, 4.3), (5.0, 4.9), (6.0, 6.2)

Hints:

Create the arrays for the dataset as follows. Then create the X_bar array and use it and y to get the exact solution for w_hat, which includes w and b. Finally, you will need to plot the predicted labels using the obtained w and b against the actual labels. You will need to check how to do inverse and transpose in NumPy when implementing the analytical solution.

```
X = np.array([[1.],[2.],[3.],[4.],[5.],[6.]])
y = np.array([0.9,2.1,2.8,4.3,4.9,6.2])
```

2. Now let us address a linear regression problem with slightly more complicated data, in which each sample has two attributes instead of one. So in the following measured data points, each data point (sample) includes two attribute values (the first two elements) and a label (the last element).

Data points: (1.0,1.0,6.58593), (1.0,2.0,8.89468), (2.0,2.0,9.20705), (2.0,2.0, 11.8968), (1.5,2.5,9.8388), (2.0,4.0,13.0987), (1.0,3.0,10.0886), (3.0,1.5,9.24608)

Hints: You will need to create X and y and then follow the same procedure as Part 1. You can use the same code (in the same files or two separate files). The major difference is the data preparation and plotting parts. For plotting, you can start from the following code (make modifications if needed):

```
fig = plt.figure()
ax = plt.axes(projection='3d')
ax.scatter(X[:,0],X[:,1],y,'b*')
ax.scatter(X[:,0],X[:,1],y_fit,'rd')
# ax.show()
```

Chapter 4
Decision Trees

4.1 Overview

Decision trees are a family of algorithms that use a treelike structure to mimic humans' decision-making process. This chapter presents knowledge that is needed to understand and practice decision trees. We will first focus on the basics of decision trees. In particular, we will see how a decision tree is generated in training and used for predictions. A clear understanding will be gained for the splitting of data, selection of attributes at nodes, and the underlying information theory criteria. Next, discussions on such criteria will help us reach three classic decision tree algorithms: ID3, C4.5, and CART. These algorithms will be explained with adequate detail for implementation. After that, the common issues of decision trees, especially overfitting, will be explained. In the end, countermeasures, including common pre-pruning and post-pruning methods, will be outlined.

4.2 Basics of Decision Trees

The idea of decision tree algorithms is to use a treelike model for structuring decisions and their possible consequences to assist decision-making [49]. One decision tree has one root node, multiple inner nodes, and several leaf nodes. The training of a classification or regression model using a decision tree involves gradual splitting of the samples. As shown in Fig. 4.1, this process starts from the root node, which thus contains all the samples. Starting from the root node, the dataset will be split into subsets assigned to inner nodes and/or leaf nodes connecting to the root node. This splitting of the datasets is carried out according to a selected sample attribute. The selection of attributes is usually made to achieve some goals, such as higher purity in the total information according to the information theory. Different attribute selection rules lead to different decision tree algorithms. Each data subset

Z. "L." Liu, *Artificial Intelligence for Engineers*,
https://doi.org/10.1007/978-3-031-75953-6_4

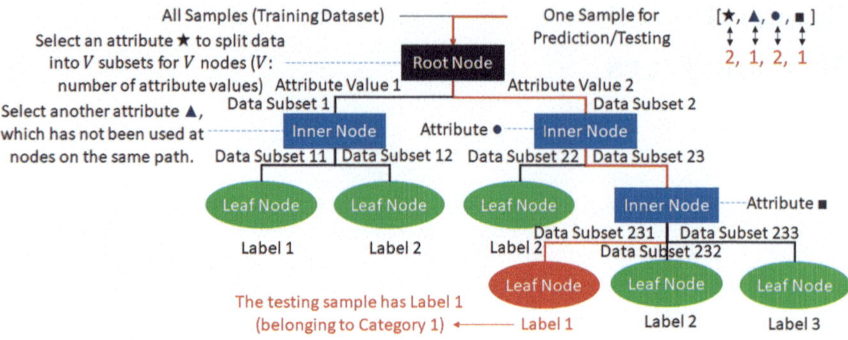

Fig. 4.1 Conceptual illustration of decision tree

or the data at secondary (second-level) inner nodes is further split until it reaches nodes where data cannot be split anymore, that is, leaf nodes. Eventually, any sample will be associated with a leaf node. In this data-splitting process, the attributes will be "consumed": an attribute that has been selected for data splitting at a node will not be selected in the remainder of the process.

Any leaf node is associated with a "decision," e.g., a category label in a classification task. Therefore, any path connecting the root node and a leaf node represents a series of conditions that leads to a decision. In a classification task, this path is associated with the values of different attributes of the sample, and each node on the path is associated with an attribute. Therefore, there are two key things in this process:

(1) When can we tell a leaf node is reached (not an inner node)?
(2) How to select the attribute at a node that determines the way to split the sample set at that node?

For the first question, the leaf node is a node at which samples cannot be separated anymore. The detailed rules are as follows:

(1) All the samples at the current node belong to one single category (data cannot be further classified).
(2) All the attributes have been adopted in the previous nodes on the path leading to this node, or all the samples at the current node have the same attribute (data cannot be further split).
(3) There are no samples at the current node (no data to split/classify).

For the second question, the key in the search for the best way of splitting the data is to identify a measure for quantifying the goodness of different ways of splitting the dataset. A widely accepted measure comes from the information theory, i.e., the information entropy. The information entropy quantifies the purity of information. A lower information entropy indicates a higher level of purity and less uncertainty. That is, there is relatively high uncertainty in the dataset before classification. As the relationship between samples and their correct classifications become clearer,

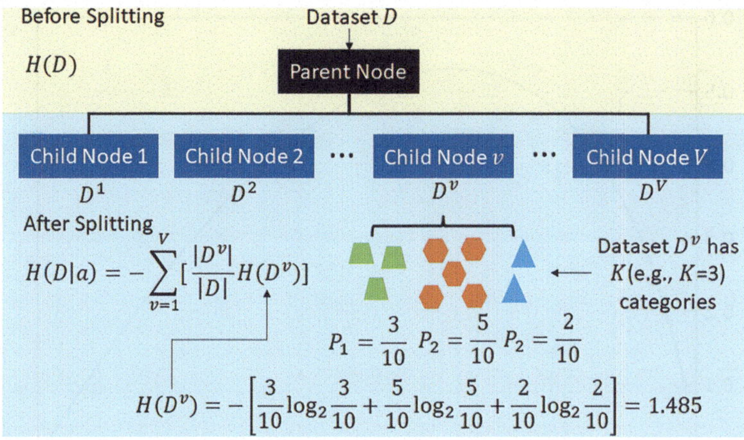

Fig. 4.2 An example for dataset splitting and information entropy change

this uncertainty drops. A decision-making process, such as classification, is to let samples with identical conditions be grouped in the same class, in which the information gets purer and uncertainty drops. Therefore, a better measure should be associated with faster drops of the information entropy in the data splitting process. Accordingly, we need to select an attribute at a node, based on which the dataset splitting processes will lead to a lower information entropy for the next split data or a higher drop in the information entropy in the splitting process. The selection of methods for quantifying this drop leads to various common decision tree algorithms, e.g., ID3 [50], C4.5 [51], and CART [52].

Figure 4.2 shows the data splitting at one node. At this (parent) node, a dataset, which consists of D samples belonging to K categories, will be split. We need to select an attribute to split the data. Let us consider an attribute a from a pool of candidate attributes: attributes that have not been used in the previous steps of forming the decision tree. This feature has V different values, which will split the dataset into V subsets, leading to V nodes on the next level. In the following introduction, we will see that the number of samples in each subset D^v, i.e., $|D^v|$, is significant in decision tree algorithms, in addition to the number of data subsets, i.e., V.

It is worthwhile to note that the meanings of "splitting" and "classification" (or "class") are different in the above introduction. An obvious distinction is that "splitting" and "classification" are related to V and "K," respectively. The splitting of data is carried out according to a selected attribute at each node on every level, leading to nodes on the next level. Consequently, each subset of data corresponds to a node on the next level. By contrast, classification also occurs in that data splitting process; however, we will only see the classification results at the leaf nodes (last

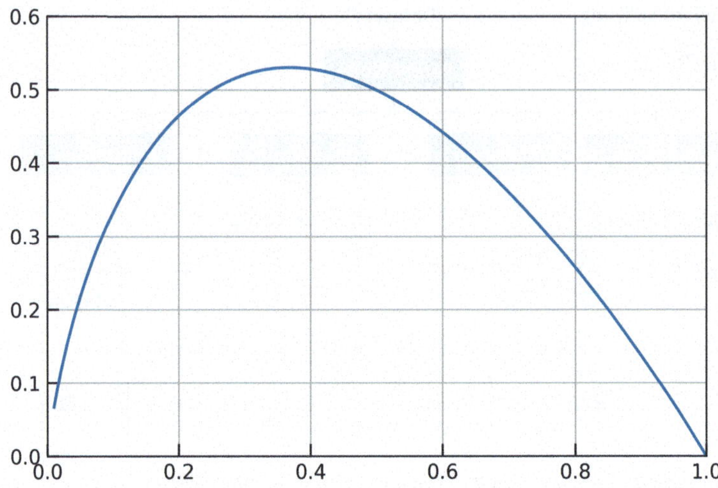

Fig. 4.3 Information entropy function

level of nodes). That means, in a subset generated by splitting a data set, there are very likely data samples belonging to multiple classes. Differentiating these two terms in decision trees is critical to the understanding of Fig. 4.2 and the following definition of information entropy.

Taking binary classification as an example, K is 2 as long as the dataset at this node is still separable—not a leaf node according to Rule 1. Therefore, each category contains $P_k = D_k/D$ of all the samples. Then, the information entropy associated with the dataset before splitting is

$$\mathrm{H}(D) = -\sum_{k=1}^{K} P_k \log_2 P_k \tag{4.1}$$

The general meaning of "entropy" is a measure of the chaos of the system. Accordingly, the lower the information entropy, $\mathrm{H}(D)$, the purer the information, and the less the uncertainty. Therefore, the purpose of splitting a dataset at one node in the process of "growing the decision tree" with more (levels of) nodes is to reduce the information entropy.

If we plot the information entropy function as Fig. 4.3, we can see that a low H function value is associated with very small or large P_k values ($0 \leqslant P_k \leqslant 1$). Thus, the decrease of H in splitting corresponds to a process in which more and more samples in the subset will be in one or a few groups (classes) and eventually belong to only one group. In this way, the information is getting purer as directed by this information entropy function.

But please keep in mind that our ultimate goal is to reduce the uncertainty or increase the purity of information. The information entropy is one way of quantifying such uncertainty and purity. Although it is possibly the most common and straightforward way, there are still other ways.

4.3 Classic Decision Tree Algorithms

4.3.1 *ID3 Algorithm*

The ID3 (short for Iterative Dichotomiser 3) algorithm is one of the earliest decision tree algorithms. This algorithm adopts the information gain as the measure to tell how well the data splitting according to an attribute can help reduce the uncertainty or increase the purity of information in the data. For this goal, the information gain, $\mathrm{Gain}(D, a)$, measures the difference between the entropy values before and after the set D is split according to an attribute a. Accordingly, $\mathrm{Gain}(D, a)$ is formulated as

$$\mathrm{Gain}(D, a) = \mathrm{H}(D) - \mathrm{H}(D|a) = \mathrm{H}(D) - \sum_{v=1}^{V}\left[\frac{|D^v|}{|D|}\mathrm{H}(D^v)\right] \tag{4.2}$$

where $\mathrm{H}(D)$ is the entropy of the set D to be split (i.e., entropy before splitting), K is the number of subsets generated by splitting the data according to attribute a, k is the kth subset, $\mathrm{H}(D|a)$ is the entropy of the split data (after splitting), and $|D|$ is the number of samples in set D. Therefore, we have $D = \bigcup_{v-1}^{V} D^v$.

As seen in the above equation, the information gain is the difference between the information entropy at the current node and the weighted sum of the entropy values at the nodes on the next level, which are generated by splitting the data according to a selected feature a. The information gain $\mathrm{Gain}(D, a)$ is positive; thus, the information entropy decreases in the data splitting (growth of tree) process. Accordingly, the amount of information is reduced, the purity of information increases, and the level of uncertainty decreases. The term "information gain" is used to refer to the information that we "gain" from the system by splitting the data, which equals the decrease in the total amount of information in the system or data.

4.3.2 *C4.5 Algorithm*

The measure adopted in ID3, i.e., information gain, tends to choose an attribute that leads to more subsets (or more nodes on the next level), which produces a higher information gain value. In an extreme case, the unique identification (ID) may be selected as the attribute for data splitting in ID3, because this selection leads to more subsets. Under this condition, each subset contains only one sample, and the classification is finished as no data subset can be further split. Apparently, a decision tree generated in this way is weak in generalization, or, in other words, makes no sense for the intended classification purpose. This is like "we usually do not classify people according to their unique passport number." To overcome this issue, the information gain ratio, as formulated in the following equation, is proposed in the C4.5 decision tree algorithm.

$$\text{Gain\_ratio}(D, a) = \frac{\text{Gain}(D, a)}{\text{IV}(a)} \tag{4.3}$$

A major change from the information gain in ID3 to the gain ratio in C4.5 is $\text{IV}(a)$, which is a value pertaining to attribute a. $\text{IV}(a)$ is defined as follows:

$$\text{IV}(a) = -\sum_{v=1}^{V} \frac{|D^v|}{|D|} \log_2 \frac{|D^v|}{|D|} \tag{4.4}$$

The above definition generates higher values of $\text{IV}(a)$ when the use of attribute a can lead to more subsets of data (number of nodes): a greater V. Therefore, the use of $\text{IV}(a)$ in the denominator of the equation for the information gain ratio results in the tendency of selecting an attribute that has fewer attribute values (or leads to less of nodes). This tendency can also threaten the generalization of the model. To overcome this issue, C4.5 does not use the information gain ratio directly for selecting a feature for data splitting. Instead, this algorithm first seeks attributes that can generate information gain that is higher than the average among all the candidate attributes. Then, among all the "above-average" attributes, it picks the one associated with the highest gain ratio.

4.3.3 CART Algorithm

CART (classification and regression tree) algorithm is a newer and more widely accepted decision tree algorithm. For example, the decision tree classifier in the Scikit-learn package uses an optimized version of CART by default. In the CART algorithm, the Gini index is adopted for selecting attributes.

Before we introduce the Gini index, let us first take a look at the Gini value, which is used to construct the Gini index. The Gini value, $\text{Gini}(D)$, quantifies the probability that the labels of two samples that are randomly selected from a dataset D are different from each other. Therefore, the lower the Gini value, the purer the data, and the lower the uncertainty. Therefore, the Gini value provides another way to quantify the amount and purity of information and the amount of uncertainty in the system (data) in addition to the information entropy. According to the definition, the Gini value is mathematically formulated as

$$\text{Gini} = \sum_{k=1}^{K} \sum_{k' \neq k} (P_k P_{k'}) \tag{4.5}$$

The above equation can be reformulated as below to obtain a more common equation of the Gini value:

$$\begin{aligned}
\text{Gini} =& \sum_{k=1}^{K} \sum_{k' \neq k} (P_k P_{k'}) \\
=& P_1 \cdot (P_1 + P_3 + \cdots + P_K) + \cdots + P_k \cdot (P_1 + P_2 + \cdots + P_{K-1}) \\
=& P_1(1 - P_1) + P_2(1 - P_2) + \cdots + P_K(1 - P_K) \\
=& (P_1 + P_2 + \cdots + P_K) - (P_1^2 + P_2^2 + \cdots + P_K^2) \\
=& 1 - \sum_{k=1}^{K} P_k^2
\end{aligned} \tag{4.6}$$

The Gini value is calculated for each subset D^v. Then the weighted sum of the Gini values of all the generated subsets is obtained as the Gini index. Therefore, for a dataset D that is split according to an attribute a, the Gini index $\text{Gini\_index}(D, a)$ is formulated as

$$\text{Gini\_index}(D, a) = \sum_{v=1}^{V} \frac{|D^v|}{|D|} \text{Gini}(D^v) \tag{4.7}$$

In CART, we select the attribute that will lead to split data, i.e., a collection of the data subsets, with the lowest Gini index. Figure 4.4 shows the data splitting process for training with the well-known Iris dataset using the CART function in Scikit-learn. The Gini index values, number of samples, size of subsets, and classification labels are given out for each node.

4.3.4 Implementation

The coding of decision trees is relatively complicated compared to other basic machine learning algorithms. In this subsection, we provided example code for a key component of training decision tree models: the selection of a feature for splitting data. This code exemplifies the development of dataset splitting measures for ID3, C4.5, and CART algorithms.

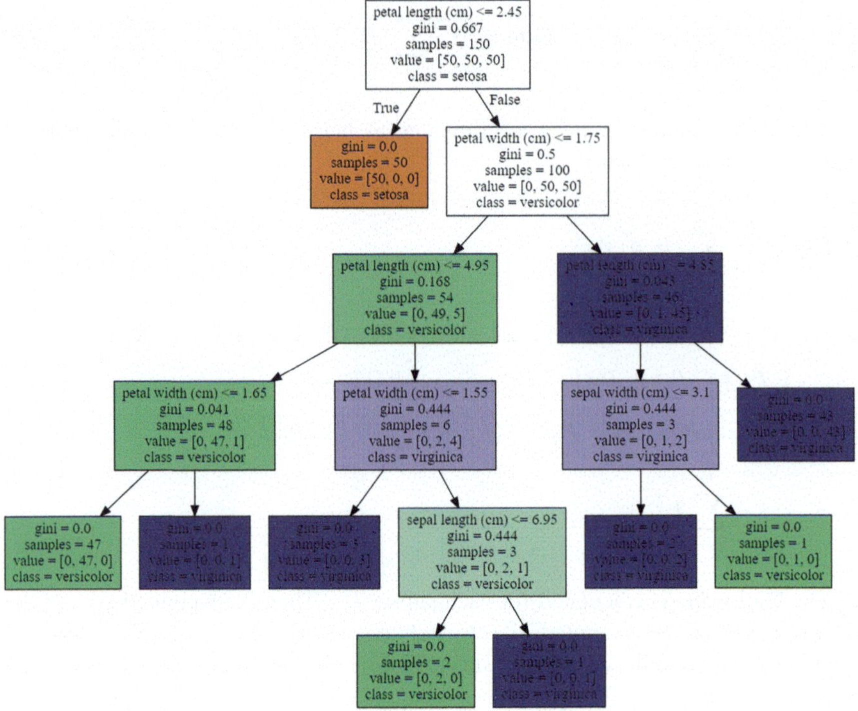

Fig. 4.4 Tree generated by fitting the Iris dataset with CART

```
# Calculate the Information Entropy
def Entropy(dataSet):
    N_instances = len(dataSet)
    Category_sizes = {} # A dictionary to store the numbers of
    samples for each category like {''yes'': 2, ''no'': 3 }
    for Instance in dataSet:
        currentLabel = Instance[-1]
        if currentLabel not in Category_sizes:
            Category_sizes[currentLabel] = 0
        Category_sizes[currentLabel] += 1
    Ent = 0.0
    Gini = 1.0
    for key in Category_sizes:
        prob = float(Category_sizes[key])/N_instances
        Ent -= prob * log2(prob)
        Gini -= prob**2
    return Ent, Gini

# Select which feature will be used for splitting data at the
    current node
def chooseBestFeatureToSplit(dataSet, method): # method = ID3,
    C4_5, or CART
```

```
N_features = len(Features) # len(dataSet[0]) - 1 # The last
element of every instance is not a feature but a label thus
needs to be excluded
baseEntropy,baseGini = Entropy(dataSet) # BaseGini is not
used
bestMeasure = 0.0
bestFeature_num = -1
for i in range(N_features):
    feature_i_values = [example[i] for example in dataSet] #
Create a list of all possible values of feature i
    feature_i_unique_values = set(feature_i_values) # Create
a list of all unique values of feature i
    newEntropy = 0.0
    IV_a = 1e-5
    Gini_index = 0.0
    for feature_value in feature_i_unique_values:
        subDataSet = splitDataSet(dataSet, i, feature_value)
        prob = len(subDataSet) / float(len(dataSet)) # Dv/D
        Ent, Gini = Entropy(subDataSet)
        newEntropy += prob * Ent
        IV_a -= prob * np.log2(prob)
        Gini_index += prob * Gini
    infoGain = baseEntropy - newEntropy # Information gain
for ID3
    GainRatio = infoGain / IV_a

    if method == 'ID3':
        Measure = infoGain
    elif method == 'C4_5':
        Measure = GainRatio
    elif method == 'CART':
        Measure = Gini_index

    if Measure > bestMeasure:
        bestMeasure = Measure
        bestFeature_num = i

return bestFeature_num
```

4.4 Issues and Techniques: Overfitting and Pruning

Like other machine learning methods, decision tree algorithms are also susceptible to overfitting issues [53]. A tree that is too large turns to overfit the training data and inadequately generalize to new samples. By contrast, a tree that is too small may not capture important structural information about the sample space, leading to underfitting.

In fact, such issues could be more serious due to the nature of the decision tree. A decision tree partitions data into smaller and smaller subsets until those final subsets are homogeneous in terms of the outcome variable (belonging to the same class). In practice, this often means that the final subsets, i.e., leaves of the tree,

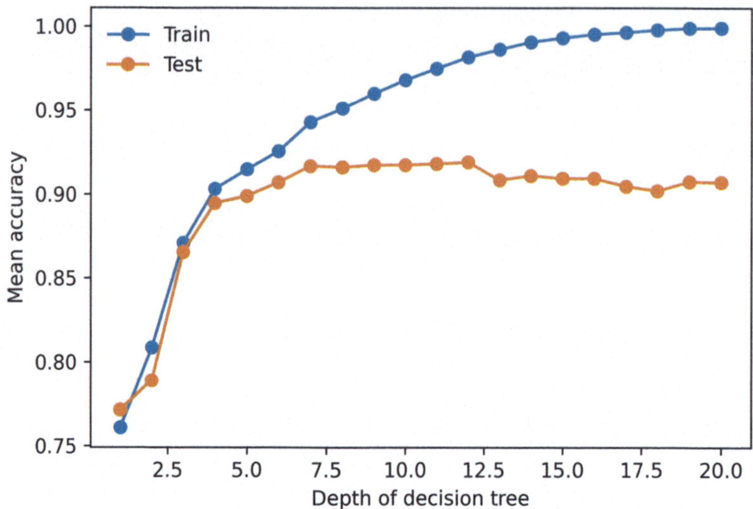

Fig. 4.5 Overfitting in the training of a decision tree (10,000 samples and each with 30 attributes)

may each consist of only one or a few data points. The tree has learned the training data exactly including both the general features and the undesired features such as noise. Due to this reason, the performance of a trained decision tree may not work well with new testing data, leading to the poor generalization capacity of this type of algorithm.

As illustrated in Fig. 4.5, when a decision tree grows in depth, the accuracy of the predictions made with the tree on the training dataset increases, whereas that on the testing data decreases as the number of nodes increases. That means the tree gets unnecessarily complicated and learns features specific to the training data. This compromises the performance of the tree when it is applied to the testing data. This is attributed to the following facts in the data:

(1) Mismeasurements: for a variety of reasons, the value of an attribute or class may be incorrectly measured or missing. This may happen because of incorrect perception, measurement, recording, or transcription.
(2) Incomplete attributes: The selected attributes may not be the perfect criterion of classification. Also, it is possible that extraneous factors that significantly affect the results are not included as attributes.
(3) Coincidence: When the dataset is small, the data may coincidentally form patterns that do not reflect general rules.

Different methods are proposed to overcome overfitting [54]. In decision trees, due to the unique nature of this type of method, we usually apply some extra rules to control or/and modify the growth of the tree to actively adjust the complexity of the model. This can be done with fixed rules without assessing the data or with adaptive rules that can be adjusted according to feedback from the data analysis or training process. Two major tricks for improving the generalization of the decision tree algorithms are pre-pruning and post-pruning. The difference between them is that pre-pruning is performed before the growth of the tree is finished, while post-pruning is usually carried out after a tree is obtained. Each of these two tricks can be implemented in different ways, which will be introduced in the following subsections.

4.4.1 Pre-Pruning

Pre-pruning is proposed to stop the growth of some parts of the tree before the training or the growth of the tree is finished. Therefore, it is similar to the "early-stopping" technique used in many other machine methods for addressing overfitting issues. There are different ways of determining when to stop the growth of the tree:

(1) The simplest way is to stop training or the tree growth when the tree grows into a specific size such as a certain number of node layers.
(2) Another way is to stop the growth when the number of samples at a node is smaller than a specific value.
(3) A more widely accepted way is to check the information gain for the growth at each node, and the growth will be stopped if the information gain is lower than a threshold.

Pre-pruning has its pros and cons. The pros are that it is relatively simple to implement and has good efficiency. Due to this reason, pre-pruning can be applied to large-scale problems. Despite its straightforwardness, it is difficult to find a suitable rule to stop the growth of the trees. In particular, the situation varies from node to node. Besides, pre-pruning is implemented at the current step; thus, the situations of the following steps are unknown at the time of pruning. Due to this reason, it could be short-visioned. For instance, a pre-pruning operation may bring benefits at the current node, i.e., in the short term, but hurt improvements in the tree's performance in the long run.

4.4.2 Post-Pruning

Post-pruning is performed after a decision is generated. It simplifies the tree by pruning some parts that learn specific features of the training data. These specific features do not belong to the general data and thus should not be learned. The following are common ways of post-pruning a decision tree.

Cost-Complexity Pruning (CCP)

CCP is a post-pruning method frequently adopted in the CART algorithm. For example, the decision tree classifier in Scikit-learn uses an optimized version of CART, which includes CCP for post-pruning. Because of this, this optimized CART is also called the CART pruning algorithm in some places. The use of CCP follows the following steps.

First, CCP generates a series of trees, i.e., $T_0, \ldots, T_m$, in m steps via pruning, where T_0 is the initial tree and T_m is the root alone. At step i, the tree is created by replacing a subtree in tree $i-1$ with a leaf node. This subtree is chosen in the following way:

(1) Define the error rate of tree T over dataset S as $\text{err}(T, S)$.
(2) The subtree t that minimizes $\frac{\text{err}(T,S)-\text{err}(\text{prune}(T,t),S)}{|\text{leaves}(T)|-|\text{leaves}(\text{prune}(T,t))|}$ is chosen for removal.

where $\text{leaves}(T)$ is the number of the leaves in the untrained tree. The function $\text{prune}(T, t)$ is the tree obtained by pruning the subtrees t from the tree T.

Next, once such a series of trees has been created, the tree with the best performance is selected. This accuracy or error of each tree in the family may be estimated in two ways: cross-validation and testing with an independent training (pruning) set.

Reduced Error Pruning (REP)

REP is one of the simplest post-pruning methods. The idea is to replace every node (excluding leaf nodes) with a leaf represented by the most popular class in the data. If the new tree generated after this replacement has an error that is equal to or smaller than the error of the original tree, then we will execute the pruning via this replacement. Such pruning is performed by evaluating nodes in a bottom-up fashion. Also, a pruning or validation dataset is needed to assist the pruning operations.

Thus, REP has the advantage of simplicity and speed. In particular, REP has linear computing complexity, as each node is only visited once to evaluate the possibility of trimming it. However, it is noted that REP is usually not selected when the dataset is small. This is because REP has a proclivity towards over-pruning due to the fact that all the evidence contained in the training set, which was used to construct a fully grown tree, is ignored during the pruning process. This issue is particularly obvious when the pruning set is significantly smaller than the training set, but it becomes less significant as the percentage of instances in the pruning set grows.

Pessimistic Error Pruning (PEP)

PEP was proposed to overcome a weakness of other post-pruning methods—an independent dataset is needed for trimming. Thus, this method is distinct in that the same training set is utilized for both growing and trimming a tree. The apparent error rate, i.e., the error rate on the training set, is optimistic and cannot be used to select the best-pruned tree. Due to this consideration, the continuity correction for the binomial distribution is proposed, which may give "a more realistic error rate."

The PEP approach is accepted to be one of the most accurate decision tree pruning algorithms. In addition, due to its top-down nature, each subtree in the tree only has to be consulted once. Besides, the time complexity in the worst case is linear with the number of non-leaf nodes in the decision tree. However, PEP suffers from over-pruning because the mechanism for traversing PEP is similar to pre-pruning.

Minimum Error Pruning (MEP)

MEP adopts a bottom-up strategy that seeks a single tree with the lowest "anticipated error rate on an independent dataset." For this purpose, for every node (excluding leaf notes), we will first calculate the error $Er(t)$. Next, we calculate the errors of the branches (sub-notes) of these nodes. The errors of these branches will be added according to weights determined by the number of samples (percentages) at these branches. If this weighted sum of errors, $Er(Tt)$, is smaller than $Er(t)$, we will keep the subtree and otherwise delete it.

This post-pruning method does not require an independent pruning dataset. No matter in the original or improved versions of MEP, most information comes from the training set. Besides, its computational complexity has a linear relationship with the total number of non-leaf nodes in the unpruned tree.

Comparison and Summary

Table 4.1 is a summary of the major characteristics of the four methods introduced above.

Overall, a few general rules can be drawn regarding the selection of post-pruning methods. MEP is less accurate than PEP in general. Both of them do not require

Table 4.1 Comparison of post-pruning methods

Characteristics	CCP	REP	PEP	MEP
Independent pruning dataset	Not needed	Needed	Not needed	Not needed
Pruning direction	Botom-up	Botom-up	Top-down	Botom-up
Error estimate	Cross-validation or standard error	Pruning set	Continuous correction	Probability estimate based on a m parameter
Computation complexity	$O(n^2)$	$O(n)$	$O(n)$	$O(n)$

extra datasets, so they may suit the conditions of small datasets. REP is simple and accurate, but it requires extra datasets. CCP is comparable to REP in terms of accuracy.

4.5 Practice: Decision Trees in Scikit-learn—Training, Tree Plot, and Testing

>>> More and Up-to-Date Course Materials including Practices @ AI-engineer.org <<<

Please use the decision tree algorithm in Scikit-learn for classification in the following two steps:

1: Load the Iris dataset in Scikit-learn. Using the decision tree (e.g., tree.DecisionTreeClassifier()) in Scikit-learn to fit the data. Plot the tree using tree.plot_tree().

Note: tree.DecisionTreeClassifier() uses an optimized version of CART.

2: The Iris dataset contains 150 samples. Please use 30 samples for training. Then test the trained decision tree model with the remaining 120 samples and tell the classification results (e.g., use score() method in tree.DecisionTreeClassifier()

Chapter 5
Support Vector Machines

5.1 Overview

The concept underlying the support vector machine (SVM) is to find a "plane" that can separate samples belonging to different categories. Some of the most important concepts in SVM are margin, dual problem, and kernel trick (methods). The concept of margin is where the support vectors originate. The original SVM is defined with a hard margin, while a soft margin can enable SVMs to consider problems with imperfect data: data with noise and mislabeled points. The dual problem is an optimization technique to facilitate the search for solutions to SVM problems. Kernel methods generalize the basic SVM by introducing a kernel function so that a hyperplane can be used to separate different categories of data points without linear boundaries (planes). The kernel trick allows the SVM to consider nonlinear problems by converting the data to a higher-dimensional space, in which the problem becomes linear.

This chapter will start a tour to SVM with an introduction to the basic SVM. The idea of SVM will be explained using mathematical functions and graphical illustrations. We will see how to derive the optimization formulation of a machine learning problem using SVM. To facilitate solution, the conversion of this original formulation of the optimization problem into a dual problem will be explained. Based on that, we will show how to generalize the basic SVM for linear problems via kernel functions to obtain nonlinear SVMs. After that, soft margin as a common technique for addressing overfitting issues in SVM applications will be introduced. Like the basic SVM (with soft margin), the introduction to SVM with hard margin will also include formulations of the original optimization problem and dual problem. In the final, extra skills about SVM, including the SMO algorithm and the use of SVM for multi-class classification and regression, will be briefly discussed.

Z. "L." Liu, *Artificial Intelligence for Engineers*,
https://doi.org/10.1007/978-3-031-75953-6_5

5.2 Basics of SVM: Hard-Margin SVM

5.2.1 Basic Formulation

Let us first take a look at typical data in a binary classification problem to see how SVM [55] can be constructed. In a space of samples, as shown in Fig. 5.1, the plane separating different categories of data points can be described using the following linear function:

$$\vec{w}^T \cdot \vec{x} + b = 0 \tag{5.1}$$

where $\vec{w}$ is the normal vector of the plane, which specifies the direction of the plane, and b is the intersect, which determines the distance between the plane and the origin.

In this space, the distance from any point $\vec{x}$ to the plane $(\vec{w}, b)$ is

$$r = \frac{\left|\vec{w}^T \cdot \vec{x} + b\right|}{\|\vec{w}\|} \tag{5.2}$$

The magnitudes of $\vec{w}$ and b can be adjusted by multiplying them with the same constant so that the two margin lines (or called sides or boundaries) in Fig. 5.1 have the functions of

$$\begin{cases} \vec{w}^T \cdot \vec{x} + b = +1 \\ \vec{w}^T \cdot \vec{x} + b = -1 \end{cases} \tag{5.3}$$

It is noted that, in SVM, "margin" usually refers to the distance between the separating hyperplane and any of the boundaries of the separation zone or the

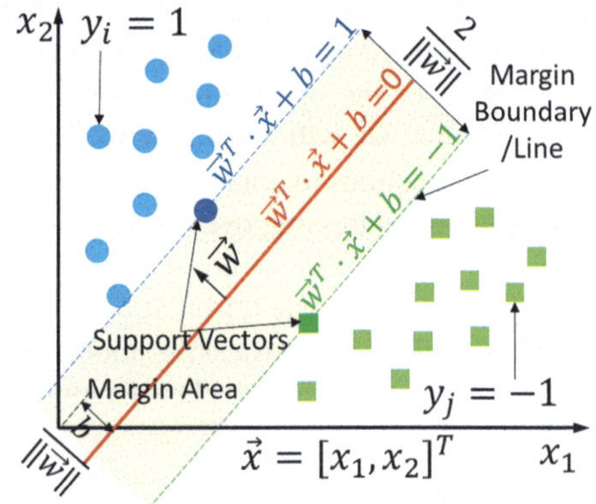

Fig. 5.1 Conceptual illustration of SVM

distance between the two boundaries. However, it is also seen that "margin" is used to refer to the separation zone. To avoid confusion, we reserve the term "margin" for the distance and use "margin area" for the separation zone. The two boundaries of the separation zone, which are also called positive and negative hyperplanes in some literature, are called margin lines or margin boundaries.

These two margin boundaries are the two (dashed) lines that are parallel to the separating hyperplane and passing through the points closest to the hyperplane. The points on these boundaries are the support vectors. Therefore, the data points that are out of the margin area should satisfy the following relations:

$$\begin{cases} \vec{w}^T \cdot \vec{x} + b \geqslant +1, & y = +1 \\ \vec{w}^T \cdot \vec{x} + b \leqslant -1, & y = -1 \end{cases} \tag{5.4}$$

where $y = +1$ assigns a label of $+1$ to this point.

The separating hyperplane, which is a plane in this 2D case, is in the middle of the margin area. Accordingly, the margin, or the width of the margin area, is $\frac{2}{\|\vec{w}\|}$. The two regions on two sides of the margin area mark the two categories. In a binary classification task using the SVM, the two categories are generally labeled as $+1$ and -1 (for y_i). Based on the definitions of the margin and label, we can see that valid points, which are on two sides of the margin area (including points on the margin boundaries), must satisfy the following condition:

$$y_i \cdot (\vec{w}^T \cdot \vec{x}_i + b) \geqslant 1, i = 1, 2, \cdots, I \tag{5.5}$$

The above equation can be written in a more general array format as

$$\vec{y} \odot (\bar{X} \cdot \vec{w} + b) \geqslant \vec{1} \tag{5.6}$$

where $\odot$ is the element-wide product and $\vec{1}$ is the array that has the same size as $\vec{y}$ and all the elements as 1.

If data points belonging to two classes are distributed as shown in Fig. 5.1, then the two categories of points can be separated by an infinite number of planes. In the SVM, the plane associated with the best classification performance is believed to be the plane that can achieve the greatest (widest) margin. Accordingly, the mathematical description of the basic SVM for a binary classification task is

$$\begin{cases} \max_{\vec{w},b} \frac{2}{\|\vec{w}\|} \quad \left(\text{or } \max_{\vec{w},b} \frac{2}{\|\vec{w}\|^2}\right) \\ \text{s.t. } y_i \cdot (\vec{w}^T \cdot \vec{x}_i + b) \geqslant 1, i = 1, 2, \cdots, I \end{cases} \tag{5.7}$$

The above problem can be reformulated to utilize the minimization function in optimization.

$$\begin{cases} \min_{\vec{w},b} \frac{1}{2}\|\vec{w}\|^2 \\ \text{s.t. } y_i \cdot (\vec{w}^T \cdot \vec{x}_i + b) \geqslant 1, i = 1, 2, \cdots, I \end{cases} \tag{5.8}$$

5.2.2 *Dual Formulation*

The above optimization problem is a convex quadratic programming problem and can be solved with some optimization packages. However, a more widely accepted way is to convert this problem into a dual problem by multiplying two sides of the equation with the Lagrange multipliers ($\alpha_i \geqslant 0$). This will bring two benefits: (1) making the optimization problem easier to solve and (2) facilitating the introduction of kernel functions for tackling nonlinear problems.

The Lagrange function of the above problem can be written as

$$\ell(\vec{w}, b, \vec{\alpha}) = \frac{1}{2}\|\vec{w}\|^2 + \sum_{i=1}^{I} \alpha_i \left[1 - y_i \cdot \left(\vec{w}^T \cdot \vec{x}_i + b\right)\right] \tag{5.9}$$

where $\vec{\alpha} = [\alpha_1, \alpha_2, \cdots, \alpha_I]^T$ are Lagrange multipliers. The optimal (minimal) value of the function locates where the partial derivatives of the equation with respect to $\vec{w}$ and b ($\frac{\partial l}{\partial \vec{w}}$ and $\frac{\partial l}{\partial b}$, respectively) are zero, that is,

$$\begin{cases} \vec{w} = \sum_{i=1}^{I} \alpha_i y_i \vec{x}_i = \bar{X}^T \cdot (\vec{\alpha} \odot \vec{y}) \\ 0 = \sum_{i=1}^{I} \alpha_i y_i = \vec{\alpha}^T \cdot \vec{y} \end{cases} \tag{5.10}$$

Substituting the above equations into the Lagrange function yields the dual problem of the SVM:

$$\begin{cases} \min_{\vec{\alpha}} \left[\frac{1}{2} \sum_{i=1}^{I} \sum_{j=1}^{I} (\alpha_i \alpha_j y_i y_j \vec{x}_i^T \cdot \vec{x}_j) - \sum_{i=1}^{I} \alpha_i\right] = \\ \min_{\vec{\alpha}} \left\{\frac{1}{2}[\bar{X}^T \cdot (\vec{\alpha} \odot \vec{y})]^T \cdot [\bar{X}^T \cdot (\vec{\alpha} \odot \vec{y})] - \vec{1}^T \cdot \vec{\alpha}\right\} \\ \text{s.t. } \sum_{i=1}^{I} \alpha_i y_i = \vec{\alpha}^T \cdot \vec{y} = 0, \qquad 0 \leqslant \alpha_i, i = 1, 2, \cdots, I \end{cases} \tag{5.11}$$

where $\vec{1}$ is a column array, which has the same size as $\vec{\alpha}$ and all the elements equaling 1.

Solving the above dual problem yields $\vec{\alpha}$. The optimal $\vec{w}$ and b correspond to the plane that separates the two categories the best in the SVM. According to the Karush-Kuhn-Tucker conditions (KKT), the original and dual problems will have the same optimal values. The following optimal solution to the dual problem can be obtained:

$$w_j^* = \sum_{i=1}^{I} \alpha_i y_i x_{ij} \tag{5.12}$$

Or equivalently,

$$\vec{w}^* = \sum_{i=1}^{I} \alpha_i y_i \vec{x}_i \tag{5.13}$$

The calculation of b^* requires the use of support vectors. Let us recall the following equation (for margin boundary) that the support vector needs to satisfy:

$$y^* \cdot (\vec{w}^{*T} \cdot \vec{x}^* + b^*) = 1 \tag{5.14}$$

where $\vec{x}^*$ and y^* are a support vector (vector/array for a point on the margin boundaries) and its label, respectively. The symbol "*" is used for such a sample and its label because the point also corresponds to the optimal margin. Due to the same reason, $\vec{w}$ and b also carry the "*" sign.

The support vectors are associated with positive Lagrange multipliers. This is because any inequality constraint is defined as $g_i(x^*) \geqslant 0$. In optimization, we have the complementary slackness condition as part of the KKT conditions. This condition states that $g_i(x^*) \cdot \alpha_i = 0$. For points that are not support vectors, we have $g_i(x^*) > 0$; hence, we must have $\alpha_i = 0$.

Multiplying two sides of Eq. 5.13 with y^*, we get

$$y^{*2} \cdot (\vec{w}^{*T} \cdot \vec{x}^* + b^*) = y^* \tag{5.15}$$

We know $y^{*2} = 1$ because y can only be ± 1, so we have

$$b^* = y^* - \vec{w}^{*T} \vec{x}^* = y^* - \left(\sum_{i=1}^{I} \alpha_i y_i \vec{x}_i \right)^T \cdot \vec{x}^* \tag{5.16}$$

After the $\vec{w}^*$ and b^* are obtained, the function of the optimal hyperplane separating the two classes is

$$\vec{w}^{*T} \cdot \vec{x} + b^* = 0 \tag{5.17}$$

Accordingly, we can use the following equation for predictions in a binary classification task with the SVM:

$$f(\vec{x}) = \text{sign}(\vec{w}^{*T} \cdot \vec{x} + b^*) \tag{5.18}$$

where "sign" is the sign function, which outputs 1 if the input is greater than 0, output 0 if the input is 0, and output -1 if the input is lower than 0.

5.3 Generalization of SVM: Kernel Methods

In reality, many problems are not linearly classifiable. This means that, in a sample space, the points in the two categories cannot be separated by a plane or a hyperplane. In order to classify such samples, we can map the samples from the original space, i.e., $\vec{x}$, into a space with higher dimensions, i.e., $\phi(\vec{x})$. It is possible that a hyperplane can be sought for the data points in the high-dimensional space. The function of the hyperplane in the new space can be formulated as

$$f(\vec{x}) = \vec{w}^T \cdot \phi(\vec{x}) + b \tag{5.19}$$

The optimization problem becomes

$$\begin{cases} \min_{\vec{w},b} \frac{1}{2}\|\vec{w}\|^2 \\ \text{s.t. } y_i \cdot (\vec{w}^T \cdot \phi(\vec{x}_i) + b) \geqslant 1, i = 1, 2, \cdots, I \end{cases} \tag{5.20}$$

The corresponding dual problem is

$$\begin{cases} \min_{\vec{\alpha}} \left\{ \frac{1}{2} \sum_{i=1}^I \sum_{j=1}^I \left[\alpha_i \alpha_j y_i y_j \phi(\vec{x}_i)^T \cdot \phi(\vec{x}_j)\right] - \sum_{i=1}^I \alpha_i \right\} \\ \text{s.t. } \sum_{i=1}^I \alpha_i y_i = 0, \qquad 0 \leqslant \alpha_i, i = 1, 2, \cdots, I \end{cases} \tag{5.21}$$

The solution to the above problem involves the calculation of $\phi(\vec{x}_i)^T \cdot \phi(\vec{x}_j)$. This is the inner product of the mappings of $\vec{x}_i$ and $\vec{x}_j$. Because the mapping leads to a space with a higher dimension, the calculation of the inner product is usually challenging. Therefore, we usually introduce the following kernel function [56]:

$$k(\vec{x}_i, \vec{x}_j) =< \phi(\vec{x}_i), \phi(\vec{x}_j) >= \phi(\vec{x}_i)^T \cdot \phi(\vec{x}_j) \tag{5.22}$$

The function of the hyperplane with a kernel function, which is called the support vector expansion, is written as

$$f(\vec{x}) = \vec{w}^T \cdot \phi(\vec{x}) + b = \sum_{i=1}^{I} \alpha_i y_i \phi(\vec{x}_i)^T \cdot \phi(\vec{x}_i) + b = \sum_{i=1}^{I} \alpha_i y_i k(\vec{x}_i, \vec{x}_i) + b \tag{5.23}$$

The following are the common kernel functions:

(1) Linear kernel: $k(\vec{x}_i, \vec{x}_j) = \vec{x}_i^T \cdot \vec{x}_j$

(2) Polynomial kernel: $k(\vec{x}_i, \vec{x}_j) = (\vec{x}_i^T \cdot \vec{x}_j)^d$

(3) Gauss kernel: $k(\vec{x}_i, \vec{x}_j) = \exp\left(-\frac{\|\vec{x}_i - \vec{x}_j\|^2}{2\sigma^2}\right)$

(4) Laplace kernel: $k(\vec{x}_i, \vec{x}_j) = \exp\left(-\frac{\|\vec{x}_i - \vec{x}_j\|}{\sigma}\right)$

(5) Sigmoid kernel: $k(\vec{x}_i, \vec{x}_j) = \tanh(\beta \vec{x}_i^T \cdot \vec{x}_j + \theta)$

It is worthwhile to point out that SVM is relatively sensitive to the selection of kernels in addition to missing data. Like the applications of kernels in other types of machine learning, the selection of kernels in real problems still heavily relies on experience. There is a lack of theories for guiding and explaining the selection of suitable kernels to address specific problems with SVM.

5.4 Soft-Margin SVM

Overfitting is a major potential issue in the application of SVMs. To overcome overfitting, we learned from previous chapters that linear models include regularization and decision trees use pruning techniques. SVM adopts a very unique soft-margin technique. This technique can prevent us from learning from noise data points.

The mapping of data to a higher-dimensional space helps consider nonlinearity in data. However, this still does not address issues caused by the noise data points or outliers. For example, there may be some "bad" data points that are located in the wrong groups. In this case, we can use a "soft" margin, which will not be strictly enforced as a "hard" margin, to overlook these noise or bad points. With a soft margin, the learning agent can avoid learning from these points. Soft-margin SVM allows data points to fall within the margin or even on the wrong side of the decision boundary [57].

5.4.1 *Basic Formulation*

To understand the concept of soft margin, let us first revisit the equation for hard margin:

$$y_i \cdot (\vec{w}^T \cdot \vec{x}_i + b) \geqslant 1, i = 1, 2, \cdots, I \tag{5.24}$$

In order to allow a point, e.g., Point i, to sit inside the margin area, we can reduce the value of 1 with a relaxation or slack variable ξ_i for this point:

$$y_i \cdot (\vec{w}^T \cdot \vec{x}_i + b) \geqslant 1 - \xi_i \tag{5.25}$$

The value $1 - \xi_i$ gives out the minimum required margin for Point i. If a point is inside the margin area are while on the correct side (the side of its true label) of the center line ($\vec{w}^T \cdot \vec{x} - b = 0$), then $0 < \xi_i < 1$. If this point surpasses the center line but remains inside the margin area, the relaxation variable needs to be $\xi_i < 0$. Usually, we require $\xi_i \geqslant 0$. We can assign different ξ values to different points to allow different degrees of relaxation at different data points. For points that do not violate the hard-margin requirement, we have $\xi = 0$.

In addition to the relaxation variable for different points, we also frequently introduce a penalty parameter C. This regularization parameter is used to multiply the sum of all the relaxation variables as a term in the loss function. The loss function is combined with the constraints to obtain the following mathematical description of the SVM with the soft margin:

$$\begin{cases} \min_{\vec{w},b} \frac{1}{2}\|\vec{w}\|^2 + C\sum_{i=1}^{I} \xi_i \\ \text{s.t. } y_i \cdot (\vec{w}^T \cdot \vec{x}_i + b) \geqslant 1 - \xi_i, \xi_i \geqslant 0, i = 1, 2, \cdots, I \end{cases} \tag{5.26}$$

where the penalty parameter C tells how much the violation of the hard-margin constraint will be penalized. A greater C value poses a higher preference for points that less violate the hard-margin constraint. That is, we will have fewer points violating the hard-margin requirement. Accordingly, the model will need to become more complicated (less smooth) to allow this to happen.

5.4.2 *Dual Formulation*

The loss function of the corresponding dual problem is

$$\ell(\vec{w}, b, \vec{\xi}, \vec{\alpha}, \vec{\mu}) = \frac{1}{2}\|\vec{w}\|^2 + \sum_{i=1}^{I} C\xi_i + \sum_{i=1}^{I} \alpha_i[-y_i \cdot (\vec{w}^T \cdot \vec{x}_i + b) + 1 - \xi_i] - \sum_{i=1}^{I} \mu_i \xi_i \tag{5.27}$$

where $\alpha_{I\times 1}$ and $\mu_{I\times 1}$ are Lagrangian multipliers. The second and third terms on the right-hand side are arranged in this way (e.g., the negative sign in front of the third term) because we should enforce inequality constraints as $f \leqslant 0$ so that the Lagrange multipliers are non-negative. However, the Lagrange multipliers for equality constraints as $f = 0$ can be positive or negative depending on the problem and adopted conventions.

To obtain the optimal value, the gradient of the above loss equation with respect to $\vec{w}$, b, and $\vec{\xi}$ must be zero:

$$\nabla_{\vec{w}} \ell(\vec{w}, b, \vec{\xi}, \vec{\alpha}, \vec{\mu}) = \vec{w} - \sum_{i=1}^{I} \alpha_i y_i \vec{x}_i = 0 \tag{5.28}$$

$$\nabla_b \ell(\vec{w}, b, \vec{\xi}, \vec{\alpha}, \vec{\mu}) = -\sum_{i=1}^{I} \alpha_i y_i = 0 \tag{5.29}$$

$$\nabla_{\xi_i} \ell(\vec{w}, b, \vec{\xi}, \vec{\alpha}, \vec{\mu}) = C - \alpha_i - \mu_i = 0 \tag{5.30}$$

The solution to the above three equations are

$$\vec{w}^* = \sum_{i=1}^{I} \alpha_i y_i \vec{x}_i \tag{5.31}$$

$$\sum_{i=1}^{I} \alpha_i y_i = 0 \tag{5.32}$$

$$C - \alpha_i - \mu_i = 0 \tag{5.33}$$

Substituting the above equation into the Lagrange function yields the dual problem that is identical to the hard-margin problem except for the upper bound (i.e., $\alpha_i \leqslant C$) for the Lagrange multipliers α_i:

$$\begin{cases} \min_{\vec{\alpha}} \left[\frac{1}{2} \sum_{i=1}^{I} \sum_{j=1}^{I} (\alpha_i \alpha_j y_i y_j \vec{x}_i^T \cdot \vec{x}_j) - \sum_{i=1}^{I} \alpha_i \right] \\ \text{s.t. } \sum_{i=1}^{I} \alpha_i y_i = 0, \qquad 0 \leqslant \alpha_i \leqslant C, i = 1, 2, \cdots, I \end{cases} \tag{5.34}$$

The parameter b^* then can be calculated as

$$b^* = y^* - \vec{w}^{*T} \cdot \vec{x}^* = y^* - \left(\sum_{i=1}^{I} \alpha_i y_i \vec{x}_i \right)^T \cdot \vec{x}^* \tag{5.35}$$

After the w^* and b^* are obtained, the function of the hyperplane separating the two classes is determined as

$$\vec{w}^{*T} \cdot \vec{x} + b^* = 0 \tag{5.36}$$

Accordingly, we can use the following equation for SVM in a binary classification task:

$$f(x) = \text{sign}(\vec{w}^{*T} \cdot \vec{x} + b^*) \tag{5.37}$$

5.5 More About SVM

A few issues with SVM need to be mentioned to better understand and apply SVM.

5.5.1 *SMO Algorithm*

First, the training of SVM can encounter problems with large datasets. As shown in the solutions, e.g., equations for calculating $\vec{w}^*$ and b^*, the implementation of these equations will involve multiplications and additions between arrays. This will consume large amounts of memory and time when the dataset is large. A popular method to address this problem is the sequential minimal optimization (SMO) method [58]. Essentially, SMO breaks an SVM optimization problem into a series of smallest possible subproblems, which are then solved analytically. SMO is an optimization topic and thus not introduced in detail here.

5.5.2 *SVM for Multiclass Classification and Regression*

Second, the classic SVM was developed to address binary classification problems. Additional effort is needed to extend such SVM to multiclass classification problems and regression problems. For multiclass classification, this can be accomplished both directly and indirectly, while for regression, we will need to slightly modify the way of using the margin for constructing the loss function.

The direct way of extending SVM for multiclass classification is to search for multiple hyperplanes instead of one to separate the three or more categories of data. This can be done directly by adding the loss associated with additional hyperplanes to the loss function. Though this extension appears straightforward, the implementation may require extensive computing. The indirect method involves the combination of multiple binary classifiers to generate multiclass classifiers, either one-vs-all (or rest) and one-vs-one. In one-vs-one, we create K binary classifiers for a dataset with K classes, for which each classifier tells whether a sample belongs to one class or the other classes. Then we compare the results of the K samples and determine which class is the most likely one, e.g., based on distance or probability. This method usually leads to relatively poor results with high biases. In one-vs-one, we will need to train $C_2^K = \frac{P_2^K}{2!} = \frac{K!}{2!(K-2)!} = \frac{K*(K-1)}{2}$ binary classifiers to consider all binary classification possibilities. Then, a sample will be tested with all these binary classifiers, and the class with the most votes will be determined as the label of the sample in the multiclass classification problem. It is not difficult to see that this method requires more computation due to the involvement of more binary classifiers.

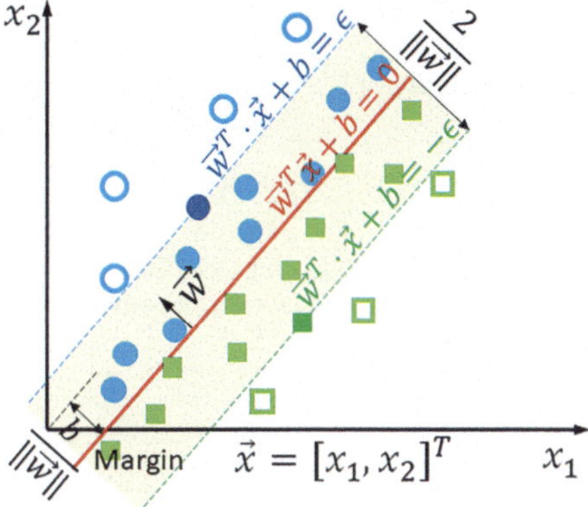

Fig. 5.2 Conceptual illustration of SVR

SVM can be applied to regression, which can lead to support vector regression (SVR) as an extension or application of the classic SVM. As illustrated in Fig. 5.2, when using the SVR, we will also need to find a hyperplane. The difference is that we now want the total of distances between all the data points and this plane to be as small as possible. In practice, we usually set up a soft margin or threshold (ϵ), so that all the data points within the margin area are considered to have no loss. Then, we just need to calculate the distances between the points out of the margin area and the hyperplane to obtain the total loss.

5.6 Practice: Use of SVMs in Scikit-Learn for Classification and Regression

>>> More and Up-to-Date Course Materials including Practices @ AI-engineer.org <<<

The following code presents an implementation of the SVM. Please modify the code to implement the SVM using the dual problem formulation.

Hints: You will need to change some lines between "# Prepare optimizer function" and "# Generate data for the SVM separation curve." You will need to modify the objective (loss) function (variables will be changed from "wb0" to "alpha" (for $\vec{\alpha}$)) and the associated constraints (the dual problem has both equality and inequality constraints). After the Lagrange multipliers, i.e., alpha, are obtained, you will need to obtain $\vec{w}^*$ and b^*, for which you will need to identify a support vector for calculating b^*.

```
import numpy as np
import matplotlib.pyplot as plt
from scipy.optimize import minimize
```

```
X = np.array([[-1, -1.5], [0, -1], [1, -0.5], [-0.5, 0], [1, -1],
     [1.5, 1], [1, 2], [0.5, 1.5], [2.5, 1.5], [2, 2]])
y = np.array([1, 1, 1, 1, 1, -1, -1, -1, -1, -1])


# Prepare optimizer function
tole = 1e-8 # Tolerance
fun = lambda wb: wb[:-1]@wb[:-1] # Optimize that work with lambda
     function with one argument (general function can have more
    arguments)
# Add equality constraints
cons = (#{'type': 'eq', 'fun': lambda alpha: alpha @ y}, # func
    =0
{'type': 'ineq', 'fun': lambda wb: y*(X@wb[:-1]+wb[-1])-1} # func
     > 0
)
# Optimization
w0 = np.ones(X.shape[1])/X.shape[1] # Initial value; alpha0 must
    satisfy the above constraints
b0 = 1
wb0 = np.append(w0,b0)
res = minimize(fun, wb0, method='SLSQP',constraints=cons,tol=tole
    ) # SLSQP BFGS
wb = res.x
print(wb)

# Obtain the model
w_star = wb[:-1]
# alpha[alpha>1e-4] yields the indices for the positive elements
    (2 or more) in Lagrange multipliers (support vectors)
b_star = wb[-1] # [0] gives out the first support vector
# margin_star = 2/np.abs(np.sum(w_star**2))
print('The parameter for getting the minimum loss is:',w_star,
    b_star)

# Generate data for the SVM separation curve
a = -w_star[0] / w_star[1]
xx = np.linspace(X[:,0].min(), X[:,0].max(),100)
yy = a * xx + (b_star) #/ w_star[1]

fig = plt.figure()
# Plot original data with label
plt.scatter(X[:,0],X[:,1],marker='o',color='b')
for X1,y1 in zip(X,y):
label = "{:.2f}".format(y1)
plt.annotate(label, # this is the text
(X1[0],X1[1]), # these are the coordinates to position the label
textcoords="offset points", # how to position the text
xytext=(0,10), # distance from text to points (x,y)
ha='center') # horizontal alignment can be left, right or center
# Plot the SVM separation curve
plt.scatter(xx,yy,marker='.',color='r')
```

Chapter 6
Bayesian Algorithms

6.1 Overview

This chapter introduces Bayesian methods, which are also called Bayesian algorithms, Bayesian machine learning, and probabilistic machine learning in the literature. Because such Bayesian methods were initially proposed for and closer to classification tasks by nature, the use of the "Bayesian classifier" is also very common or even predominant in some technical publications. However, it is worthwhile to mention that such Bayesian classifiers, which belong to parametric machine learning methods, can also be extended to regression tasks. Other nonparametric Bayesian methods like Bayesian process are proposed more for regression and may appear much different from traditional Bayesian classifiers.

In this chapter, we will first provide a general background for statistics-based machine learning, which contains statistical inference adopted by both frequentists and Bayesians. The frequentists' inference method, i.e., maximum likelihood estimation, is used by many other machine learning methods like artificial neural networks, while Bayesians' method, i.e., Bayesian estimation, is adopted as the basis of Bayesian methods in this chapter. Then, major parametric Bayesian methods, e.g., naive Bayes classifier, Bayesian networks, and Markov processes, will be discussed. Next, one nonparametric Bayesian method, i.e., Gaussian process, will be introduced.

This chapter differs from the other chapters in a few ways, which will need close attention to avoid possible confusion. First, the diverse, incomplete, and conflicting statistics contents in different machine learning literature, especially Bayesianism-related topics, can cause severe learning difficulties. We will start the chapter with an introduction to extra statistics knowledge considering the overlap between machine learning and applied statistics. You can skip this section if you just want to know how specific Bayesian methods work. Second, a few other criteria for classifying machine learning methods like discriminative versus generative and parametric

Z. "L." Liu, *Artificial Intelligence for Engineers*,
https://doi.org/10.1007/978-3-031-75953-6_6

versus nonparametric can also be hard to understand. These will also be explained when appropriate as we move through the technical content.

6.2 Statistics Background for Machine Learning

6.2.1 Statistics and Machine Learning

As can be found in the appendix for statistics, the role and significance of statistics in the general context of machine learning can vary depending on the background, goal, and focus of machine learning practitioners. Despite this fact, we need to be aware that, for some machine learning methods, especially the Bayesian methods in this chapter, a strong background in statistics is required in order to better understand and utilize such methods. Due to this consideration, statistics knowledge is presented in this chapter in addition to what is provided in the appendices. As a result, this section will offer a big picture for the role of statistics in machine learning, which is both needed in this chapter and desired or helpful for other machine learning topics.

6.2.2 Frequentists and Bayesians

In statistics, there are two mainstreams: frequentism and Bayesianism. Frequentists build the probability from past events. In this way, the probability does not depend on our beliefs, but instead, objective measurements of what have occurred. For example, if we toss a coin 100 times and heads show up 38 times, then the frequency/probability of heads is $38/100 = 0.38$. By contrast, Bayesians view probability as a measure of belief, and thus, the probability is subjective and refers more to the future. Due to this reason, the workflow of Bayesians starts with a belief, or called a prior, and then obtains some data to update our belief for an outcome called a posterior. As more data is obtained, the posterior becomes a new prior, and this process can be repeated to improve the model with more data.

When using statistics to address problems, such as classification and regression problems in machine learning, we essentially try to understand how different samples are distributed in the sample space, which determines their discrete (classification) or continuous (regression) labels/targets. We can understand the learning (or narrowly, training) process as an effort to search for the best statistical model to describe the data, so that such a model can be used for future predictions. Such a model can be defined with parameters represented by $\vec{\theta} = [\theta_1, \theta_2, \cdots]$. So machine learning for both classification and regression essentially aims to infer the parameters $\vec{\theta}$ for defining the model.

The difference between the two ways of treating probability also leads to the reality that frequentists mostly use the maximum likelihood estimation, while Bayesians adopt Bayesian estimation. In the following, these two statistical inference methods will be introduced first. The former may be brought up again when introducing other machine learning methods. The introduction to the latter will recall and further elaborate on some statistical concepts introduced in the appendix for statistics as well as other basic statistics literature.

Here, let us first recall Bayes' theorem, which will be used in different ways in this chapter:

$$P(A \mid B) = \frac{P(B \mid A)P(A)}{P(B)}, \text{ if } P(B) \neq 0. \tag{6.1}$$

where

A and B are events and $P(B) \neq 0$.
$P(A \mid B)$ is a conditional probability: the probability of event A occurring given that B is true. It is also called the posterior probability of A given B.
$P(B \mid A)$ is also a conditional probability: the probability of event B occurring given that A is true. It can also be interpreted as the **likelihood** of A given a fixed B.
$P(A)$ and $P(B)$ are the prior probabilities (also called marginal probabilities) of observing A and B, respectively.

If we substitute the law of total probability into the Bayes' theorem, we can obtain the format of the Bayes' theorem that we typically use in Bayesian classifiers:

$$P(A \mid B) = \frac{\sum_{n=1}^{N} P(B_n \mid A)P(A)}{P(B)} \tag{6.2}$$

where $\{B_n : n = 1, 2, 3, \ldots, N\}$ is a finite or countably infinite partition of a sample space (in other words, a set of pairwise disjoint events whose union is the entire sample space), in which each event B_n is measurable, and A is an event in the same probability space.

6.2.3 *Overview of Statistical Inference*

Statistical inference aims to learn characteristics of a population, i.e., the underlying probability distribution of all the possible data, from observed data sampled from the population [59]. The goal of statistical inference or development of statistical models in machine learning, no matter with maximum likelihood estimation (MLE), maximum a posteriori estimation (MAP), or Bayesian inference, is to infer the optimal model parameterized by $(\vec{\theta})$ from data $(\bar{X})$.

The Bayes' theorem can be reformulated as follows for statistical inference:

$$P(\vec{\theta}|\bar{X}) = \frac{P(\bar{X}|\vec{\theta}) \cdot P(\vec{\theta})}{P(\bar{X})} \tag{6.3}$$

$P(\bar{X})$ is something we generally cannot compute. However, the value of $P(\bar{X})$ does not matter that much because it is just a normalization constant. When comparing models, we are mainly interested in expressions containing $\vec{\theta}$, because $P(\bar{X})$ stays the same for each model.

To explain MLE, MAP, and Bayesian using the above equation, it is better to first recall the fact that an algorithm is usually selected prior to an estimation. This algorithm defines a hypothesis space as explained in previous chapters. In other words, this algorithm provides a collection (a space) of possible models (hypothesis). Next, let us use the following analogy to easily understand the thing we are trying to do. What we have is some data ($\bar{X}$) generated by one or multiple machines (analogous to models). Our aim is to look for such a machine(s), related to or defined by $\vec{\theta}$, so that we can best reproduce the same data. In this analogy, selecting an algorithm is analogous to finding a manufacturer who has a collection of such machines.

MLE aims to maximize the likelihood as follows:

$$\vec{\theta}^* = \arg\max_{\vec{\theta}} P(\bar{X}|\vec{\theta}) \tag{6.4}$$

Thus, MLE tries to identify one model parameterized by $\vec{\theta}$ that mostly likely generates $\bar{X}$. That is, the probability of generating $\bar{X}$ under the condition of $\vec{\theta}$ is maximized in MLE.

MAP differs from MLE in that MAP includes prior knowledge. In the same analogy, we can understand this as that different machines have different probabilities of delivering the claimed performance. $P(\vec{\theta})$ can be viewed as prior knowledge provided by the manufacturer about such probabilities. If we include this term in the estimation, we can obtain the following equation based on the MLE equation (Eq. 6.4) and the above Bayes' equation (Eq. 6.3):

$$\vec{\theta}^* = \arg\max_{\vec{\theta}} P(\bar{X}|\vec{\theta})P(\vec{\theta}) = \arg\max_{\vec{\theta}} P(\vec{\theta}|\bar{X}) \tag{6.5}$$

As can be seen, the major difference between MLE and MAP is whether to include prior knowledge $P(\vec{\theta})$. In a more strict description, $P(\vec{\theta})$ is (prior knowledge about) the probability of the model parameterized by $\vec{\theta}$ in the hypothesis space. This difference actually explains why MLE is more susceptible to overfitting than MAP on datasets with outliers. This is because MLE is more inclined to fit to outliers without prior knowledge for excluding such outliers. Two simple algorithms that represent MLE and MAP are logistic regression with and without L2 regularization. Without regularization, logistic regression uses MLE and would contain all data points including outliers. With regularization, logistic regression can filter out outliers that do not comply with the regularization term.

Bayesian inference is much different because it searches for a possible combination of different models, or ensemble, instead of one model. Therefore, instead of the optimal model $\vec{\theta}^*$, Bayesian methods aim to obtain the possibilities of different models $P(\vec{\theta}|\bar{X})$, or more accurately, $P(\vec{\theta}_t \bar{X})$. The number of models can be large or even infinite. In the latter case, we can perform a certain number of sampling, e.g., T $(t \in \{1, \cdots, T\})$. Then, the prediction made by the overall model (or ensemble) is the average weighted by $P(\vec{\theta}_t \bar{X})$.

Compared with MLE and MAP, Bayesian methods have the following advantages:

- Bayesian methods perform better on small datasets. This is because, as mentioned, they are by nature ensemble learning. Due to the same reason, they are less susceptible to overfitting.
- It is more convenient to incorporate prior knowledge into the Bayesian estimate. Therefore, extra constraints or assumptions about data or model distributions can be easily integrated.
- Bayesian methods can better handle uncertainty. MLE and MAP only obtain $\vec{\theta}$ via learning, e.g., weights of linear models and ANNs. Such deterministic predictions may not be adequate or reasonable in many conditions. For example, a classification model of this type outputs labels corresponding the classes with the highest probabilities. However, no information is available about the confidence level of such a prediction. Bayesian can provide such confidence information, which can prompt us to intervene when the confidence level is too low.

In the following, we will first provide more information about the use of MLE. After that, the general Bayesian estimation will be introduced, followed by detailed parametric and nonparametric Bayesian methods.

6.2.4 *Maximum Likelihood Estimation (MLE)*

Suppose we have a training dataset $\bar{X}$, which has I independent and identically distributed (called i.i.d. in many places) samples. Let us assume the distribution of the samples can be described using a statistical model characterized by $\vec{\theta}$. The basic idea of MLE in frequentism is that $\vec{\theta}$ is unknown but exists objectively, and $\vec{\theta}$ as a constant(s) can be estimated by ensuring that a given dataset (for training) has the maximum likelihood to appear (or be observed). Following this understanding, we can obtain the mathematical formulation of MLE:

$$\theta_{MLE} = \arg\max_{\vec{\theta}} O(\bar{X}|\vec{\theta}) = \arg\min_{\vec{\theta}} \ell(\bar{X}|\vec{\theta}) \tag{6.6}$$

where $P(\vec{x}_i|\vec{\theta})$ is the probability that the occurrence of sample $\vec{x}_i$ is predicted by the model parameterized by $\vec{\theta}$.

The MLE function $O(\bar{X}|\vec{\theta})$ is the objective function constructed as the total probability of all the samples. The loss function ℓ is the opposite of the objective function O. These equations can be constructed as follows if we assume the attributes are independent of each other:

$$\begin{aligned} O(\bar{X}|\vec{\theta}) &= \ell(\bar{X}|\vec{\theta}) = P(\bar{X}|\vec{\theta}) = P(\vec{x}_1, \cdots, \vec{x}_I|\vec{\theta}) \\ &= P(\vec{x}_1|\vec{\theta}) \cdot \cdots P(\vec{x}_i|\vec{\theta}) \cdots P(\vec{x}_I|\vec{\theta}) = \prod_{i=1}^{I} P(\vec{x}_i|\vec{\theta}) \end{aligned} \tag{6.7}$$

In a general multiclass classification problem, what we try to maximize is the expectation of the total probability. When different samples and different attributes are not equally weighted, the objective is calculated using $O(\vec{x}|\vec{\theta}) = \sum_{j=1}^{J} \lambda_{i,j} P(x_{ij}|\vec{\theta})$, in which $\lambda_{i,j}$ is used to represent the weights. Accordingly, we obtain

$$\theta_{MLE} = \arg\max_{\vec{\theta}} O(\bar{X}|\vec{\theta}) = \arg\max_{\vec{\theta}} \left[\sum_{j=1}^{J} \lambda_{i,j} P(x_{ij}|\vec{\theta}) \right] \tag{6.8}$$

In practice, the value of $P(x_i|\vec{\theta})$ can be very small because J can be large in a large dataset. The product of such small values may cause numerical issues. To allow for such problems, we usually use the log of the above function as the loss function, i.e., $\ell \leftarrow \ln(\ell)$, for optimization. Then, the mathematical formulation of MLE excluding weights $\lambda_{i,j}$ becomes

$$\ell = \sum_{i=1}^{I} \ln[-P(\vec{x}_i|\vec{\theta})] \tag{6.9}$$

$$\theta_{MLE} = \arg\max_{\vec{\theta}} \ln[P(\bar{X}|\vec{\theta})] \tag{6.10}$$

Let us employ a simple example to understand MLE. Here, we have a bag containing black and white balls, for which we do not know the ratio of black balls to white balls. We took a ball from the bag and put it back. We performed 100 ball-taking operations, and we got 75 white balls in total. What is the proportion of white balls?

MLE can be used to estimate the above parameters $\vec{\theta}$, that is, the proportion of white balls P in this example:

$$O(\bar{X}|\vec{\theta}) = P(\vec{x}_1|\vec{\theta}) \cdot P(\vec{x}_2|\vec{\theta}) \cdots P(\vec{x}_{100}|\vec{\theta}) = P^{75} \cdot (1 - P)^{25} \tag{6.11}$$

Then the natural log of the MLE likelihood function is

$$\ln P(\bar{X}|\vec{\theta}) = -75 \cdot \ln P - 25 \cdot \ln(1 - P) \tag{6.12}$$

The maximum value appears when the derivative of the above function is 0:

$$\frac{d(\ln P(\vec{x}|\vec{\theta}))}{dP} = -\frac{75}{P} + \frac{25}{1 - P} = 0 \tag{6.13}$$

Hence, we can get $P = 0.75$.

This method is relatively simple. However, its validity lies in a strict assumption that the dataset can represent the sample space. In other words, the distribution of the training samples needs to be the same or close to the distribution in the sample space.

6.2.5 *Bayesian Estimation*

As mentioned, Bayesian estimation is different from MLE in that it assumes the models or model parameters, i.e., $\vec{\theta}$, have their probability distribution and need to be estimated for given data. Therefore, the estimation of the model parameters can differ in different training data. Due to the same reason, the model estimation can also be updated as more data is obtained.

In the following, we can use classification tasks as an example to illustrate how Bayesian estimation models for classification, or Bayesian classifiers, are built up.

Let us consider a multiclass classification problem, in which we need to classify a dataset $\bar{X}$ consisting of I samples $\vec{x}_1, \vec{x}_2, \cdots, \vec{x}_i, \cdots, \vec{x}_I$ and each sample has J attributes $x_{i1}, x_{i2}, \cdots, x_{ij}, \cdots, x_{iJ}$. We will need to classify the samples into K categories: $y_i \in \{c_1, c_2, \cdots, c_k\}$. The goal is to predict the probability of any sample $\vec{x}_i$ belonging to the kth category: $P(c_k|\vec{x}_i)$.

Next, we slightly modified the above Bayes' theorem to illustrate its use in Bayesian estimation for classification purposes. We will replace $\vec{\theta}$ with c_k, which represents "belonging to class c_k," and replace $\bar{X}$ with sample "$\vec{x}_{ij}$." Thus, we are now considering individual models and samples instead of all the models (ensemble) and all the samples. Also, to maintain a consistent understanding with the previous subsection, we can understand c_k as a model that solely generates or predicts Class k samples:

$$P(c_k \mid \vec{x}_i) = \frac{P(\vec{x}_i, c_k)}{P(\vec{x}_i)} = \frac{P(\vec{x}_i|c_k)P(c_k)}{P(\vec{x}_i)} = \frac{P(x_{i1}, x_{i2}, \cdots, x_{iJ}|c_k)P(c_k)}{P(\vec{x}_i)} \tag{6.14}$$

where the joint probability $P(x_{i1}, x_{i2}, \cdots, x_{iJ}|c_k)$ is the probability that all the attribute values of sample $\vec{x}_i$ appear simultaneously. That is, each attribute of Sample

$\vec{x}_i$ has multiple attribute values, and the value taken by Sample $\vec{x}_i$ will have a probability. Such probabilities, as well as the joint probability, can be estimated using the training data.

The above formulation of Bayes' theorem casts a theoretical basis for all the Bayes classifiers. In fact, in a more general sense, it presents a framework for understanding all the classification methods or even all the supervised learning methods. Based on the way of using the above equation, algorithms can be classified into discriminative models and generative models.

Discriminative models assume a function (implicit or explicit) for $P(c_k|\vec{x}_i)$ directly. A majority of the classic machine learning algorithms like linear models, support vector machines, traditional ANNs, and KNNs are discriminative models. By contrast, generative models assume some function forms for $P(\vec{x}_i, c_k)$ and $P(\vec{x}_i)$—estimating the parameters of $P(\vec{x}_i, c_k)$ and $P(\vec{x}_i)$ directly from training data—to calculate $P(c_k \mid \vec{x}_i)$ indirectly. Bayesian methods like naive Bayes and Bayesian networks, as well as less common approaches like Markov random fields and hidden Markov models (HMM), produce generative models.

Intuitively, we can see that discriminative models are more straightforward and thus can be simpler than generative models due to the above fact. However, generative models may provide better insights into the data. Using classification tasks as an example, we can simply understand that discriminative models identify boundaries in the data space, whereas generative models attempt to model how data is distributed throughout the space.

6.3 Parametric Bayesian Methods

6.3.1 Naive Bayes Classifier

One major difficulty in applying the above framework (Eq. 6.14) for tasks like classification is that the joint probability, i.e., $P(x_{i1}, x_{i2}, \cdots, x_{iJ}|c_k)$, is not easy to deal with. This is because it is hard to tell the relationship between attributes or their corresponding random variables. In particular, some of these variables may be mutually dependent. In that case, the calculation of $P(x_{i1}, x_{i2}, \cdots, x_{iJ}|c_k)$ will be really challenging because it requires us to identify such interdependencies first.

The naive Bayes classifier simplifies the problem by assuming that all the attributes (the corresponding random variables) are independent of each other. As a result, the joint probability becomes the product of the probabilities of all the attributes: $P(x_{i1}, x_{i2}, \cdots, x_{iJ}|c_k) = \prod_{j=1}^{J} P(x_{ij}|c_k)$. Substituting this into Eq. 6.14, we can obtain the major equation for the naive Bayes classifier:

$$P(c_k \mid \vec{x}_i) = \frac{\prod_{j=1}^{J} P(x_{ij}, c_k)}{P(\vec{x}_i)} = \frac{\prod_{j=1}^{J} P(x_{ij} \mid c_k)P(c_k)}{P(\vec{x}_i)} \tag{6.15}$$

where $P(x_{ij}|c_k)$ is the probability of the attribute value of the jth attribute of Sample $\vec{x}_i$ in all the samples in Class c_k. Thus, x_{ij} only provides a specific attribute value out of all possible V values for this attribute, i.e., a_v, in which $v \in \{1, 2, \cdots, V\}$.

It is noted that, in some literature, the subscripts i and k are omitted to simplify the use of the Bayes' theorem in Bayesian estimation for multiclass classification problems. Accordingly, $\vec{x}$ is a vector representing any sample, and c is any class. This equation is presented in the following but will not be adopted to ensure a more accurate in the description:

$$P(c \mid \vec{x}) = \frac{\prod_{j=1}^{J} P(x_j \mid c)P(c)}{P(\vec{x})} = \frac{P(c)}{P(\vec{x})} \prod_{j=1}^{J} P(x_j|c) \tag{6.16}$$

The use of Eq. 6.15 requires us to find a way of calculating $P(x_{ij} \mid c)$ and $P(c_k)$. For $P(c_k)$, it can be easily computed as the percentage of Class c_k samples in all the samples. Thus, the following equation can be used:

$$P(c_k) = \frac{|D_{c_k}|}{|D|} \tag{6.17}$$

where $|D_{c_k}|$ and $|D|$ are the numbers of the samples in Class c_k and in the whole dataset, respectively.

The calculation of $P(x_{ij}|c_k)$ will be much different for discrete (nonnumeric) and continuous (numeric) attribute values. When we have a finite number of discrete values for an attribute, then $P(x_{ij}|c_k)$ can be calculated as the ratio of samples with the attribute value x_{ij}, i.e., $D_{c_k,x_{ij}}$, to the number of all the samples in this class, D_{c_k}:

$$P(x_{ij}|c_k) = \frac{|D_{c_k,x_j}|}{|D_{c_k}|} \tag{6.18}$$

where $|D_{c_k,x_j}|$ is the number of samples with attribute value j in Class c_k. Thus, it is actually not related to Sample $\vec{x}_i$.

When we have continuous attribute values such as float numbers, the probability $P(x_{ij}|c_k)$ can be computed by assuming that the attribute (random variable) follows a probability distribution. A predominant assumption is the Gaussian distribution: $P(\vec{x}_{ij}|c_k) \sim \mathcal{N}(\mu_{k,j}, \sigma_{k,j}^2)$. Based on this assumption, $P(x_{ij}|c_k)$ can be computed as

$$P(x_{ij}|c_k) = \frac{1}{\sqrt{2\pi}\sigma_{k,j}} \exp\left[-\frac{(x_{ij} - \mu_{k,j})^2}{2\sigma_{k,j}^2}\right] \tag{6.19}$$

The above equation can be used to calculate the probability of any attribute of any sample. Parameters about the variable distribution, i.e., $\mu_{k,j}$ and $\sigma_{k,j}$, can be obtained from the training data. For example, we can find all the samples in Class

c_k and assess the distribution of the values of attribute j to compute the mean and the standard deviation for the probability density function: $\mu_{k,j}$ and $\sigma_{k,j}$.

When all the attributes have continuous values, we will need to use a Gaussian distribution for each of them. The naive Bayes classifier, in this case, is also called the Gaussian naive Bayes classifier in some places.

The implementation of the naive Bayes classifier is very straightforward. What can be confusing is that the training and testing processes are lumped together in some way. Let us assume that we want to predict the classification of a sample $\vec{x}_i$ based on a given labeled training dataset $\bar{X}$. Then we will just need to calculate $P(c_k)$ and $P(x_{ij}|c_k)$ with the training data using Eq. 6.17, Eq. 6.18, and Eq. 6.19. In particular, $P(x_{ij}|c_k)$ can be calculated directly for discrete attribute values, while for continuous attribute values, the computation of μ and σ needs to be carried out with the training data before the computation of the probability. Next, we just need to use Eq. 6.15 for predicting the labels of new samples.

As can be seen, naive Bayes classifiers need us to assess the distribution of the data. Thus, to ensure a smooth implementation, we need enough data to compute the probabilities. For those probabilities calculated using percentages, we will need to have samples in the numerators and denominators or the ratios. Usually, we modify the equations for calculating $P(c_k)$ and $P(x_{ij}|c_k)$ for discrete attribute values. The following Laplace correction is commonly adopted to address this issue, which can help smoothen the predicted probability distributions:

$$P(c_k) = \frac{|D_{c_k}| + 1}{|D| + K} \tag{6.20}$$

$$P(x_{ij}|c_k) = \frac{|D_{c_k,x_j}| + 1}{|D_{c_k}| + N_j} \tag{6.21}$$

where K is the number of classes in the training dataset and N_j is the total number of attribute values in Attribute j.

For probabilities calculated using distribution functions, i.e., $P(x_{ij}|c_k)$ for discrete attribute values, we will need to make sure we have enough training samples to obtain a distribution function. This can help avoid the unintended removal of information about attributes due to inadequate or missing attribute values.

The following code shows how to train a regression model using the Gaussian naive Bayes algorithm explained in this subsection. The Iris dataset from Scikit-learn was adopted for training. The performance of the trained model was compared with that obtained using the Gaussian naive Bayes from Scikit-learn.

```
from sklearn.datasets import load_iris
from sklearn.model_selection import train_test_split
import numpy as np
from sklearn.naive_bayes import GaussianNB

# ----------------- Load data ----------------- #
x, y = load_iris(return_X_y=True)
```

```
x_train, x_test, y_train, y_test = train_test_split(x, y,
    random_state=20190308, test_size=0.3)

# Calculate the probabilities of every attribute value P(x_j|c)
    for P(x|c)
def gaussion_pdf(x_test, x):
    return np.exp(-(x_test-x.mean(0))**2 / (2 * x.std(0)**2)) /
    np.sqrt(2 * np.pi * x.std(0)**2)

# -------------- Define Naive Bayes Model -------------- #
classes = np.unique(np.concatenate([y_train, y_test], 0))
pred_probs = []

# Loop over all classes to obtain the probabilities of all
    testing samples belonging to different classes
for i in classes:
    idx_i = y_train == i # idx_i stores all indices of samples in
     Class c (y_train == i)
    p_c = len(idx_i) / len(y_train) # Calculate P(c)
    p_x_c = np.prod(gaussion_pdf(x_test, x_train[idx_i]), 1) #
    Use Gaussian to calculate P(x|c) for continuous attribute
    values
    prob_i = p_c * p_x_c # Joint probability prob_i is the
    probability of sample i belonging to Class c
    pred_probs.append(prob_i)
# Array for probabilities of samples belonging to different
    classes: N_samples x N_classes
pred_probs = np.vstack(pred_probs).T
# Index of the class with the highest probability as the
    classification for each sample
label_idx = pred_probs.argmax(1)
y_pred = classes[label_idx] # Select the class based on the index

score = np.count_nonzero(y_test == y_pred)/len(y_test) # Accuracy
    : correct when the predicted class is the same as the actual
    label
print('The accuracy of the self-developed Gaussian Naive Bayes is
    : ',score)

# ------------------ Apply Gaussian Naive Bayes from Scikit-learn
     ------------------ #
model = GaussianNB()
model.fit(x_train, y_train)
print('The accuracy of Gaussian Naive Bayes from Scikit-learn is:
     ', model.score(x_test, y_test))
```

The execution of the above code produced the following results:

```
The accuracy of self-developed Gaussian Naive Bayes is:
    0.9333333333333333
The accuracy of Gaussian Naive Bayes from Scikit-learn is:
    0.9333333333333333
```

6.3.2 Semi-Naive Bayesian Classifier

Naive Bayesian classifiers are fairly simple to understand and implement. However, the independence assumption of attributes adopted by naive Bayesian classifiers may be difficult to hold in many cases. To address this issue, semi-naive Bayesian classifiers are proposed by relaxing the assumption of the conditional independence of attributes to some extent. This "semi-relaxation" of the independence requirement explains the origin of "semi-naive."

Semi-naive Bayesian classifiers still consider the relatively strong attribute dependencies but overlook weak interdependence to reach a compromise between accuracy and convenience. The different ways of selecting dependencies lead to different semi-naive Bayesian classifiers. In the following, common semi-naive Bayesian classifiers, i.e., one-dependent estimator (ODE), tree augmented naive Bayes (TAN) [60], and averaged ODE (AODE) [61], are introduced.

One-Dependent Estimator (ODE)

ODE is one of the most popular categories of semi-naive Bayesian classifiers. ODE is distinct in that it assumes each attribute depends on at most one another attribute:

$$P(\vec{x}_i, c_k) = P(c_k) \cdot \prod_{j=1}^{J} P(x_{ij}|c_k, pa_j) \tag{6.22}$$

where pa_j is the attribute that the jth attribute of Sample $\vec{x}_i$, i.e., x_{ij}, depends on. The above equation is formulated for Sample $\vec{x}_i$, so i is carried in the corresponding terms. However, such dependency is not related to the sample being considered. It is also called the parent attribute of attribute j.

Let us use the following example to illustrate the use of ODE. This example is for the selection of aggregates for road construction. There are two classes, "Y(1)" and "N(0)," based on three attributes, i.e., size, color, and shape. Table 6.1 presents a training dataset. Let us show how to predict the label of a testing sample: [Big, Gray, Round].

To use ODE, we will first need to specify the interdependencies between attributes. In Fig. 6.1, "$x_{*1} \leftarrow x_{*2}$" means x_{*2} depends on x_{*1}, or equivalently, attribute x_{*2}'s pa is x_{*1}. Thus, the attribute "size" depends on the attribute "shape": (1) size = big depends shape = round, (2) the attribute "color" depends on no other attribute, and (3) the attribute "shape" depends on the attribute "size": shape = round depends on size = big.

Then, the calculation for ODE is performed as follows. First, we will need to calculate the prior probabilities.

$$P(c = 1) = \frac{4 + 1}{10 + 2} = \frac{5}{12} \tag{6.23}$$

Table 6.1 Training dataset for an example of aggregate selection

Size	Color	Shape	Label
Small	Gray	Irregular	0
Big	Green	Irregular	1
Big	Green	Round	1
Big	Gray	Round	0
Big	Gray	Irregular	0
Small	Green	Round	1
Big	Gray	Irregular	0
Small	Green	Irregular	0
Small	Gray	Round	0
Big	Green	Round	1

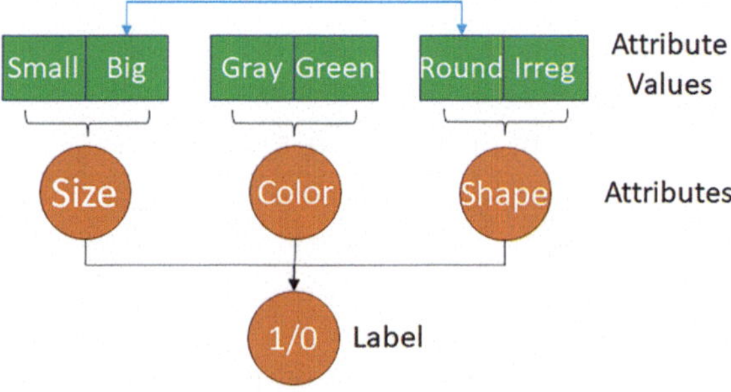

Fig. 6.1 Interdependencies between attributes for aggregation example

$$P(c=0) = \frac{6+1}{10+2} = \frac{7}{12} \tag{6.24}$$

Next, the likelihoods considering the above interdependencies between attributes are calculated as

$$P(Size = \text{Big}|c = 1, Shape = \text{Round}) = \frac{2+1}{3+2} = \frac{3}{5} \tag{6.25}$$

$$P(Color = \text{Gray}|c = 1) = \frac{0+1}{4+2} = \frac{1}{6} \tag{6.26}$$

$$P(Shape = \text{Round}|c = 1, Size = \text{Big}) = \frac{2+1}{3+2} = \frac{3}{5} \tag{6.27}$$

$$P(Size = \text{Big}|c = 0, Shape = \text{Round}) = \frac{1+1}{2+2} = \frac{2}{4} \tag{6.28}$$

$$P(Color = \text{Gray}|c = 0) = \frac{5+1}{6+2} = \frac{6}{8} \tag{6.29}$$

$$P(Shape = \text{Round}|c = 0, Size = \text{Big}) = \frac{1+1}{3+2} = \frac{2}{5} \tag{6.30}$$

Then, we will calculate the posterior probabilities based on the above prior probabilities and joint probabilities:

$$\begin{aligned} &P(c = 1|\vec{x} = [\text{Big, Gray, Round}]) \\ &= \frac{P(c = 1) \cdot P(\text{Big}|c = 1, \text{Round}) \cdot P(\text{Gray}|c = 1) \cdot P(\text{Round}|c = 1, \text{Big})}{P(\vec{x})} \\ &= \frac{0.025}{P(\vec{x})} \end{aligned} \tag{6.31}$$

$$\begin{aligned} &P(c = 0|\vec{x} = [\text{Big, Gray, Round}]) \\ &= \frac{P(c = 0) \cdot P(\text{Big}|c = 0, \text{Round}) \cdot P(\text{Gray}|c = 0) \cdot P(\text{Round}|c = 0, \text{Big})}{P(\vec{x})} \\ &= \frac{0.0875}{P(\vec{x})} \end{aligned} \tag{6.32}$$

The above results show that the probability of this sample belonging to $c = 0$ is greater than that of $c = 1$. Thus, this sample is predicted to be the Class of 0.

Variations of ODE

SPODE

The super-parent ODE, or written as SPODE, is a special type of ODE classifier. ODE has gained great popularity due to this simplicity in selecting the parent attribute. We know that, in ODE, one attribute can only depend on up to one other attribute, which is called the parent attribute. As illustrated in Fig. 6.2, SPODE simplifies the search for the parent attribute by specifying that there is one parent attribute and all the attributes except the parent attribute depend on it.

AODE

AODE (averaged one-dependent estimator) is an ensemble learning mechanism that integrates multiple ODEs to create a more powerful ODE. Unlike SPODE, which selects a super-parent attribute, AODE tests the use of each selected attribute as a super-parent to build SPODE and then combines those with enough samples to generate the ensemble.

Not all the attributes need to be selected. For example, in a typical example with J attributes, usually J' ($J' < J$) attributes are selected. This is performed by checking the attribute values of this testing sample being considered, then counting the number of samples with the same attribute values for this attribute

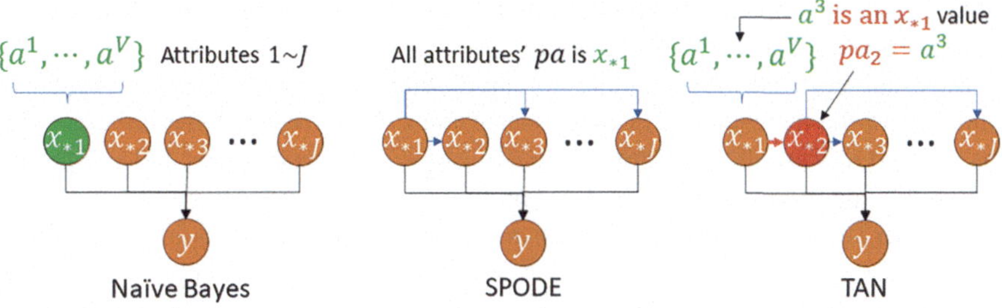

Fig. 6.2 Naive Bayes and semi-naive Bayes methods

under consideration. If the number exceeds a threshold value, e.g., 30, then this attribute will be selected to generate an SPODE for constructing the ensemble. The ensemble can be formulated as follows:

$$P(\vec{x}_i, c_k) = P(c_k) \sum_{j=1}^{J'} P(c, x_{ij}) \prod_{n=1}^{J} P(x_{in}|c_k, x_{ij}) \tag{6.33}$$

Tree Augmented Naive Bayes (TAN)

TAN further loosens the interdependency requirement. Instead of limiting the number of parent attributes to up to 1, TAN tests the interdependency between any two attributes. Then, only those pairs with significant interdependencies will be kept. Conditional mutual information is adopted in TAN to measure the significance of interdependencies. After selecting the dependencies to keep, we will also need to select the root attribute/variable and select the direction of dependencies so that the tree can be generated.

TAN is between the two extremes, i.e., naive Bayes and ODE for high simplicity and the Bayesian network for high flexibility, which will be introduced later. The detailed procedure for TAN will not be discussed here considering this fact and the limited space.

6.3.3 Bayesian Network

Bayesian network, also known as Bayes network, belief network, Bayes net, or decision network, further loosens the assumption of interdependencies [62]. For this purpose, a directed acyclic graph (DAG) is adopted to describe the dependency relationship between attributes. Thus, a Bayesian network is a probabilistic graphical model that represents a set of variables and their conditional dependencies via

DAG. In addition, conditional probability tables (CPT) are employed to describe the joint probability distribution of attributes.

Bayesian networks are suitable for predicting the probabilities of events as outcomes of the factors that can cause such events. For example, Bayesian networks can be used to learn the probabilistic relationships between diseases and symptoms and predict the diseases based on the symptoms.

We will show how to use a Bayesian network to predict the joint probability, i.e., $P(\vec{x}_i|c) = P(x_{i1}, x_{i2}, \cdots, x_{iJ}|c)$. For this purpose, we will first introduce the basic structure of the Bayesian network. Then, we will use one example to show how to predict the joint probability. Next, we will show the implementation of the Bayesian network. Coding Bayesian networks, as well as using such code, could be complicated and trivial. Possibly due to this reason, it is not provided in common machine packages like Scikit-learn. We will show the implementation of Bayesian using another package: pgmpy [63]. The implementation of the Bayesian network on three levels will be introduced: (1) manually constructing the network structure and entering the parameters (CPTs), (2) autonomously obtaining the parameters by training the network with training data, and (3) autonomous search for the optimal network structure.

Structure

The structure of Bayesian networks needs to be understood from two relevant aspects: basic units and interdependencies. Figure 6.3 shows the three basic units in typical Bayesian networks: tail-to-tail, head-to-tail, and head-to-head.

These three basic units can indicate the following interdependencies:

- Common cause: If C is known, A and B are independent. This is written as $(A \perp B|C)$.
- Causal/evidential: If C is known, A and B are independent. This is written as $(A \perp B|C)$.
- Common evidence: If C is unknown, A and B are independent. Or, if C is known, A and B are not independent. This is written as $(A \not\perp B|C)$.

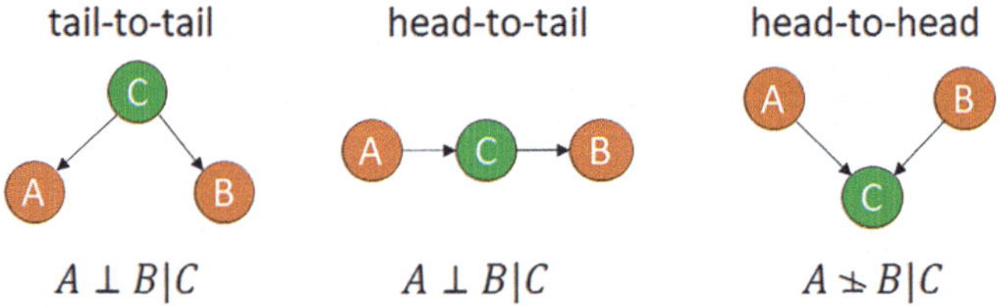

Fig. 6.3 Three basic units in typical Bayesian networks

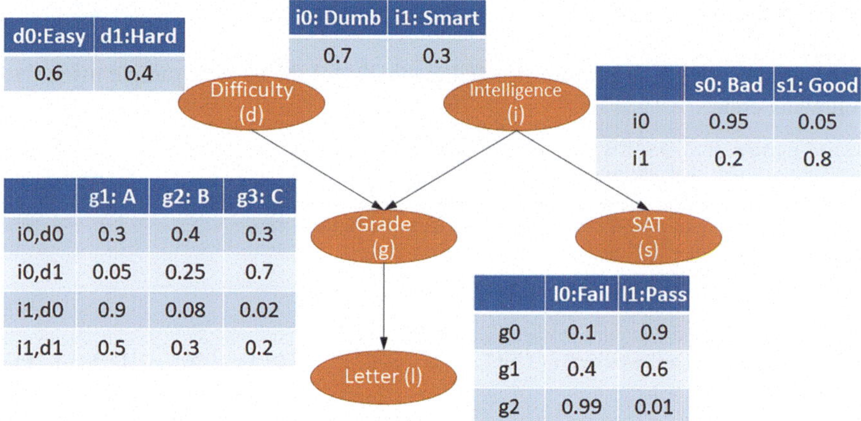

Fig. 6.4 Example for using Bayesian networks

The above interdependencies can be used to calculate the joint probabilities using the following equation:

$$P(\vec{x}_i|c) = P(x_{i1}, x_{i2}, \cdots, x_{iJ}|c) = \prod_{j=1}^{J} P(x_{ij}|c, \text{parents}(x_{ij})) \tag{6.34}$$

where parents(x_{ij}) are all the parent attributes. Thus, we need to consider the parent attributes of any attribute when computing the joint probability.

The use of this equation will be illustrated using the example in Fig. 6.4, which is frequently adopted for illustrating the application of the Bayesian network.

Without considering the interdependencies given in the above Bayesian network, the joint probability can be formulated as

$$P(D, I, G, L, S) = P(L|S, G, D, I) \cdot P(S|G, D, I) \cdot P(G|D, I) \cdot P(D|I) \cdot P(I) \tag{6.35}$$

Applying the local independence conditions in the above equation, we will get

$$P(D, I, G, L, S) = P(L|G) \cdot P(S|I) \cdot P(G|D, I) \cdot P(D) \cdot P(I) \tag{6.36}$$

$$\begin{aligned} P(G) &= \sum_{D,I,L,S} P(L|G) \cdot P(S|I) \cdot P(G|D, I) \cdot P(D) \cdot P(I) \\ &= \sum_D \sum_I \sum_L \sum_S P(L|G) \cdot P(S|I) \cdot P(G|D, I) \cdot P(D) \cdot P(I) \\ &= P(G|D, I) \cdot \sum_D P(D) \cdot \sum_I P(I) \cdot \sum_L P(L|G) \cdot \sum_S P(S|I) \end{aligned} \tag{6.37}$$

Implementation

A Bayesian network can be constructed manually for the above example using the following code.

```
from pgmpy.models import BayesianNetwork # BayesianModel in old
    versions of pgmpy
from pgmpy.factors.discrete import TabularCPD
from pgmpy.inference import VariableElimination
import networkx as nx
import matplotlib.pyplot as plt

# Define the Bayesian model structure using edges
model = BayesianNetwork([('D', 'G'), ('I', 'G'), ('G', 'L'), ('I'
    , 'S')])

#----------------------Enter probabilities manually
    ----------------------------

# Define CPD
cpd_d = TabularCPD(variable='D', variable_card=2, values=[[0.6],
    [0.4]])
cpd_i = TabularCPD(variable='I', variable_card=2, values=[[0.7],
    [0.3]])

cpd_g = TabularCPD(variable='G', variable_card=3,
                   values=[[0.3, 0.05, 0.9,  0.5],
                           [0.4, 0.25, 0.08, 0.3],
                           [0.3, 0.7,  0.02, 0.2]],
                  evidence=['I', 'D'],
                  evidence_card=[2, 2])

cpd_l = TabularCPD(variable='L', variable_card=2,
                   values=[[0.1, 0.4, 0.99],
                           [0.9, 0.6, 0.01]],
                   evidence=['G'],
                   evidence_card=[3])

cpd_s = TabularCPD(variable='S', variable_card=2,
                   values=[[0.95, 0.2],
                           [0.05, 0.8]],
                   evidence=['I'],
                   evidence_card=[2])

# Correlate DAG with CPDs
model.add_cpds(cpd_d, cpd_i, cpd_g, cpd_l, cpd_s)
#
    ------------------------------------------------------------------


model.get_cpds()
# Check the CPDs of different nodes
```

```
for cpd in model.get_cpds():
    print("CPD of {variable}:".format(variable=cpd.variable))
    print(cpd)

# Check the network structure and CPDs: valid if the total of CPD
     equals 1
model.check_model()

# Plot the Bayesian network
nx.draw(model,
        with_labels=True,
        node_size=1000,
        font_weight='bold',
        node_color='y',
        pos={"L": [4, 3], "G": [4, 5], "S": [8, 5], "D": [2, 7],
    "I": [6, 7]})
plt.text(2, 7, model.get_cpds("D"), fontsize=10, color='b')
plt.text(5, 6, model.get_cpds("I"), fontsize=10, color='b')
plt.text(1, 4, model.get_cpds("G"), fontsize=10, color='b')
plt.text(4.2, 2, model.get_cpds("L"), fontsize=10, color='b')
plt.text(7, 3.4, model.get_cpds("S"), fontsize=10, color='b')
plt.title('test')
plt.show()
```

In the above code, the parameters, i.e., CPDs, were entered manually. However, such parameters should be determined by data via training. Thus, when we have training data, we will need to obtain the parameters based on the structure and the training data. These parameters can be automatically obtained by fitting the model with a fixed structure to the training data. The following code illustrates this process.

```
from pgmpy.models import BayesianNetwork
from pgmpy.factors.discrete import TabularCPD
from pgmpy.inference import VariableElimination
import networkx as nx
from matplotlib import pyplot as plt
from pgmpy.estimators.MLE import MaximumLikelihoodEstimator

import numpy as np
import pandas as pd
raw_data = np.random.randint(low=0, high=2, size=(1000, 5)) #
    Create data using random numbers
data = pd.DataFrame(raw_data, columns=["D", "I", "G", "L", "S"])

# Define the Bayesian model structure using edges
model = BayesianNetwork([('D', 'G'), ('I', 'G'), ('G', 'L'), ('I'
    , 'S')])


#----------------------Generate probabilities from data
    ------------------------
# Obtain the parameters (CPDs) by fitting the model to the data
model.fit(data, estimator=MaximumLikelihoodEstimator)

model.get_cpds()
```

```
# Check the CPDs of different nodes
for cpd in model.get_cpds():
    print("CPD of {variable}:".format(variable=cpd.variable))
    print(cpd)
```

On a higher level, we will not need to stick with a network structure that we lay down manually, which may not be the optimal structure. Such structures can be sought based on scores with methods provided by pgmpy like exhaustive search and hill climb method. In addition, structural learning can also be performed with constraint or hybrid (score and constraint). The following code shows the use of two different methods with scores.

```
import pandas as pd
import numpy as np
from pgmpy.estimators.StructureScore import BDeuScore, K2Score,
    BicScore
from pgmpy.models import BayesianNetwork

# Generate samples using random numbers: there are 3 variables,
    in which Z depends on X and Y
data = pd.DataFrame(np.random.randint(0, 4, size=(5000, 2)),
    columns=list('XY'))
data['Z'] = data['X'] + data['Y']

bdeu = BDeuScore(data, equivalent_sample_size=5)
k2 = K2Score(data)
bic = BicScore(data)

# Method 1: Exhaustive Search
from pgmpy.estimators import ExhaustiveSearch

es = ExhaustiveSearch(data, scoring_method=bic)
best_model = es.estimate()
print(best_model.edges())

print("\nAll DAGs by score:")
for score, dag in reversed(es.all_scores()):
    print(score, dag.edges())


# Method 2: HillClimbSearch
from pgmpy.estimators import HillClimbSearch

hc = HillClimbSearch(data)
best_model = hc.estimate(scoring_method=BicScsore(data))
print(best_model.edges())
```

6.4 Bayesian Nonparametrics

6.4.1 Parametric Versus Nonparametric Models

Overview

Though much less common than the classification of supervised versus unsupervised machine learning, parametric versus nonparametric has also been adopted to differentiate machine learning methods. The definitions of these two types of models stem from the mapping sought in most problem-solving processes: the mapping from the input x to the output y, which is the essence/goal of both AI and traditional engineering methods for problem-solving.

$$y = f(x) \tag{6.38}$$

This mapping in Eq. 6.38 can be instantiated by a math function or, more generally, as a model consisting of many functions and/or mathematical operations. To estimate the mapping, we need to fit a candidate model to the data (more accurately, training data). This candidate model is untrained because it contains random or nonoptimal parameters. These parameters are fixed in the fitting process. Or, in other words, the model is trained.

The form of the function, or more generally, the structure of the model, needs to be determined before the fitting (or training) process. Thus, to find the model with the optimal fitting results, we may have to make some assumptions about the mapping f. For example, we need to select whether to use a first-order or a second-order polynomial function before we attempt to find the best model by fixing the coefficients in the selected polynomial function.

The model mentioned above can be parametric or nonparametric. Parametric models, or more accurately, parametric methods/algorithms, refer to models (or the algorithms generating such models) in which the form of the mapping, such as the type of the mathematical function, is pre-assumed. By contrast, nonparametric models (or their algorithms) exclude such assumptions in a strict sense. As a result, a nonparametric model is mostly determined by the training data and an algorithm without predefined parameters. It is noted that the boundary between parametric and nonparametric models may not be that evident in many cases.

Parametric Models

Some examples of parametric methods in machine learning include linear models, linear discriminant analysis (LDA), naive Bayes, and perceptron (including shallow neural networks). Deep neural networks hold a relatively vague spot in this classification, but they are viewed as parametric methods if we accept that the network with multiple layers of operations (e.g., addition, multiplication, surrogate, pooling, and normalization) defines a mapping function beforehand, though this function can be really complicated and its shape needs to be optimized by updating the network weights.

The biggest disadvantage of parametric methods is that the assumptions we make may not always be true or suitable. For instance, we may assume that the form of the function is linear while the problem may have a nonlinear nature. Therefore, these methods involve less flexible algorithms and are usually used for less complex problems. However, parametric methods tend to be relatively fast and require significantly less data compared to nonparametric methods. Due to the same reason, parametric methods are more interpretable.

Nonparametric Models

On the contrary, nonparametric methods, which are also called case-based, instance-based, or memory-based methods in some places, refer to a set of algorithms that do not make assumptions about the form of the function to be estimated. Due to the exclusion of assumptions, such methods can estimate the unknown function of any form. K-Nearest neighbor, decision trees, hierarchical Dirichlet process version of LDA, RBF, and Gaussian processes are common nonparametric learning algorithms. For example, the basic support vector machine algorithm uses support vectors from the training data points to define the model, so it can be viewed as a "nonparametric model" algorithm. In such models, the number of parameters grows with the size of the training set. Thus, the number of parameters in a model can increase as more data becomes available.

Nonparametric methods can be more accurate as they seek to best fit the data points without limiting the fitting model with predefined parameters. Some of such algorithms have the potential of generating extraordinary performance due to the high flexibility stemming from the exclusion of assumptions about the model. However, this comes at the expense of more data for estimating the mapping function. Due to the same reason, such algorithms can be less computationally efficient. Besides, such algorithms are susceptible to overfitting. This is because the extra flexibility in these algorithms attributed to the unconstrained fitting function could allow learning errors and noise as well.

From Parametric to Nonparametric Bayesian Algorithms

A majority of classic Bayesian algorithms are parametric in that they are represented by a fixed, finite number of parameters. Bayesian networks use directed graphs to encode patterns of probabilistic dependencies between different random variables for mining causal relationships between these variables in data. As introduced in the previous section, Bayesian techniques can be used to learn both the parameters and structure of the network. Linear-Gaussian models are an important special case of Bayesian networks where the variables of the network are all jointly Gaussian.

Latent Dirichlet allocation adopts a set of documents (e.g., web pages) that each comprises topics. Related models include nonnegative matrix factorization and probabilistic latent semantic analysis. Linear dynamical systems are a time series model in which low-dimensional Gaussian latent states evolve over time. The sparse coding algorithm models each data point as a linear combination of elements drawn from a larger dictionary, which is learned by resembling the receptive fields of neurons in the primary visual cortex.

By contrast, as an ongoing research area within machine learning and statistics, Bayesian nonparametrics aim at models that can be infinitely complex from a Bayesian perspective [64]. That is, though we cannot explicitly represent infinite objects in their entirety (e.g., all possible samples), we can still perform posterior inference in the models while only explicitly representing a finite portion of them. For example, Gaussian processes put priors over functions such that the attribute values of any finite set of samples are jointly Gaussian. As a result, posterior inference is tractable in many cases.

6.4.2 Gaussian Processes

Gaussian processes may not be at the center of the current machine learning hype, but they are still used at the forefront of research. For instance, they were employed to automatically tune the MCTS hyperparameters for AlphaGo Zero. This type of nonparametric Bayesian algorithm can be very easy to use while providing rich modeling capacity and uncertainty estimates. However, the algorithm is not that easy to grasp, especially if you are new to nonparametric models. This section is proposed to present a more visual and intuitive introduction to Gaussian processes without totally abandoning the theory.

Introduction to Gaussian Process

A Gaussian process (GP) is a powerful model that can be used to represent a "distribution (Gaussian) of functions" [65]. In a 2D regression problem, this distribution of functions can be visualized as a Gaussian distribution of curves, as shown in Fig. 6.5. That is, after the "training process" is finished, we would obtain an infinite number of fitting curves that are associated with different probabilities at a finite number of data points. Among them, we have a curve with the highest probability, i.e., the mean curve or the curve connecting all the mean values at the data points. We can also obtain confidence intervals, e.g., a range where a certain percentage of curves will fall in.

Usually, we do the regression at a finite number of given data points (training and testing). In GP, the prediction or the label of each data point can be viewed as a random variable that obeys the Gaussian distribution. Accordingly, all the random

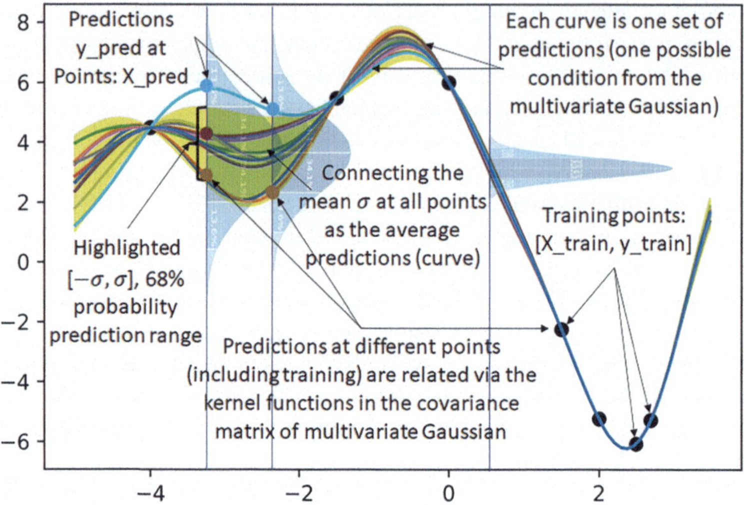

Fig. 6.5 Conceptual illustration of the Gaussian process

variables corresponding to all the points form a multivariate Gaussian distribution, as illustrated in the following equation:

$$\begin{bmatrix} y_0 \\ y_1 \\ \vdots \\ y_I \end{bmatrix} \sim \mathcal{N}\left(\vec{\mu}, \bar{\Sigma}\right) \sim \mathcal{N}\left(\begin{bmatrix} 0 \\ 0 \\ \vdots \\ 0 \end{bmatrix}, \begin{bmatrix} \sigma_{11} & \sigma_{12} & \dots & \sigma_{1I} \\ \sigma_{21} & \sigma_{22} & \dots & \sigma_{2I} \\ \vdots & \vdots & \ddots & \vdots \\ \sigma_{I1} & \sigma_{I2} & \dots & \sigma_{II} \end{bmatrix}\right) \tag{6.39}$$

where $\sim \mathcal{N}(\vec{\mu}, \bar{\Sigma})$ means following a Gaussian (normal) distribution characterized by the vector of means $\vec{\mu}$ and a covariance matrix $\bar{\Sigma}$. Usually, the vector of means is assumed to be a vector of zeros, which is proven adequate to give the "fitting function" or model enough flexibility. The overall location and trend of the curve are determined by the training data points.

The covariance matrix of the multivariate distribution is determined by a kernel function that defines the covariance coefficient between any two random variables in the matrix. The kernel function or the covariance coefficients calculated with it determine how the values between neighboring points are related to each other—the smoothness of the curve. A typical option for such a kernel function is a squared exponential function:

$$k(x, x') = \exp\left[-\frac{(x - x')^2}{2}\right] \tag{6.40}$$

Gaussian process is different from many supervised learning methods, because it does not contain the typical training and testing steps, which is called eager learning. By contrast, it is similar to KNN, which is also a nonparametric algorithm, in that both of them belong to lazy learning. In such lazy learning algorithms, there is no obvious training step/stage, and most computing occurs during the testing/prediction. Because of this, this type of machine learning algorithm has a high demand for memory for storing the training data. This difference is very obvious in the implementation of such methods.

Using a 2D regression problem as an example, we have X_train, y_train, and X_test and need to predict y_test. The covariance matrix can then be obtained with these given data and the equations for multivariate Gaussian distributions. With the covariance matrix and a zero vector of means, we can predict the possible curves consisting of the predicted values of the random variables at the testing data points. With a number of predicted curves, we can get the most possible (mean) curve and confidence intervals. Thus, there are training and testing data but no distinct steps for training and testing.

Also, as described above, the predictions were obtained based on the analytical equations from the training data (X_train, y_train) and X_test directly without assuming any specific math functions for the model. Instead, we only assume the values at all the data points follow a multivariate Gaussian function with a specific kernel function for the covariance coefficients. This is why Gaussian process is usually believed to be a nonparametric machine learning method.

Modeling Functions Using Multivariate Gaussian

The key idea behind GP is that a mapping function can be modeled using an infinite dimensional multivariate Gaussian distribution. In other words, every data point in the input space is associated with a random variable, and the joint distribution of these variables is modeled as a multivariate Gaussian. This may not be easy to understand for people who are not familiar with multivariate Gaussian distributions. Let us use a simpler case, i.e., a unit 2D Gaussian, to see what it looks like.

First, let us take a look at a very simple multivariate Gaussian distribution that consists of multiple univariate standard Gaussian distributions. Such a function has a math formulation as follows:

$$\begin{bmatrix} y_0 \\ y_1 \\ \vdots \\ y_I \end{bmatrix} \sim \mathcal{N}\left(\vec{\mu}, \bar{\Sigma}\right) \sim \mathcal{N}\left(\begin{bmatrix} 0 \\ 0 \\ \vdots \\ 0 \end{bmatrix}, \begin{bmatrix} 1 & 0 & \dots & 0 \\ 0 & 1 & \dots & 0 \\ \vdots & \vdots & \ddots & \vdots \\ 0 & 0 & \dots & 1 \end{bmatrix}_{I \times I}\right) \tag{6.41}$$

Next, let us implement such a function. It is a probability distribution. So, we can sample 10 curves at 20 data points. Accordingly, I is 20 in this example.

```
import numpy as np
import matplotlib.pyplot as plt

def plot_unit_gaussian_samples(D):
  p = plt.figure()

  xs = np.linspace(0, 1, D)
  for color in range(10): # 10 for 10 curves
    ys = np.random.multivariate_normal(np.zeros(D), np.eye(D))
    plt.plot(xs, ys)
  return p

plot_unit_gaussian_samples(20)
```

As can be seen in Fig. 6.6, 10 curves sampled using this simple multivariate Gaussian distribution are illustrated using different colors. Each predicted value at any of the 20 data points is one sample from a univariate Gaussian distribution. One curve consisting of 20 random variable values was generated at one time. However, the curve is very "noisy" or "rough," which is much different from smooth math functions that we typically use for machine learning models. In other words, any two predicted values in each curve are not related. This is caused by the zero non-diagonal values in the covariance matrix, making the 20 predicted values in one curve not different from 20 predicted values from 20 separate standard Gaussian distributions.

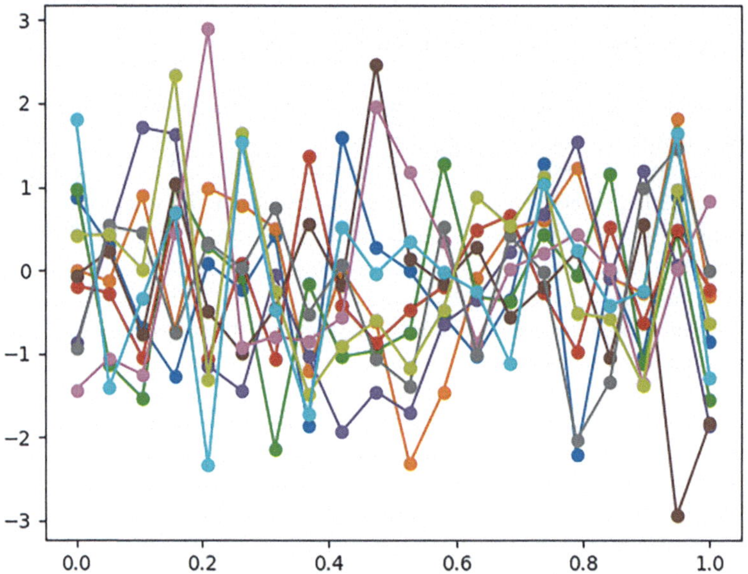

Fig. 6.6 Samples from a 20D Gaussian without kernel smoothing

In order to obtain smooth functions, we can generate a complicated covariance matrix using a squared exponential kernel equation. This kernel function, together with the zero array for the mean vector, can be defined using the following code:

```
def kernel(X1,X2):
    Sigma = np.empty((X1.shape[0],X2.shape[0]))
    for i in range(X1.shape[0]):
        for j in range(X2.shape[0]):
            Sigma[i,j] = np.exp(-(X1[i]-X2[j])**2/2)
    return Sigma

def mean(X):
    Mu = np.zeros(X.shape[0])#np.mean(X,axis=0)
    return Mu
```

So, to get the smoothness we want, we will consider two random variables y_i and y_j plotted at x_i and x_j to have a covariance as $cov(y_i, y_j) = k(x_i, x_j)$. This dictates that the closer the two points, the higher their covariance.

Using the kernel function from above, we can get this matrix with "kernel(xs, xs)." Now, let us plot another 10 samples from the 20D Gaussian with the new covariance matrix. For this purpose, we will need to use mean(xs) and kernel(xs, xs) to replace np.zeros(D) and np.eye(D) in the above code, respectively. Then, plot the curves again, and we get Fig. 6.7.

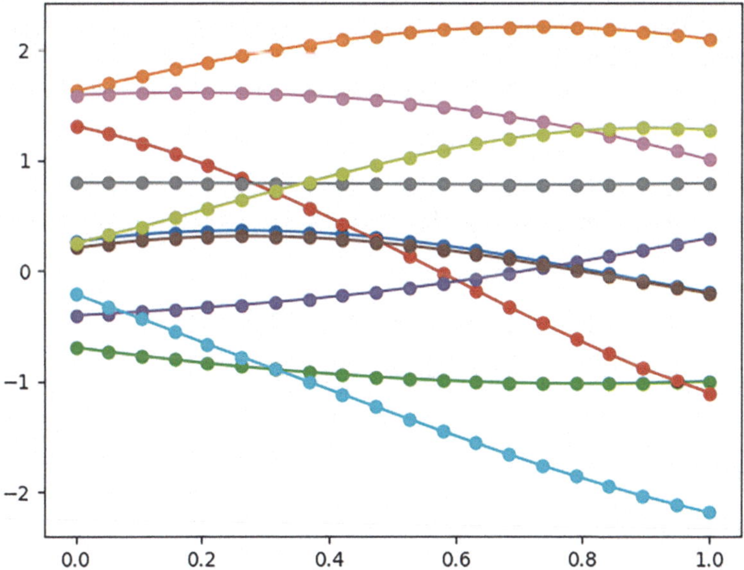

Fig. 6.7 Samples from a 20D Gaussian with kernel smoothing

Making Predictions Using a Prior and Observations

Now, we get to the heart of GPs. As mentioned, though there are no explicit training and testing steps, we still need to predict the target values (y_test) at the testing data points (X_test). In GP, the mapping $y = f(x)$ is represented by a probabilistic model $p(\vec{y}|\vec{x})$ using a multivariate normal:

$$p(\vec{y}|\vec{x}) = \mathcal{N}(\vec{y}|m(\vec{x}), \bar{\mathbf{K}}) \tag{6.42}$$

where $\bar{\mathbf{K}} = k(\vec{x}, \vec{x})$, in which the kernel matrix $\bar{K}$ is calculated with the kernel function introduced above, and $m(\vec{x}) = \vec{0}$, in which $m(\vec{x})$ is the function for obtaining mapping $\vec{x}$ to its corresponding vector of means.

We have some training data with inputs $\vec{x}$ and outputs $\vec{y} = f(\vec{x})$. Now let us say we have some new points $\vec{x}_*$ where we want to predict $\vec{y}_* = f(\vec{x}_*)$. Accordingly, we assume all the data, including both the training and testing data, follows the same distribution. Therefore, we assume these two types of data from one multivariate Gaussian distribution, $p(\vec{y}, \vec{y}_*|\vec{x}, \vec{x}_*)$, as follows:

$$\begin{pmatrix} \vec{y} \\ \vec{y}_* \end{pmatrix} \sim \mathcal{N}\left(\begin{pmatrix} m(\vec{x}) \\ m(\vec{x}_*) \end{pmatrix}, \begin{pmatrix} \bar{\mathbf{K}} & \bar{\mathbf{K}}_* \\ \bar{\mathbf{K}}_0^T & \bar{\mathbf{K}}_{**} \end{pmatrix}\right) \tag{6.43}$$

where $\bar{\mathbf{K}} = k(\vec{x}, \vec{x})$, $\bar{\mathbf{K}}_* = k(\vec{x}, \vec{x}_*)$ and $\bar{\mathbf{K}}_{**} = k(\vec{x}_*, \vec{x}_*)$. As before, we stick with a zero mean. Thus, we can use what we know, $\vec{x}$ (X_train) and $\vec{x}_*$ (X_test), to calculate the covariance matrix.

Next, we will need to predict $p(\vec{y}_*|\vec{x}_*)$. This requires us to calculate $\vec{\mu}_*$ and $\bar{\Sigma}_*$:

$$p(\vec{y}_*|\vec{x}_*, \vec{x}, \vec{y}) = \mathcal{N}(\vec{y}_*|(\vec{\mu}_*, \bar{\Sigma}_*)) \tag{6.44}$$

This is conditioning multivariate Gaussian. Here, we will skip the deductions and show the equations directly:

$$\vec{\mu}_* = m(\vec{x}_*) + \bar{\mathbf{K}}_*^T \bar{\mathbf{K}}^{-1}(\vec{y} - m(\vec{x})) \tag{6.45}$$

$$\bar{\Sigma}_* = \bar{\mathbf{K}}_{**} - \bar{\mathbf{K}}_*^T \bar{\mathbf{K}}^{-1} \bar{\mathbf{K}}_* \tag{6.46}$$

Now we have a posterior distribution over $\vec{y}_*$ using a prior distribution and some observations!

Example

In this subsection, we will use one example to illustrate the above process for implementing Gaussian process. Let us first assume we want to model data that follows a 5th-order polynomial function as given in Eq. 6.47:

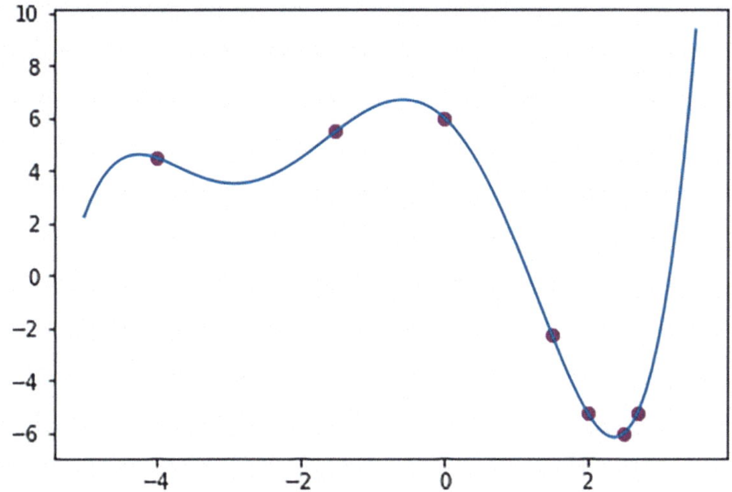

Fig. 6.8 True solution and training data for Gaussian process example

$$f(x) = 6 - 2.5x - 2.4x^2 - 0.1x^3 + 0.2x^4 + 0.03x^5 \tag{6.47}$$

```
import numpy as np
import matplotlib.pyplot as plt

# Input
def input_fun(X):
  Coefs = np.array([6,-2.5,-2.4,-0.1,0.2,0.03])
  y = np.sum(np.tensordot(X , np.ones(Coefs.size), axes=0 )** np.
    arange(Coefs.size) * Coefs,axis=1)
  return y

X_true = np.linspace(-5,3.5,100)
y_true = input_fun(X_true)
plt.plot(X_true,y_true)
```

Next, let us generate a few scatter data points as the training data points.

```
X_train = np.array([-4,-1.5,0,1.5,2,2.5,2.7])
y_train = input_fun(X_train)
plt.figure(1)
plt.scatter(X_train,y_train,color='m')
```

Figure 6.8 gives the true solution (curve) and training data (dots).

The mean vector and covariance matrix can be generated for this problem using the following code.

```
# Kernel
def kernel(X1,X2):
  # Sigma = np.exp(-(X1.reshape(X1.shape[0],-1) - X2)**2/2)
  Sigma = np.empty((X1.shape[0],X2.shape[0]))
```

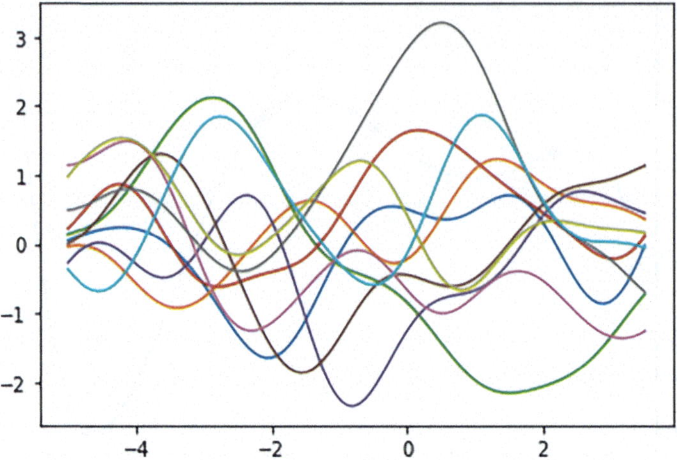

Fig. 6.9 Multivariate Gaussian sampling without relating training data to testing data

```
    for i in range(X1.shape[0]):
        for j in range(X2.shape[0]):
            Sigma[i,j] = np.exp(-(X1[i]-X2[j])**2/2)
    return Sigma

def mean(X):
    Mu = np.zeros(X.shape[0])#np.mean(X,axis=0)
    return Mu

Sigma = kernel(X_train,X_train)
Mu = mean(X_train)
```

The following is not a necessary step for problem-solving. We just want to see what we will get from multivariate Gaussian sampling without relating the training data to the testing data (Fig. 6.9).

```
# Sampling for showing the Gaussian samples
Sigma_sampling = kernel(X_true,X_true)
Mu_sampling = mean(X_true)
for n_sampling in range(10):
    y_sample = np.random.multivariate_normal(Mu_sampling,
      Sigma_sampling) #with smoothing (Sigma with kernel)
    plt.figure(2)
    plt.plot(X_true,y_sample)

    y_sample = np.random.multivariate_normal(Mu_sampling,np.eye(
      Mu_sampling.shape[0])) #without smoothing (Sigma without
      kernel)
    plt.figure(0)
    plt.plot(X_true,y_sample)
```

The next step is the key step for conditioning multivariate Gaussian, which will obtain the coefficients for the testing data.

```
# Prediction
X_pred = X_true

K_whole = kernel(np.hstack((X_train,X_pred)),np.hstack((X_train,
    X_pred))) # Sigma matrix for the combined Gaussian
    distribution of [X_train, X_pred]
K = K_whole[0:X_train.shape[0],0:X_train.shape[0]]
K_ast = K_whole[:X_train.shape[0],X_train.shape[0]:]
K_ast_ast = K_whole[X_train.shape[0]:,X_train.shape[0]:]
M = Mu
M_ast = mean(X_pred)

Mu_pred = M_ast + ( K_ast.T @ np.linalg.inv(K) @ (y_train - Mu).
    reshape((y_train.size,1)) ).reshape(M_ast.size)
Sigma_pred = K_ast_ast - K_ast.T @ np.linalg.inv(K) @ K_ast
```

With the GP model, we can then sample ten curves and plot the result.

```
plt.figure(3)
for n_sampling_pred in range(10):
  y_pred = np.random.multivariate_normal(Mu_pred,Sigma_pred)
  plt.plot(X_pred,y_pred) # Plot n (number of iterations)
    predicted curves
```

Finally, we can obtain the average curve and confidence intervals by assessing the curves that we sample.

```
plt.plot(X_pred,Mu_pred) # Predict average prediction
# sigma = np.std(X_pred) # Incorrect
sigma = Sigma_pred.diagonal()**0.5 # Correct
plt.fill_between(X_pred,Mu_pred-2*sigma,Mu_pred+2*sigma,color='
    yellow',alpha=0.9) # Show the region with mu +- sigma (68%)
    (2*sigma is 95%, 3*sigma is 99.7%)
```

The final results are shown in Fig. 6.10.

Summary

After going through the above theory and implementation, we can draw the following key points about Gaussian process:

(1) At each point for prediction, which can be viewed as the points for interpolation, the prediction at this point is treated as a random variable whose value can be drawn from a multivariate Gaussian distribution.
(2) One set of predicted target values at all the prediction data points represents one possibility from the multivariate Gaussian distribution.
(3) The average curve of all the predicted curves is the mean (most possible) prediction. The ranges of different confidence can also be obtained. Thus, GP does not give out one definite prediction, but, instead, gives out different sets

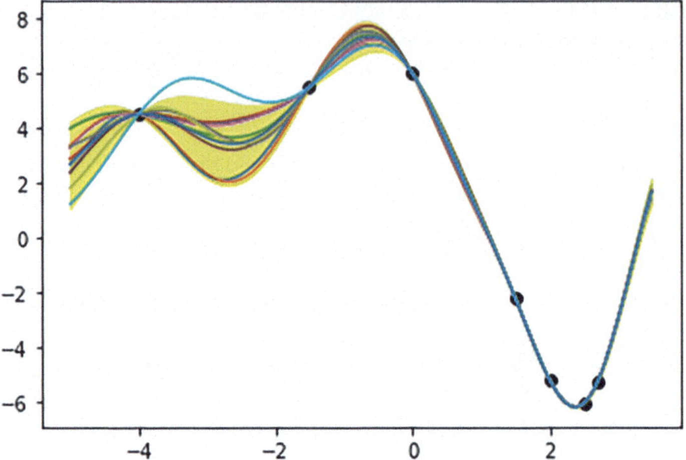

Fig. 6.10 Final results of Gaussian process analysis

of predictions (corresponding to different models) that correspond to different probabilities.

(4) Prediction at different points (including the training points) in a set of predictions (one curve) is related to each other via the kernel function in the covariance matrix of multivariate Gaussian. This kernel function, which can use different math functions, is given to define how to smoothen the curve. Usually, the closer the prediction point is to a training point, the smaller the difference in their target values.

(5) The training points form a multivariate Gauss distribution with the prediction points. Thus, their values determine the general trends and means of predictions at the neighboring prediction points.

Besides, please note that the selection of the kernel is made by human experts while the determination of the parameters can be done automatically by minimizing a loss term. This is the realm of Gaussian process regression. Finally, the handling of noisy data also deserves close attention, especially when we cannot get perfect samples of the hidden function. In this case, we need to factor this uncertainty into the model for better generalization.

In summary, the locations (X_train) and target values (y_train) of the training points, the selection of the kernel function, and the locations of the prediction data points (X_pred) determine what the predictions look like. The predictions are different curves with different probabilities.

6.5 Practice: Code Gaussian Naive Bayes Classifier, Try Bayesian Network, and Apply Gaussian Process

>>> More and Up-to-Date Course Materials including Practices @ AI-engineer.org <<<

1. Please use the Gaussian Naive Bayes Classifier from Scikit-learn to classify the Iris dataset as follows:

```
from sklearn.datasets import load_iris
from sklearn.model_selection import train_test_split
from sklearn.naive_bayes import GaussianNB

# ----------------- Load data ----------------- #
x, y = load_iris(return_X_y=True)
x_train, x_test, y_train, y_test = train_test_split(x, y,
    random_state=20190308, test_size=0.3)
```

2. Please develop your own code to implement the Gaussian naive Bayes classifier algorithm. Use the code to analyze the Iris testing dataset in Part 1.

3. *Please install pgmpy and run the Bayesian network code in the book (three script files). Try to understand how this code addressed the given example problem.

4. Please apply the Gaussian process code provided in the book to the training and true data. Then, obtain the average curve and mark the region corresponding to 99.7% confidence intervals.

```
import numpy as np
from matplotlib import pyplot as plt

def f(x):
    # True function to be approximated.
    return x * np.sin(x)

X_true = np.linspace(0, 10, 100)
y_true = f(X_true)
plt.plot(X_true, y_true)

X_train = np.array([1,2.5,4,4.5,7,9.5])
y_train = f(X_train)
```

Chapter 7
Artificial Neural Networks

7.1 Overview

This chapter presents essential knowledge about the artificial neural network (ANN), or more specifically, shallow ANNs. We will first check the structure of biological neurons to understand how ANN can mathematically imitate biological NN's behavior. Next, the mathematical formulation of the well-known M-P neuron model and other significant concepts, such as activation function, will be discussed. Then, we will see how perceptron, as one of the simplest and most typical NNs, works from a mathematical perspective. In particular, the training of such a network via backpropagation will be detailed. The yielded equations will be generalized for NNs with more than three layers to guide implementations. Based on that, extra skills about NN implementations, the illustration of the implementation procedure with an example, and the transformation of data as it moves through a network will be discussed in detail. Finally, a few other shallow ANN issues will be briefly mentioned.

7.2 Basics of Artificial Neural Networks

7.2.1 From Biological Neural Network to ANN

The artificial neural network (ANN) was inspired by the way that biological neural networks work [66]. It conceptualizes such working mechanisms and then mimics the functions with mathematical formulations [67]. Thus, prior to an introduction to ANN, let us first take a look at the structure of a biological neural network, especially how different units, i.e., neurons and their components, work with each other so that information can be processed.

Z. "L." Liu, *Artificial Intelligence for Engineers*,
https://doi.org/10.1007/978-3-031-75953-6_7

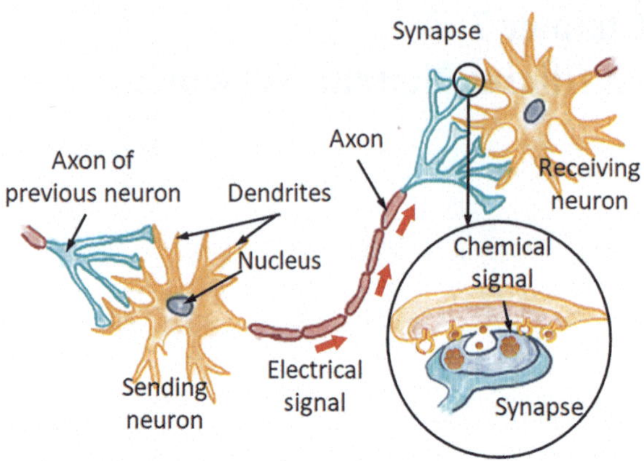

Fig. 7.1 Architecture and working mechanisms of neurons

As one of the two basic types of cells, neurons serve the role of transmitting information using electrical impulses and chemical signals. As illustrated in Fig. 7.1, a typical neuron consists of a cell body and two extensions: an axon and several (e.g., 5–7) dendrites. A dendrite has the shape of tree branches and receives information from another cell. An axon has a long-tail shape and transmits information out of the cell. Information is transmitted and processed within the cell as electrical impulses. Neighboring neurons communicate with each other via chemicals across a small contact region called a synapse, which is between an axon and a dendrite of the next neuron.

In biological neural networks, every neuron is connected to other neurons. When the electrical potential of a neuron exceeds a threshold value, it will be activated. Once activated, this neuron transmits chemicals to the connected neurons to change their electrical potential. In 1943, McCulloch and Pitts conceptualized the above model [9]. This model is the well-known M-P model, which is widely adopted for building ANNs in machine learning. Mathematically, the M-P model for a neuron can be described as

$$y = f\left(\sum_{j=1}^{J} w_j x_j - \theta\right) \tag{7.1}$$

where the output y is formulated as a function of the input x_j in which j is one of the J input variables (or attributes), w_j is the corresponding weight, and θ is the threshold for the neuron.

Thus, as shown in Fig. 7.2, a sample or the input (column) array for Neuron k, which is marked as $\vec{x}_i$, will first be added up according to the corresponding weights: $\vec{w}_k^T \cdot \vec{x}_i = \sum_{j=1}^{J} w_{kj} x_{ij}$. The weighted sum as the result of the transfer function marked as "$\sum$" in the figure will be treated by the activation function $f(\cdot)$ to produce the output y_k.

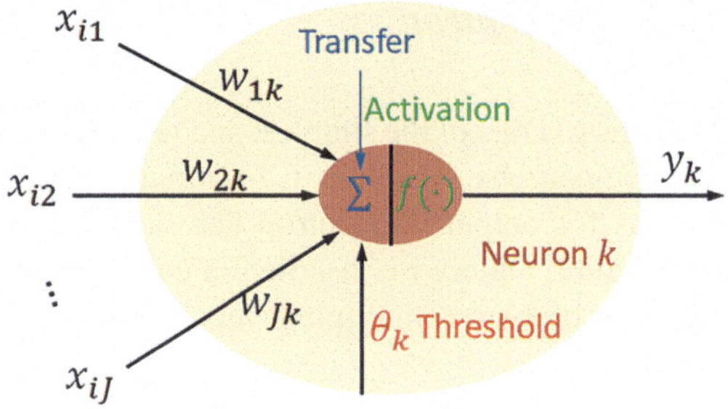

Fig. 7.2 M-P formulation of the working mechanism of neurons

7.2.2 *Activation Function*

The function f in Eq. 7.1 is the activation function. The ideal activation function is the step function. But we usually adopt continuous and smooth functions such as the Sigmoid and tanh functions as the activation function. The major purpose of using an activation function is to squash the input, which can span over a wide range of values, into a small range, e.g., 0 to 1, as the range for output. As a result, an activation function is also called a squashing function. The neural network as a network of connected neurons thus can be viewed as the combination of many functions such as $y = f\left(\sum_{j=1}^{J} w_j x_j - \theta\right)$.

The Sigmoid function in the following is a classic activation function:

$$y = \frac{1}{1 + e^{-x}} \tag{7.2}$$

This function converts any value into the range between 0 and 1. In addition to squeezing the input into the designated range of output, this function has another significant property that promotes its widespread adoption in the backpropagation for training neural networks. That is, the derivative of the Sigmoid function can be conveniently expressed using this function itself:

$$y' = y \cdot (1 - y) \tag{7.3}$$

Another popular activation function is the tanh function:

$$y = \frac{e^x - e^{-x}}{e^x + e^{-x}} \tag{7.4}$$

This function squeezes the input into a range between -1 and 1. Similar to the Sigmoid function, the derivative of tanh can also be expressed using the function itself as

$$y' = 1 - y^2 \tag{7.5}$$

7.2.3 Perceptron

Perceptron is one of the simplest and most typical neural networks. It consists of two layers of neurons. One layer is for input, while the other layer consists of functional neurons for output. Perceptron can only learn linear problems due to its simple architecture. Otherwise, oscillation can occur in the learning process. To consider nonlinear problems, multiple layers of function neurons need to be adopted.

7.2.4 Multiple-Layer Feedforward Neural Network

A feedforward neural network is an ANN wherein connections between the nodes do not form a cycle. As such, it is different from its descendant: recurrent neural networks. The most common ANNs are multiple-layer feedforward neural networks, which can be obtained by stacking multiple layers of perceptrons. In such a model, every layer of neurons is fully connected with the neurons in the next layer. Neurons from the same layer do not connect with each other. In a typical setting, we have an input layer, one to multiple hidden layers, and an output layer. While the input layer only takes the input (no other math operations), hidden layers and the output layer are composed of functional neurons like the above M-P neuron.

7.3 Training with Backpropagation

7.3.1 Concepts

Backpropagation is the most common learning algorithm for training ANNs. In a simple way, the weights, which determine the behavior of the network, are updated based on the difference between the prediction and true solution (labels). The algorithm thus involves two steps: forward propagation (forward pass) and backward propagation (backward pass or backpropagation). Forward propagation refers to the calculation and storage of the outputs of individual neuron layers, layer by layer from the input layer to the output layer.

Backward propagation, which will be termed backpropagation thereafter for consistency, strictly refers only to the calculation of the gradients and the updates on the weights using the obtained gradients. In each backward pass, the gradients are first obtained from the difference between the predicted and true solution in the output layer and then propagated backward layer by layer to the first hidden layer. Such gradients or corrections, as the derivatives of the difference between predicted and actual labels with respective to different weights, can be viewed as the contributions of these weights to the errors. Therefore, the weights are updated according to their contributions to the errors. For example, weights or their corresponding neurons that cause an overestimate of the output will be modified so that they will lead to lower predictions in the next forward pass.

The term "backpropagation" is often used loosely to refer to the entire training process or algorithm, including both of the two passes and how the gradient is used, such as by stochastic gradient descent. In this book, to avoid confusion, "forward pass" and "backward pass" will be used for the two specific processes with different directions, while "backpropagation" will be used to refer to the algorithm including both passes and the optimization method (solver).

We can view the threshold θ as a dummy node. However, more frequently, θ is counted as part of the weights, i.e., one extra w. Then, in each step of the backward pass, the weights $w's$ and $\theta's$ will be updated as

$$w_j + \Delta w_j \longrightarrow w_j \tag{7.6}$$

As mentioned above, the increment or correction, Δw_j, is determined with the difference between the predicted and actual values, which is included in the loss function ℓ:

$$\Delta w_j = -\eta \frac{\partial \ell}{\partial w_j} \tag{7.7}$$

where η is the learning rate. One backward pass consists of updating all the weights associated with connections between any two neurons in the network. It usually takes place from the last layer of neurons connecting to the output and then propagates backward to the previous layers and ultimately to the input layer. That is why it is called backpropagation. A learning process consists of multiple rounds of alternate forward and backward processes. The forward process obtains the predictions, and the backward process calculates the error as the difference between predictions and labels, based on which backpropagation is carried out. The whole learning process is, therefore, an optimization process whose goal is to minimize the difference between the predictions and labels.

7.3.2 Backpropagation in a 3-Layer Network

In the following, we will use a three-layer feed forward network as an example to show the detailed process of backpropagation. Equations for calculating the increments in weights will be derived, which are essential to coding a neural network. Illustrated in Fig. 7.3 is a representative neural network consisting of three layers of neurons, in which the bottom is the input layer containing J neurons. Therefore, the forward direction in the figure is the upward direction, which converts the input $\vec{x}$ (or x_j) into the output $\vec{y}$ (or y_k). Except for the input layer, all the layers have input and output together with weights and their associated math operations.

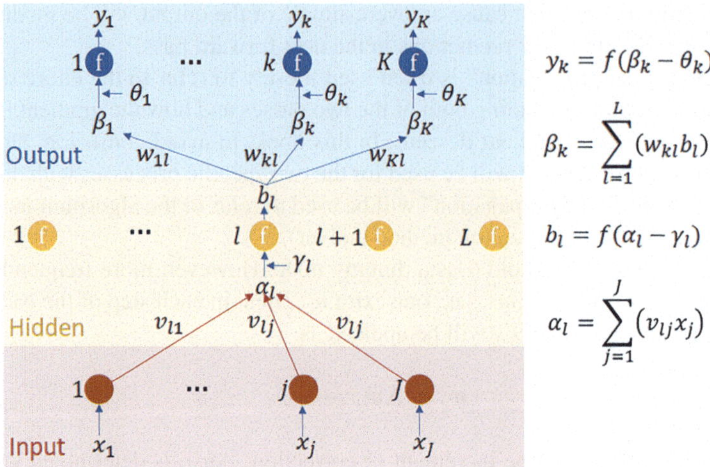

Fig. 7.3 Forward propagation in a 3-layer ANN

Taking the output layer, for example, the input is β and the output is y. The weights include ws and θs between the hidden and output layer, and their associated math operations are the linear combination and activation. In this network, any neuron in the hidden layer, $l \in \{1, \cdots, L\}$, is connected to all the neurons in the previous layer (J) including Neuron j and all the neurons in the next layer (K) including Neuron k. A weight w_{kl} connects Neuron l in the hidden layer to Neuron k in the next (output) layer.

As shown in Fig. 7.3, four types of operations are involved in converting the input of the network to its output. Among them, two generates α and β as linear combinations of weights:

$$\alpha_l = \sum_{j=1}^{J}(v_{lj}x_j) \tag{7.8}$$

$$\beta_k = \sum_{l=1}^{L}(w_{kl}b_l) \tag{7.9}$$

Thus, α and β can be viewed as intermediate weights. In fact, α and β are inputs for the hidden and output layers, respectively. The other two types of operations convert the input of each layer into the output of that layer. For the hidden layer, that is

$$b_l = f(\alpha_l - \gamma_l) \tag{7.10}$$

and for the output, we have

$$y_k = f(\beta_k - \theta_k) \tag{7.11}$$

Next, let us differentiate the predictions of the networks, $\vec{\tilde{y}}$, from the labels, $\vec{y}$. There are K output nodes; thus, both $\vec{\tilde{y}}$ and $\vec{y}$ should have K elements. For example, $\vec{\tilde{y}}$ should be $[\tilde{y}_1, \tilde{y}_2, \cdots, \tilde{y}_k, \cdots, \tilde{y}_K]$. Each of the elements can also be arrays with one or more dimensions. After each forward pass, $\vec{\tilde{y}}$ is calculated. Then, the error can be calculated with the given labels. Most commonly, the mean square root is used for constructing the error or loss:

$$\ell = \frac{1}{2}\sum_{k=1}^{K}(\tilde{y}_k - y_k)^2 \tag{7.12}$$

Then, the backward pass can start from the above loss/error function. As introduced above, the purpose is to minimize the above error using gradient descent. There are four types of weights in the above neural network. Therefore, we need to calculate the four types of gradients in each iteration: Δw, $\Delta\theta$, Δv, and, $\Delta\gamma$. Let us see how to derive equations for calculating each of them using primarily the chain rule. First, Δw is the derivative of the error function ℓ with respect to w:

$$\Delta w_{kl} = -\eta\frac{\partial \ell}{\partial w_{kl}} \tag{7.13}$$

As seen in Fig. 7.3, multiple operations and variables are involved to connect w to $\tilde{y}$. Thus, we can apply the chain rule:

$$\begin{aligned}\Delta w_{kl} &= -\eta\frac{\partial \ell}{\partial \tilde{y}_k}\frac{\partial \tilde{y}_k}{\partial \beta_k}\frac{\partial \beta_k}{\partial w_{kl}} \\ &= -\eta(\tilde{y}_k - y_k)[\tilde{y}_k(1-\tilde{y}_k)](b_l)\end{aligned} \tag{7.14}$$

Please note that the above deduction assumed the use of the Sigmoid activation function, whose derivative is $f' = f \cdot (1 - f)$ (Eq. 7.3).

We can use a new symbol as follows to simplify the formulation:

$$g_k = \frac{\partial \ell}{\partial \beta_k} = (\tilde{y}_k - y_k)[\tilde{y}_k(1-\tilde{y}_k)] \tag{7.15}$$

Then, Δw_{kl} can be simply written as

$$\Delta w_{kl} = -\eta g_k b_l \tag{7.16}$$

Likewise, the partial derivative of the error function with respect to θ is derived as

$$\begin{aligned}\Delta\theta_k =& -\eta\frac{\partial\ell}{\partial\tilde{y}_k}\frac{\partial\tilde{y}_k}{\partial\theta_k}\\ =& -\eta(\tilde{y}_k - y_k)[-\tilde{y}_k(1-\tilde{y}_k)] = \eta(\tilde{y}_k - y_k)[\tilde{y}_k(1-\tilde{y}_k)]\\ =&\eta g_k\end{aligned} \tag{7.17}$$

Next, let us move to the previous layer for Δv and $\Delta\gamma$:

$$\begin{aligned}\Delta v_{lj} =& \frac{\partial\ell}{\partial v_{lj}}\\ =& -\eta\frac{\partial\ell}{\partial\beta_k}\frac{\partial\beta_k}{\partial b_l}\frac{\partial b_l}{\partial\alpha_l}\frac{\partial\alpha_l}{\partial v_{lj}}\\ =& -\eta g_k w_{kl}[b_l(1-b_l)]x_j\end{aligned} \tag{7.18}$$

Similarly, let us assume

$$e_l = \frac{\partial\ell}{\partial\alpha_l} = \frac{\partial\ell}{\partial\beta_k}\frac{\partial\beta_k}{\partial b_l}\frac{\partial b_l}{\partial\alpha_l} = g_k w_{kl}[b_l(1-b_l)] \tag{7.19}$$

Then, Δv_{kl} can be simply written as

$$\Delta v_{lj} = -\eta g_k w_{kl}[b_l(1-b_l)]x_j = -\eta e_l x_j \tag{7.20}$$

Accordingly, $\Delta\gamma$ can be obtained as

$$\Delta\gamma_l = \eta e_l = \eta g_k w_{kl}[b_l(1-b_l)] \tag{7.21}$$

When switching to a different activation function, e.g., tanh, we just need to replace all the $[\tilde{y}_k(1-\tilde{y}_k)]$ with $(1-\tilde{y}_k^2)$ and replace all the $[b_l(1-b_l)]$ with $(1-b_l^2)$.

7.3.3 Backpropagation in Neural Networks with 3+ Layers

The expressions for the weights in the above 3-layer network can be easily extended to obtain expressions for the gradients of weights in a neural network consisting of more layers, e.g., $N > 3$. To simplify the deduction, we will adopt the tensor notation rather than the index notation. We will also need to start from the weights connecting to the last layer of neurons. As shown in Fig. 7.4, we will get the following equation similar to deduction for the 3-layer network:

$$\Delta\bar{w}^N = -\eta\vec{g}^N \otimes \vec{O}^{N-1} \tag{7.22}$$

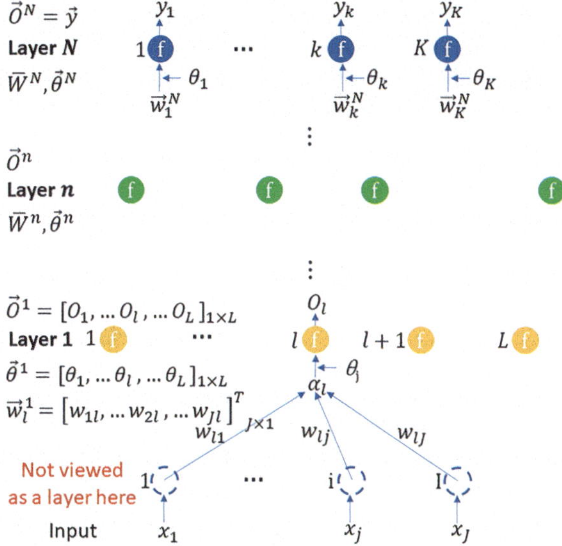

Fig. 7.4 Workflow of multiplayer NN

and

$$\Delta\vec{\theta}^N = \eta\vec{g}^N \tag{7.23}$$

where $\otimes$ is the external (or tensor) product. In the latest version of NumPy, the operator for the external product is "numpy.tensordot" without any contraction (of axes), while "@" or "numpy.dot" is for the dot/inner product, which contracts the neighboring axes of the two tensors. You can see that we use superscripts to represent the number of the layer. This is much different from the subscripts, which were used in the above equations for node numbers. The subscripts now can be excluded because of the use of the tensor notation. $\vec{O}^{N-1}$ is the tensor for the output of the second last layer (i.e., Layer $N-1$), which corresponds to $\vec{b}$ in the above 3-layer model. $\vec{g}$ is written as follows when Sigmoid is used as the activation function:

$$\vec{g}^N = \frac{\partial \ell}{\partial \vec{I}^N} = (\tilde{\vec{y}} - \vec{y}) \odot [\tilde{\vec{y}} \odot (1 - \tilde{\vec{y}})] \tag{7.24}$$

where $\odot$ is the element-wise product, and I is the tensor of input of the last layer. In NumPy, we simply use the symbol "*" for the element-wise product.

Then, we can move to the second last layer, and the weight tensor $\bar{w}^{N-1}$ is expressed as

$$\Delta\bar{w}^{N-1} = -\eta\vec{g}^N \cdot \bar{w}^N \odot [\vec{O}^{N-1} \odot (1 - \vec{O}^{N-1})] \otimes \vec{O}^{N-2} = -\eta\vec{g}^{N-1} \otimes \vec{O}^{N-2} \tag{7.25}$$

where $\vec{g}^{N-1}$ is used to replace e_l in the deduction for the 3-layer network. This is because the g_k and e_l in the deduction for the 3-layer network have a similar meaning: the derivative of the loss function with respect to the input of the corresponding layer. Therefore, $\vec{g}^{N-1}$ is formulated as

$$\vec{g}^{N-1} = \frac{\partial \ell}{\partial \vec{I}^{N-1}} = \vec{g}^{N} \cdot \bar{w}^{N} \odot [\vec{O}^{N-1} \odot (1 - \vec{O}^{N-1})] \tag{7.26}$$

and the weight θ^{N-1} is

$$\Delta\vec{\theta}^{N-1} = \eta\vec{g}^{N-1} \tag{7.27}$$

If we continue the deduction to the next layer, we will see that the equations for Layer $N-2$ will be the same as those for Layer $N-1$. The only difference is the substitution of N with $N-1$, and as a result, $N-1$ becomes $N-2$. That is to say, any hidden layer n, i.e., a layer between Layer 2 and Layer $N-1$ (including these two layers), has the following equations for weight updates:

$$\vec{g}^{n} = \frac{\partial \ell}{\partial \vec{I}^{n}} = \vec{g}^{n+1} \cdot \bar{w}^{n+1} \odot [\vec{O}^{n} \odot (1 - \vec{O}^{n})] \tag{7.28}$$

$$\Delta\bar{w}^{n} = -\eta\vec{g}^{n} \otimes \vec{O}^{n-1} \tag{7.29}$$

$$\Delta\vec{\theta}^{n} = \eta\vec{g}^{n} \tag{7.30}$$

The above three equations can be used in a loop to update the weights in all the hidden layers.

7.4 Implementation

The theory of ANN is not difficult to understand. However, the implementation of a shallow neural network, especially the implementation of the forward and backward passes, could be trivial and susceptible to mistakes even with the details offered in the previous section. What is particularly troublesome is the treatment of the dimensions or sizes of data, in the form of arrays or tensors, as the data moves through the network layer by layer. This can be viewed as a flow of tensors as coined in "TensorFlow."

This section will provide enough information for you to quickly put together your own code for implementing ANNs. For this purpose, we will first get familiar with some skills and concepts that have been widely used for training ANNs. Next, a typical procedure illustrated by pseudo-code will be provided to show the detailed steps for ANN training. Finally, based on one example, more details will be presented about the tensor flow, especially how data changes in shape as it moves through a typical network.

7.4.1 Practical Skills

We will need to understand how data moves through the neural network as well as several critical skills and concepts before we can implement the training of ANNs and understand the relevant literature well. The input, $\vec{x}$, is the data without labels, while the output, $\vec{y}$, is the corresponding label. Each time, one sample or multiple samples are fed into the network in the format of an array. This data array moves forward through the network to obtain a prediction that can be compared with the corresponding label. The difference between the predicted and actual labels is propagated backward in terms of gradients to adjust the weights. The new weights are used in the forward pass of the next iteration. Therefore, the model represented by the weights of the network can be updated every forward-backward process. This process is called an iteration of the training process.

The real implementation can be more complicated due to the use many extra skills and terms. The first is about the use of batches, epochs, and related optimizers. If one sample is used for each (gradient descent) iteration, then the training method (optimizer) is called stochastic gradient descent. Here, "stochastic" is used because of the fact that updating the model after processing each sample usually leads to high randomness/stochasticity due to the variability in data points (samples). In this case, each update will take a relatively short time, but lots of iterations may be needed to identify a good direction of optimization. The other extreme is to process the whole training set for each iteration, and updates on the model are made in the backward pass with gradients of all the samples. This is called batch gradient descent.

More frequently, it is not possible or efficient to process the whole dataset within one iteration. Thus, we frequently break the training dataset into multiple "mini batches" of samples. Then, updates on the model are made in each iteration for one mini batch. This is called the mini-batch gradient descent, which is also loosely called batch gradient descent in many places. We can calculate the loss or gradient for samples within one mini batch one by one before adding the loss or gradient up for updating the model. Or alternatively, we can treat all samples together as one array to obtain gradients for the model update. The use of mini batches also prompts us to use another concept, "epoch." One epoch means one dataset or the same number of samples as one dataset have been processed once, depending on the sampling method.

The selections of batch size and epoch number (or iteration number) need to consider the available resources and goals and to perform a trial in some cases. As the batch sizes increase, the use of memory and the consequent calculation time for each batch will increase dramatically. Compared with the stochastic and mini-batch methods, batch gradient descent method is more susceptible to the traps of local optima. This is because the randomness associated with small batch sizes adds noise to the training process, which helps the training to get out of local optima. But such randomness may also make it hard to find a good direction within an affordable amount of time.

7.4.2 Procedure for An Example

To simply illustrate the implementation of the neural network, we will start with a shallow neural network. Let us use the Boston House Prices (BHP) dataset from Scikit-learn as the example data. The dataset consists of 506 samples; each sample has 13 numbers (attribute values) as the input data and 1 number as the label/output. The 13 numbers of each sample will be fed as input to the network, processed layer by layer, and finally outputted as the prediction for that sample. The difference between the prediction and the label is the error, which needs to be minimized. To avoid confusion and facilitate coding, we will not count the input layer as one layer, because its input and output are identical. Accordingly, the first hidden layer will be viewed as the first layer. The following is the pseudo-code for implementing the neural network. The BHP dataset and a 3-layer (excluding the input layer) network are used to mark the sizes of arrays in the training process to better present the implementation details.

Simple Neural Network Training

Prepare data, e.g., get $\bar{X}$ and $\vec{y}$ from the BHP dataset.

Define network architecture, e.g., use a list [4,8,1] to represent a net with 3 layers, which have 4, 8, and 1 neurons. The number of neurons in the last layer is determined by the size of the label for each sample, i.e., 1 for BHP data.

Set up hyperparameters, e.g., η, and variables to be stored during training, e.g., $\bar{w}$,$\vec{\theta}$, $\vec{O}$, and loss, etc.

Repeat a certain number of iterations or until a certain rule is satisfied (e.g., loss improvements are smaller than a tolerance).

Initialize the network, e.g., filling the weight arrays of $\bar{w}$ and $\vec{\theta}$ with random numbers.

Repeat for every sample in a batch.

Repeat for every layer in a network in the forward pass using equation $\vec{O}^n = f(\bar{w}^n @ \vec{O}^{n-1} + \vec{\theta}^n)$.

Repeat for every layer in a network in the backward pass using equations $\vec{g}^n = \vec{g}^{n+1} @ \bar{w}^{n+1} \cdot [\vec{O}^n \cdot (1 - \vec{O}^n)]$, $\Delta\bar{w}^n = -\eta \cdot \vec{g}^n \otimes \vec{O}^{n-1}$, $\Delta\vec{\theta}^n = \eta \cdot \vec{g}^n$. Here, the weights of the whole batch are updated but will not be used until the next batch.

Calculate the loss.

The following code shows the training of a 3-layer feedforward NN following the above procedure. Please note that only the key steps in the forward and backward passes are given.

```
# Main loop for training
while True:
    # Forward
    Alpha = V @ X_train.T #X_train[i].reshape(y_train[i].size,1)
    B = 1/(1 + np.exp(-(Alpha - Gamma)))
    Beta = W @ B
```

```
    Y_hat = 1/(1 + np.exp(-(Beta - Theta)))
    y_hat = Y_hat.T
    Ek = 0.5*np.sum((y_hat - y_train)**2,axis=1)

    # Backward
    G = (y_train.T-Y_hat)*Y_hat*(1-Y_hat)
    Delta_W = eta * G @ B.T /y_train.shape[0]
    Delta_Theta = np.sum(eta*G,axis=1) /y_train.shape[0]
    Delta_Theta = Delta_Theta.reshape(Delta_Theta.shape[0],1)
    E = -(W.T @ G) * B * (1-B)
    Delta_V = eta * E @ X_train /y_train.shape[0]
    Delta_Gamma = - eta * np.sum(E,axis=1) /y_train.shape[0]
    Delta_Gamma = Delta_Gamma.reshape(Delta_Gamma.shape[0],1) #
    Change the size from (Node,) to (Node,1)

    # Improve with the learning rate and the overall descent
    direction
    W = W + Delta_W
    Theta = Theta + Delta_Theta
    V = V + Delta_V
    Gamma = Delta_Gamma

    error = np.sum(Ek)/np.sum(y_train**2)
    Iteration = Iteration + 1
    print("Iteration:",Iteration,"Error is", error)
    if error < tol:
        break

# Prediction
Alpha = V @ X_test.T
B = 1/(1 + np.exp(-(Alpha - Gamma)))
Beta = W @ B
Y_hat_test = 1/(1 + np.exp(-(Beta - Theta)))
y_hat_test = Y_hat_test.T
Ek_test = 0.5*np.sum((y_hat_test - y_test)**2)
y_pred = y_hat_test
print('The loss (score) of training is', Ek_test)
```

7.4.3 *Shape and Arrangement of Arrays for Data

To better understand what the data, i.e., different arrays for input, weights, thresholds, and output, looks like, we first write out these arrays using the way that we used in the other chapters of the book.

$$\bar{X} = [\vec{x}_1, \vec{x}_2, \cdots, \vec{x}_I]^T = \begin{bmatrix} x_{11} & x_{12} & \cdots x_{1J} \\ \vdots & \ddots & \vdots \\ x_{I1} & x_{I2} & \cdots x_{IJ} \end{bmatrix}_{I \times J} \tag{7.31}$$

$$\bar{V} = [\vec{v}_1, \vec{v}_2, \cdots, \vec{v}_L] = \begin{bmatrix} v_{11} & v_{12} & \cdots v_{1L} \\ \vdots & \ddots & \vdots \\ v_{J1} & v_{J2} & \cdots v_{JL} \end{bmatrix}_{J \times L} \tag{7.32}$$

$$\vec{\gamma} = [\gamma_1, \gamma_2, \cdots, \gamma_L] \tag{7.33}$$

$$\bar{W} = [\vec{w}_1, \vec{w}_2, \cdots, \vec{w}_K] = \begin{bmatrix} w_{11} & w_{12} & \cdots w_{1K} \\ \vdots & \ddots & \vdots \\ v_{L1} & v_{L2} & \cdots v_{LK} \end{bmatrix}_{L \times K} \tag{7.34}$$

$$\vec{\theta} = [\theta_1, \theta_2, \cdots, \theta_K] \tag{7.35}$$

$$\bar{y} = [\vec{y}_1, \vec{y}_2, \cdots, \vec{y}_I]^T = \begin{bmatrix} y_{11} & y_{12} & \cdots y_{1K} \\ \vdots & \ddots & \vdots \\ y_{Ii} & y_{I2} & \cdots y_{IK} \end{bmatrix}_{I \times K} \tag{7.36}$$

The above are the arrays for the data in a typical 3-layer NN. For NNs with more layers, we will use $\bar{W}$ for all weights of all the layers. Then $\bar{W}^n$ is the weights for Layer n can be calculated using the following general equation:

$$\begin{aligned} &\bar{X}_{sample\_number \times sample\_shape} \cdot \bar{W}^1_{sample\_shape \times neuron\_number\_1} \cdot \cdots \\ &\cdot \bar{W}^n_{neuron\_number\_n-1 \times neuron\_number\_n} \cdot \cdot \bar{W}^{n+1}_{neuron\_number\_n \times neuron\_number\_n+1} \cdot \cdots \\ &\cdot \bar{W}^N_{neuron\_number\_N-1 \times neuron\_number\_N} = \vec{y}_{sample\_number \times neuron\_number\_N} \end{aligned} \tag{7.37}$$

where *sample_shape* in simple cases like shallow ANNs are usually a scalar, which indicates that each sample is represented by a 1D array. However, in more complicated examples, like deep learning, which deals with image data, one image is usually a 2D or 3D array. In this case, the shape is a tuple like $length \times width$ (or written as tuple $(length, width)$) if not flattened. We can either flatten the image, e.g., turning an array with the shape of $(24, 24)$ for an image into an array with the shape of $(576,)$, to feed the image into the first hidden layer directly. But for the latter case, the ANN layer accepts the input should have the same shape, i.e., $(24, 24)$. Also, the number of neurons in the last layer should be the same as the number of output (labels).

As can be seen, the arrangement of the arrays is consistent with the other chapters in this book. However, we can notice that the position of the tensors for the tensor product in Eq. 7.37 is different from what we typically have in other machine learning topics (chapters). Other chapters mostly go with $\vec{w}^T \cdot \vec{x}$ (Convention 1: regular machine learning), while the above data flow adopts $\vec{x}^T \cdot \vec{w}$ or $\bar{X} \cdot \vec{w}$ (Convention 2: NN). Please note that they are essentially equivalent.

To better understand the above data flow, let us take a second look at the above example. Let us mark the shape of the input and the major operations, for which the arrays of all the weights and intermediate data are specified using "Convention 1." The input (one sample) has a shape of (13) or (13,). Then, the following are the variables to be calculated:

Layer 1: $\vec{w}^1_{(4,13)}$, $\vec{O}^0_{(13)} = \vec{X}_{(13)}$, $\vec{\theta}_{(4)}$

Forward: $\vec{O}^1_{(4)}$, $\vec{O}^1 = f(\vec{w}^1 @ \vec{O}^0 + \vec{\theta}^1)$
Backward: $\vec{g}^1_{(4)}$, $\vec{g}^1 = \vec{g}^2 @ \vec{w}^2 \cdot [\vec{O}^1 \cdot (1 - \vec{O}^1)]$, $\Delta\vec{w}^1_{(4,13)} = -\eta \cdot \vec{g}^1_{(4)} \otimes \vec{O}^0_{(15)}$, $\Delta\theta^1_{(4)} = \eta \cdot \vec{g}^1_{(4)}$

Layer 2: $w^2_{(8,4)}$, $\vec{O}^1_{(4)}$, $\vec{\theta}_{(8)}$

Forward: $\vec{O}^2_{(8)}$, $\vec{O}^2 = f(\vec{w}^2 @ \vec{O}^1 + \vec{\theta}^2)$
Backward: $\vec{g}^2_{(8)}$, $\vec{g}^2 = \vec{g}^3 @ \vec{w}^3 * [\vec{O}^2 * (1 - \vec{O}^2)]$, $\Delta\vec{w}^2_{(8,4)} = -\eta \cdot \vec{g}^2_{(8)} \otimes \vec{O}^1_{(4)}$, $\Delta\vec{\theta}^2_{(8)} = \eta \cdot \vec{g}^2_{(8)}$

Layer 3: $w^3_{(1,8)}$, $\vec{O}^2_{(8)}$, $\vec{\theta}_{(1)}$

Forward: $\vec{O}^3_{(1)}$, $\vec{O}^3 = f(\vec{w}^3 @ \vec{O}^2 + \vec{\theta}^3)$
Backward: $\vec{g}^3_{(1)}$, $\vec{g}^3 = (\vec{O}^3 - \vec{y}) \cdot [\vec{O}^3 * (1 - \vec{O}^3)]$, $\Delta\vec{w}^3_{(1,8)} = -\eta \cdot \vec{g}^3_{(1)} \otimes \vec{O}^2(8)$, $\Delta\vec{\theta}^3_{(1)} = \eta \cdot \vec{g}^3_{(1)}$

7.5 Other ANN Issues

A few other aspects of ANN also need to be considered, such as initialization, data normalization, and regularization. However, these topics are relatively simple for a shallow NN and can be easily handled with knowledge from previous chapters and available Python packages. Thus, they will not be extensively discussed here.

Initialization and normalization, which play a more critical role in deep learning, will be given more detail in the following chapter for deep learning.

Initialization can be easily performed with a constant number or random numbers following a specific distribution. The latter is more common, for which normal distributions are usually adopted. Data normalization can help with the computation a lot. In particular, many activation functions output values within a specific range, e.g., [0, 1] for the logistic function. If such an activation is used in the last layer of an NN, rescaling the label to the same range may be needed, though rescaling the unlabeled data can also be beneficial for improved computation efficiency and some other purposes such as comparing the losses of different models or datasets.

7.6 Practice: Modify and Assess the Architecture of an ANN

>>> More and Up-to-Date Course Materials including Practices @ AI-engineer.org <<<

Load data from two files for X (unlabeled samples) and y (labels) (for Boston House Prices). Then use MLPRegressor to solve the regression problem. Please adjust the network architecture and other parameters to obtain the best results. Observe the influence of different parameters, and think and discuss what we can learn from this process.

Hints: Please split the data for testing and training. The Boston House Prices dataset only has one label for each sample. So you can predict the actual labels versus predicted labels using a 2D plot to visualize the results.

Chapter 8
Deep Learning

8.1 From Artificial Neural Networks to Deep Learning

8.1.1 Overview

Deep learning is a subset of machine learning that uses deep artificial neural network (ANN)—algorithms inspired by the human brain—to perform human-like tasks such as speech recognition, image identification, and decision-making. The "deep" in this definition in general refers to the depth of layers in a neural network (NN). A neural network that consists of more than three layers, inclusive of the inputs and the output, can be considered a deep learning algorithm. Therefore, an in-depth understanding of deep learning cannot start without mentioning NNs, especially the work at "shallow" NNs.

In this chapter, the history of deep learning, including its root from and the preceding work in shallow NNs, will be introduced in terms of three waves first. Critical insights will be gained for the answers to some common questions for deep learning: "Is deep learning just a rebranding of neural networks?" "What are the major breakthroughs that drive the development of deep learning from all the way back in NNs to where it is?" "What deep learning can do and will be heading to?" After that, two basic elements of modern deep learning that help address vanishing gradients, i.e., activation and initialization, will be introduced. Next, the implementations of two most widely used types of deep NNs, i.e., CNN and RNN, especially the backpropagation through these networks, will be described. For CNN, detailed treatments in the backpropagation for convolution, padding and stride, ReLU, and pooling will be explained with examples. For RNN, the mathematical formulation of a typical RNN architecture will be provided. Based on that, practical deep learning skills will be first shared for various widely accepted initialization and batch normalization methods. Also, gradient descent optimizers as an essential part of deep learning will be investigated with details for common solvers including

Z. "L." Liu, *Artificial Intelligence for Engineers*,
https://doi.org/10.1007/978-3-031-75953-6_8

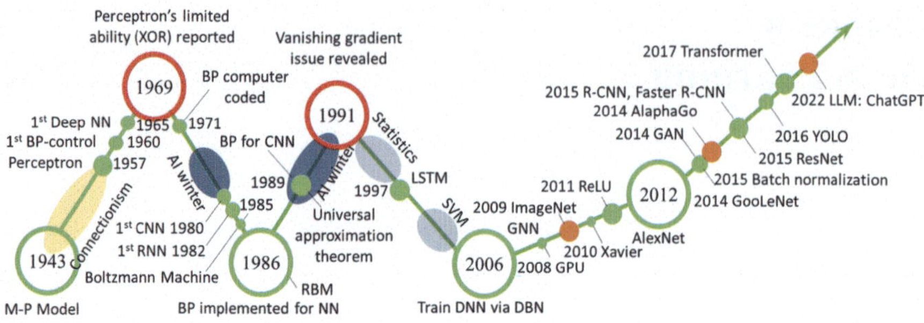

Fig. 8.1 History of deep learning development

SGD, Momentum, NAG, Adagrad, Adadelta, RMSProp, AdaMax, and Nadam. More information about data preprocessing and augmentation will be provided at the end.

8.1.2 The First Wave

As shown in Fig. 8.1, the first wave of NN or deep learning is mostly about shallow NNs, though deep NN appeared. The ANN history usually traces back to the birth of the mathematical model for the neurons of human brains, which was proposed by Walter Pitts and Warren McCulloch in 1943 [9]. This model, which was introduced in the chapter for ANNs, was usually called the McCulloch-Pitts (M-P) model and laid the foundation for ANNs and the succeeding work in deep learning. Later work based on the M-P model was among the earliest AI efforts that used networks or circuits of connected units to simulate intelligent behavior, which was called "connectionism" [68]. Most of these approaches were abandoned in the late 1950s as symbolic reasoning became the essence of AI, following the success of programs like the logic theorist and the general problem solver [13].

The next breakthrough was the advent of a new avatar of the M-P neuron in 1958, i.e., perceptron, which was shown to have learning capabilities of performing binary classification [69]. This inspired revolutions in the research of shallow neural networks and kept the field alive. In 1960, the first backpropagation model was proposed by Henry Kelley [70] in the context of control theory, which laid the foundation for training ANNs. Another step was made when Stuart Dreyfus presented a backpropagation model that used simple derivative chain rule in 1962 [71], which is used by most ANNs these days, instead of dynamic programming as used in earlier backpropagation models. Another work that needs to be mentioned is the multiplayer NN pioneered by Alexey Grigoryevich Ivakhnenko along with Valentin Grigorevich Lapa in 1965, which created a hierarchical representation of the neural network that used a polynomial activation function [72].

Mainstream perceptron studies came to an abrupt end in 1969, when Marvin Minsky and Seymour Papert published the book Perceptrons [17], which showed that Rosenblatt's perceptron cannot solve complicated functions like XOR. This fall of perceptron triggered a winter of NN research. After that, the field was still advanced with leaps in backpropagation and deep NNs. Especially, backpropagation was implemented with computer code via a general method for automatic differentiation by Seppo Linnainmaa in 1970 [73], though the implementation of backpropagation in NN came into play a decade later. In 1971, the effort towards deep NN continued such as an 8-layer NN using the group method of data handling (GMDH) by Alexey Grigoryevich Ivakhnenko [74].

After the major AI winter in 1974–1980, two major architectures for deep learning, i.e., convolutional neural network (CNN) and recurrent neural network (RNN), appeared. In 1980, the first CNN was proposed by Kunihiko Fukushima as neocognitron for recognizing visual patterns such as handwritten characters [75]. In 1982, the Hopfield network, which was essentially an RNN, was created by John Hopfield to serve as a content-addressable memory system [76]. Besides architectures, the use of backpropagation for propagating errors during the training of neural networks was proposed by Paul Werbos in 1982 [77]. Another influential study is the development of the Boltzmann machine by Ackley et al. (1985), which was a stochastic recurrent neural network [78]. Therefore, the above first wave started with the M-P model as a single neuron, rose with perceptron, and fell as a single-layer NN is constrained by their learning ability.

8.1.3 The Second Wave

The second wave of ANN can be viewed as the time when deep learning started growing and dominating NN work. In 1986, the successful implementation of backpropagation by Hinton et al. [79] opened gates for training complex deep neural networks easily, which was the main obstruction in earlier days of research in this area and propelled the rise of the second wave of deep learning. In the same year, a variation of the Boltzmann Machine where there was no intra-layer connection in input and hidden layer is proposed as restricted Boltzmann machine (RBM), which later became popular, especially for building recommender systems. Later, in 1989, backpropagation was successfully used to train CNN to recognize handwritten digits, which laid the foundation of modern computer vision using deep learning [80]. Besides implementations, a breakthrough in theories was achieved as George Cybenko published the earliest version of the universal approximation theorem: a feedforward neural network with a single hidden layer containing a finite number of neurons can approximate any continuous function [81]. This theorem added credibility to deep learning.

The second wave descended as Sepp Hochreiter reported the problem of vanishing gradient in 1991, which explained why the learning of deep neural networks could be extremely slow and almost impractical [82]. After another AI winter in

1987–1993, the development of deep learning was significantly slowed down due to the fast development of other machine learning methods such as those based on statistics in the 1990s and support vector machines in the early 2000s. This dark time of NN research also prompted researchers in the area to drive the rebranding of the frowned-upon NN research with the moniker "deep learning." Despite this fact, a milestone was still passed as an effective RNN architecture, i.e., long short-term memory (LSTM), was proposed by Sepp Hochreiter and Jürgen Schmidhuber in 1997, which helped revolutionize deep learning in the next wave [83]. As can be seen, the second wave started with the successful training of deep NN with backpropagation and fell as the vanishing gradient issue that prevents efficient training of deep NN was identified.

8.1.4 The Third Wave

The third wave marked the explosive development of deep learning, which might be attributed to three factors: improvements in architectures especially for addressing the vanishing gradient issue, an increase in computational power represented by the use of GPU, and the growth of data in a "big data" era especially image data. The wave was usually believed to start with an architecture breakthrough: the publication of the Deep Belief Networks in Science in 2006 [22]. This deep learning architecture as a stack of multiple RBMs showed the possibility and a feasible way of training deep NNs, though this way is not widely used due to the advent of other techniques for addressing vanishing gradient issues. As for computing power, for example, Andrew Ng's group at Stanford started advocating for the use of GPUs for training deep neural networks to speed up the training time by many folds in 2008 [23]. This could bring practicality of deep learning, i.e., training deep NNs on huge volumes of data efficiently. As for data, in 2009, Fei-Fei Li' team launched ImageNet, a database of 14 million labeled images, which served as a deep learning benchmark for the ImageNet competitions (ILSVRC) every year [24]. Another architecture improvement was made in 2011 as Glorot et al. [25] proposed to use ReLU to replace traditional Sigmoid and tanh functions as the activation function for addressing vanishing gradient problems. This presented another tool after GPU to deal with the issues of long and impractical training times of deep neural networks, which was later widely used and improved for prevalent deep NNs.

A victory of deep learning was made as AlexNet, a GPU-implemented CNN model designed by Alex Krizhevsky, won ImageNet's image classification contest in 2012 [26]. This victory triggered a new deep learning boom globally and attracted industry giant's attention. Apart from ReLU, some noticeable technical merits helped contribute to the success of AlexNet. Dropout, as an effective approach to regularization in neural networks (only used in the training process), helps address overfitting issues of deep learning by reducing interdependent learning amongst the neurons [84], while pooling as part of the model's architecture (used in training and testing) helped the model become less sensitive to some translations (i.e., improved translation invariance). Deep learning started gaining more momentum and making impacts in or even sweeping many disciplines such as computer vision and natural language processing.

In the fast-rising phase of this third wave, several milestones, either technical improvements or important events, need to be mentioned. The first is the development of the generative adversarial network (GAN) created by Ian Goodfellow in 2014 [27], which provided a way to synthesize real-like (high fidelity) data and consequently opened a new door for the application of deep learning in fashion, art, and science. A major victory of deep learning was announced as Deepmind's deep reinforcement learning model beats human champions in the complex game of Go in 2014 [85]. The improvements in initialization techniques further helped address the vanishing gradient issue. In fact, as early as 2010, Xavier Glorot and Yoshua Bengio proposed an initialization technique [86], which quickly became the default. This technique, which was generally referred to as "Xavier initialization" ($\pm 1/\sqrt{n}$ where n is the number of nodes in the prior layer) when using Sigmoid and tanh activation functions, was later slightly modified to suit the use of ReLU and referred to as "He initialization" ($\pm\sqrt{2/n}$).

What followed was the fast development of different deep learning architectures. In order to improve the learning ability of deep NNs, deeper networks were explored first by directly increasing the number of layers such as the VGG network in 2014 [87], which was found to be constrained by the limits in computational power and vanishing gradients. Later attempts were made in another two different directions and gained success: networks such as Inception (GoogLeNet as V1, 2014) explored "wider" deep NNs with the idea from "network in network" [88], while residual network, e.g., ResNet (2015), attempted at deeper NNs by introducing connections between layers that are far from each other in the same network—an idea evolving from "highway network" [89]. Later efforts were made to merge the two variants such as Xception and ResNeXt. Besides efforts at obtaining higher-accuracy deep learning such as Inception and ResNet, another direction is the advancement of deep learning in mobile devices, which is represented by many architectures proposed for such purposes such as SqueezeNet, MobileNet, and ShuffleNet [90]. While the above networks were primarily proposed for classification and regress tasks, advancements in NN architectures were also made for object detection tasks in computer vision. These are represented by regions with CNN and later Fast R-CNN [91], Faster R-CNN [92], YOLO [93], and GCN [94]. Graph neural networks (GNNs) also gained popularity at this time [95, 96]. More recently, large language models for generative AI like ChatGPT [28], which was enabled by the earlier transformer architecture [97], triggered another deep learning rush in both industry and academia.

8.1.5 *Summary of Enabling Innovations*

In summary, many factors contribute to the development of ANNs, especially the transition from shallow NNs to deep learning. Most of the factors have been mentioned in the above brief history of NN and deep learning, while a few others are more from the general field of machine learning such as pre-training, transfer learning, solvers, and regularizers. These factors were gathered in the following list:

- Better data: more data, preprocessing, normalization
- Better weights: initialization, pre-training, transfer learning
- Better network structure: CNN, LSTM, NIN, residual network
- Better solvers
- Better regularizers
- Better computing resources: GPU, parallel computing

The following introduction will be focused on network structures. CNN and RNN, as the two most representative deep learning architectures, will be explained first, in particular the backpropagation as well as unique structures and units in them. Then, in the rest of the chapter, more information will be provided for another three items on the must-known list: initialization, solvers, and preprocessing.

8.2 Convolutional Neural Network

A convolutional neural network (CNN or ConvNet) is a type of ANN that is usually discussed in the context of deep learning for analyzing visual imagery. It is actually good for identifying patterns between neighboring points in a sample, such as the spatial patterns in the image pixels close to each other. A modern CNN typically contains one or more of some basic elements such as convolution layers, activation functions like ReLU or its variant, pooling layers, and fully connected layers. In addition, Softmax or Sigmoid is additionally needed for classification tasks.

An understanding of the CNN, i.e., how it works, requires us to figure out how these different elements will be handled in backpropagation (including forward and backward passes as defined in the chapter for NN) in both training and testing. Considering that training involves both passes while testing only involves the forward pass, we will show how the two passes will be made in a typical training process for convolution layers, ReLU, and pooling layers in the following subsection. The treatment for fully connected layers is the same as what is introduced for shallow NNs. The introduction thus will present explanations for understanding the working mechanisms of these elements as well as equations and procedures for implementing the training of a CNN consisting of these elements.

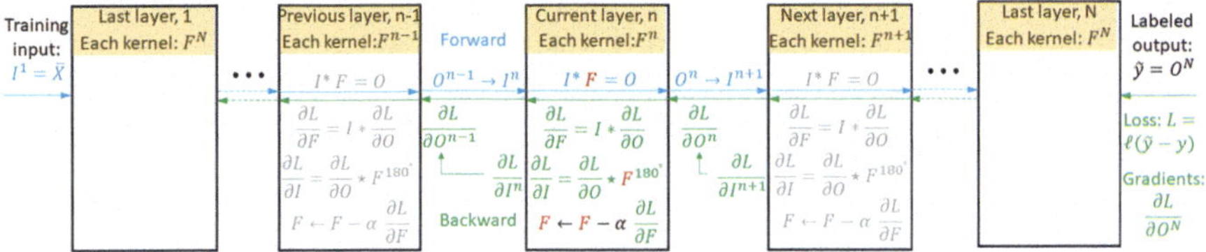

Fig. 8.2 Backpropagation through convolution layers

8.2.1 Convolution

The implementation of backpropagation including both forward and backward passes in a deep NN with exclusively convolution layers is outlined in Fig. 8.2. As illustrated, the forward pass starts from the first layer, where the input data for training, I, enters and moves along the blue arrows via the blue equations until the predicted output, O, is generated as the output of the last layer (i.e., Layer N). The difference between the prediction, O, and the labeled output, O_{true}, i.e., the loss L, will be used to calculate the gradients $\frac{\partial L}{\partial O^N}$. Then, the gradients will propagate backward through the whole network until reaching the first layer.

The main operation in the forward pass, i.e., convolution, is $I * F = O$. As a result, in the backward pass, the key steps are the calculation of $\frac{\partial L}{\partial F}$ and $\frac{\partial L}{\partial I}$ and the update on the convolution kernel F. To move backward from one layer to the previous layer, the calculated value of $\frac{\partial L}{\partial I}$ for the current layer, i.e., $\frac{\partial L}{\partial I^n}$, will be used as $\frac{\partial L}{\partial O}$ of the previous layer, i.e., $\frac{\partial L}{\partial O^{n-1}}$, because the input of the current layer I^n is the output of the previous layer O^{n-1}.

The key in the implementation of the above two-way process is the convolution operation, i.e., equation with an $*$ operator, in the forward pass and the two equations for calculating $\frac{\partial L}{\partial F}$ and $\frac{\partial L}{\partial I}$ in the backward pass, which were given in Fig. 8.2 without explanation. In the following subsection, the deduction of these equations will be presented while we explain how convolution and backpropagation work through the convolution layers. It is noted that this figure only contains convolution layers. Therefore, more operations or other modifications will be needed if other CNN elements are needed. In addition, we must notice that operations with this set of equations for using and updating F need to be performed for each convolution kernel. Usually, there is more than one kernel in each convolution layer. Therefore, such operations need to be carried out using the same equations but with different F for different kernels.

Forward Pass

Before being exposed to "convolution" in CNN, you possibly have heard about convolution in the context of general mathematics. It usually appears as a mathematical operation on two functions, e.g., y and f, generating a third function o:

$$o = (i * f)(t) = \int_{-\infty}^{\infty} i(\tau) f(t - \tau) d\tau \tag{8.1}$$

or equivalently formulated as

$$o = (i * f)(t) = \int_{-\infty}^{\infty} i(t - \tau) f(\tau) d\tau \tag{8.2}$$

where t is the axis (or independent variables) along which the three functions are defined. The operation can be understood as the area under the function $i(\tau)$ weighted by the function $f(-\tau)$ shifted by t.

The convolution in CNN is applied to data with a finite size, thus, the above equation can be slightly modified into

$$o = (i * f)(t) = \int_{0}^{t} i(\tau) f(t - \tau) d\tau \tag{8.3}$$

where t is still the axis along which the three functions are defined, and the integration bounds denote the finite size of the data.

In addition, image data, such as a photo, is usually represented as an array of numbers, in which each element or entry corresponds to a pixel in the photo. Thus, the convolution operation needs to take a discrete form so that it can be applied to discrete image data.

$$O[n] = (I * F)[n] = \sum_{m=1}^{n} I[m] \cdot F[n - m] \tag{8.4}$$

The above equation states that the element n in the generated convolution results O is obtained as a weighted sum of the (image) data, I, and the convolution kernel as the weight. The above operation needs to be repeated N times, where N is the number of elements of the generated results along the direction of the operation (shifting direction).

The schematic in Fig. 8.3 shows what happens when a convolution kernel is applied to an image. As illustrated, each convolution kernel from a convolution layer, or called a learnable filter or a neuron, is slid over the image. This starts with the red box on the left upper corner, which has the same number of elements as the kernel and can be called a receptive field. The application of the kernel to its receptive field mathematically corresponds to the computation of a dot product between the entries/elements of the filter and the array of the receptive field. This sum of this regional operation generates the $(1, 1)$ element of the output data, leading to a new element $(1, 1)$ ("4" in the red cell) in the result matrix. A stride like a step size is prescribed to move to the next receptive field. If a stride of 1 is used as illustrated in the figure, then the receptive area to the right of the red box region is the green box region. The same regional operation leads to the $(1, 2)$ element ("3"

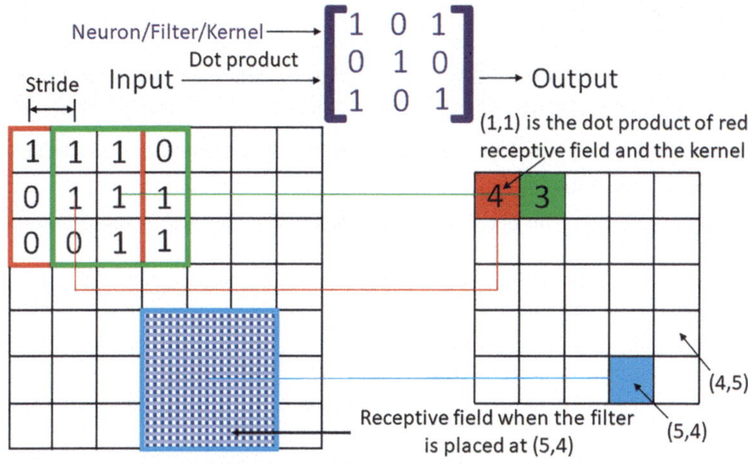

Fig. 8.3 Schematic of the application of a kernel to a 2D image (2D array)

$$\text{Convolution}(I, F) = I * F = \text{Convolution}\left(\begin{array}{|c|c|c|} \hline I_{11} & I_{12} & I_{13} \\ \hline I_{21} & I_{22} & I_{23} \\ \hline I_{31} & I_{32} & I_{33} \\ \hline \end{array}, \begin{array}{|c|c|} \hline F_{11} & F_{12} \\ \hline F_{21} & F_{22} \\ \hline \end{array} \right) = \begin{array}{|c|c|} \hline O_{11} & O_{12} \\ \hline O_{21} & O_{22} \\ \hline \end{array}$$

Fig. 8.4 An example for convolution operation

in the green cell) in the output. The same process will be repeated as the kernel moves over all the receptive fields.

To better show how different elements in the input and kernel are processed in the convolution operation, let us look at the example in Fig. 8.4. This example has a very simple 3×3 array I as input, i.e., data to which convolution is applied, and a 2×2 kernel. This example will also be employed in the following subsections for CNN to derive other equations for backpropagation.

In the above example, the following equations can be used to get the output:

$$O_{11} = I_{11}F_{11} + I_{12}F_{12} + I_{21}F_{21} + I_{22}F_{22} \tag{8.5a}$$

$$O_{12} = I_{12}F_{11} + I_{13}F_{12} + I_{22}F_{21} + I_{23}F_{22} \tag{8.5b}$$

$$O_{21} = I_{21}F_{11} + I_{22}F_{12} + I_{31}F_{21} + I_{32}F_{22} \tag{8.5c}$$

$$O_{22} = I_{22}F_{11} + I_{23}F_{12} + I_{32}F_{21} + I_{33}F_{22} \tag{8.5d}$$

The above equations can be extended for images and kernels of any size to implement the forward pass.

Backward Pass

As illustrated in Fig. 8.2, both $\frac{\partial L}{\partial F}$ and $\frac{\partial L}{\partial I}$ need to be calculated for each layer while $\frac{\partial L}{\partial O}$ is from the next layer. That is, $\frac{\partial L}{\partial O}$ is calculated as the $\frac{\partial L}{\partial I}$ of the following layer. $\frac{\partial L}{\partial F}$ is obtained to update the filter, while $\frac{\partial L}{\partial I}$ of the current layer is obtained to work as the $\frac{\partial L}{\partial O}$ of the previous layer.

The following chain rule needs to be recalled first to start deriving $\frac{\partial L}{\partial F}$ and $\frac{\partial L}{\partial I}$:

$$\frac{\partial L}{\partial F} = \frac{\partial L}{\partial O} \cdot \frac{\partial O}{\partial F} \tag{8.6a}$$

$$\frac{\partial L}{\partial I} = \frac{\partial L}{\partial O} \cdot \frac{\partial O}{\partial I} \tag{8.6b}$$

where $\frac{\partial L}{\partial O}$ can be obtained by passing the gradients layer by layer all the way back from the last layer. Thus, the above two types of gradients can be obtained as long as $\frac{\partial O}{\partial F}$ and $\frac{\partial O}{\partial I}$ can be obtained. Next, let us take a look at these two types of gradients.

As for Eq. 8.6a, in order to see what $\frac{\partial O}{\partial F}$ looks like, let us use O_{11} first:

$$O_{11} = I_{11}F_{11} + I_{12}F_{12} + I_{21}F_{21} + I_{22}F_{22} \tag{8.7}$$

Then we will have $\frac{\partial O_{11}}{\partial F_{11}} = X_{11}$, $\frac{\partial O_{11}}{\partial F_{12}} = X_{12}$, $\frac{\partial O_{11}}{\partial F_{21}} = X_{21}$, and $\frac{\partial O_{11}}{\partial F_{22}} = X_{22}$. Similarly, we can get the local gradients for O_{12}, O_{21}, O_{22}. Using these gradients, we can get the detailed equations for each element of $\frac{\partial L}{\partial F}$. Let us check two typical elements:

$$\frac{\partial L}{\partial F_{11}} = \frac{\partial L}{\partial O_{11}} \cdot X_{11} + \frac{\partial L}{\partial O_{12}} \cdot X_{12} + \frac{\partial L}{\partial O_{21}} \cdot X_{21} + \frac{\partial L}{\partial O_{22}} \cdot X_{22} \tag{8.8a}$$

$$\frac{\partial L}{\partial F_{12}} = \frac{\partial L}{\partial O_{11}} \cdot X_{12} + \frac{\partial L}{\partial O_{12}} \cdot X_{13} + \frac{\partial L}{\partial O_{21}} \cdot X_{22} + \frac{\partial L}{\partial O_{22}} \cdot X_{23} \tag{8.8b}$$

We can get the equations for all the elements in a similar way. It is not difficult to find that the above equations can be summarized as operation in Fig. 8.5:

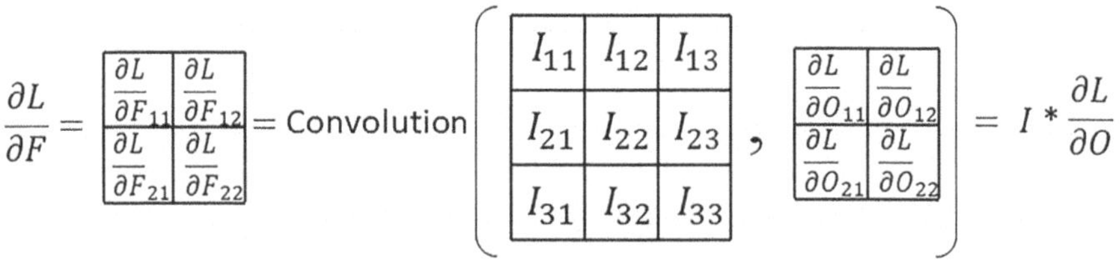

$$\frac{\partial L}{\partial F} = \begin{bmatrix} \frac{\partial L}{\partial F_{11}} & \frac{\partial L}{\partial F_{12}} \\ \frac{\partial L}{\partial F_{21}} & \frac{\partial L}{\partial F_{22}} \end{bmatrix} = \text{Convolution}\left(\begin{bmatrix} I_{11} & I_{12} & I_{13} \\ I_{21} & I_{22} & I_{23} \\ I_{31} & I_{32} & I_{33} \end{bmatrix}, \begin{bmatrix} \frac{\partial L}{\partial O_{11}} & \frac{\partial L}{\partial O_{12}} \\ \frac{\partial L}{\partial O_{21}} & \frac{\partial L}{\partial O_{22}} \end{bmatrix} \right) = I * \frac{\partial L}{\partial O}$$

Fig. 8.5 Operation for obtaining $\frac{\partial L}{\partial F}$

As for Eq. 8.6b, to get $\frac{\partial L}{\partial I}$, let us see what $\frac{\partial O}{\partial I}$ looks like: $\frac{\partial O_{11}}{\partial I_{11}} = F_{11}$, $\frac{\partial O_{11}}{\partial I_{12}} = F_{12}$, $\frac{\partial O_{11}}{\partial I_{21}} = F_{21}$, and $\frac{\partial O_{11}}{\partial I_{22}} = F_{22}$. Similarly, we can get the local gradients for O_{12}, O_{21}, O_{22}. Then, let us write out the detailed formulations of the elements of $\frac{\partial L}{\partial I}$ based on the above equations for the elements of $\frac{\partial O}{\partial I}$:

$$\frac{\partial L}{\partial I_{11}} = \frac{\partial L}{\partial O_{11}} \cdot F_{11} \tag{8.9a}$$

$$\frac{\partial L}{\partial I_{12}} = \frac{\partial L}{\partial O_{11}} \cdot F_{12} + \frac{\partial L}{\partial O_{12}} \cdot F_{11} \tag{8.9b}$$

$$\frac{\partial L}{\partial I_{13}} = \frac{\partial L}{\partial O_{12}} \cdot F_{12} \tag{8.9c}$$

$$\frac{\partial L}{\partial I_{21}} = \frac{\partial L}{\partial O_{11}} \cdot F_{21} + \frac{\partial L}{\partial O_{21}} \cdot F_{11} \tag{8.9d}$$

$$\frac{\partial L}{\partial I_{22}} = \frac{\partial L}{\partial O_{11}} \cdot F_{22} + \frac{\partial L}{\partial O_{12}} \cdot F_{21} + \frac{\partial L}{\partial O_{21}} \cdot F_{12} + \frac{\partial L}{\partial O_{22}} \cdot F_{11} \tag{8.9e}$$

$$\frac{\partial L}{\partial I_{23}} = \frac{\partial L}{\partial O_{12}} \cdot F_{22} + \frac{\partial L}{\partial O_{22}} \cdot F_{12} \tag{8.9f}$$

$$\frac{\partial L}{\partial I_{31}} = \frac{\partial L}{\partial O_{21}} \cdot F_{21} \tag{8.9g}$$

$$\frac{\partial L}{\partial I_{32}} = \frac{\partial L}{\partial O_{21}} \cdot F_{22} + \frac{\partial L}{\partial O_{22}} \cdot F_{21} \tag{8.9h}$$

$$\frac{\partial L}{\partial I_{33}} = \frac{\partial L}{\partial O_{22}} \cdot F_{22} \tag{8.9i}$$

The mathematical operation performed via the above equations can be summarized into the algebraic operation in Fig. 8.6:

In the above equation, F^{180° is obtained by rotating the array of F by 180 degrees. Taking a 2D array as an example, if there are k_h rows (height) and

$$\frac{\partial L}{\partial I} = \begin{bmatrix} \frac{\partial L}{\partial I_{11}} & \frac{\partial L}{\partial I_{12}} & \frac{\partial L}{\partial I_{13}} \\ \frac{\partial L}{\partial I_{21}} & \frac{\partial L}{\partial I_{22}} & \frac{\partial L}{\partial I_{23}} \\ \frac{\partial L}{\partial I_{31}} & \frac{\partial L}{\partial I_{32}} & \frac{\partial L}{\partial I_{33}} \end{bmatrix} = \text{Full Convolution}\left(\begin{bmatrix} F_{22} & F_{21} \\ F_{12} & F_{11} \end{bmatrix}, \begin{bmatrix} \frac{\partial L}{\partial O_{11}} & \frac{\partial L}{\partial O_{12}} \\ \frac{\partial L}{\partial O_{21}} & \frac{\partial L}{\partial O_{22}} \end{bmatrix} \right) = F^{180^\circ} \star \frac{\partial L}{\partial O}$$

Fig. 8.6 Operation for obtaining $\frac{\partial L}{\partial I}$

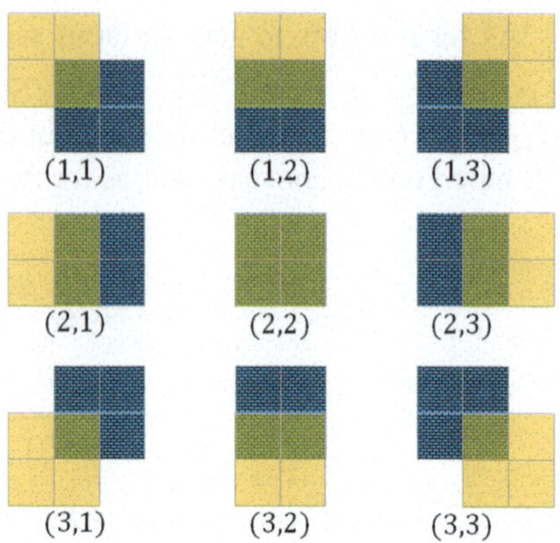

Fig. 8.7 Example of full convolution operation

k_w columns (width), then the element F_{ij} in the new array, $F^{180°}$, will be $F_{(k_h+1-i)\times(k_w+1-j)}$ in the original array, F. The "full convolution" operator represented by $\star$ in this book is different from the regular convolution, which is also called "valid convolution" in many places. The full convolution operation can be understood using the example given in Fig. 8.7. As can be seen, the array behind the operator (in yellow) is moved from the left upper corner of the array before the operator (blue), from left to right, upper to lower. The weighted sum of the overlapping area is where the local convolution occurs and generates a corresponding element in the operation result. For example, the lowest, rightmost element of the yellow array works as the weights for obtaining the sum of the highest, leftmost element (or a patch or region if a stride greater than 1 is used) of the blue array to generate the $(1, 1)$ element of the generated array. Then, the yellow array is moved one element (or patch) to the right, and the weighted sum will be the $(1, 2)$ element. The above process is repeated until all the elements (patches) of the new array are generated.

Padding and Stride

The above operations in backpropagation can be much more complicated in implementation. This is because the above illustrations were made with a very simple case: a padding of 0 and a stride of 1. However, in modern CNN practice, the use of positive padding values and strides greater than 1 is very common. When working on the implementation, such as coding a CNN ourselves, we will need to pay close attention to the sizes of different matrices and the change of these sizes as we apply different operations including convolution with a padding and a stride. A padding is used to add extra space (or elements with 0s) so that an operation such as convolution can generate an array with desired sizes. Usually, positive paddings, either identical or different paddings, can be added to different directions, i.e., height and width, to increase the size of the generated images. In common situations, the same number of padding elements are added to both ends in each direction. A stride greater than 1 is used to reduce the overlap between the neighboring receptive fields.

Let us assume I has a size of $n_h \times n_w$, F has a size of $k_h \times k_w$. In the forward pass, if we apply a convolution operation between these two arrays with paddings of $p_h \times p_w$ and strides of $s_h \times s_w$, then the size of the generated O will be $(n_h - k_h + p_h + s_h)/s_h \times (n_w - k_w + p_w + s_w)/s_w$. Accordingly, in the backward pass, we will apply a full convolution between F and $\frac{\partial L}{\partial O}$. In this case, $\frac{\partial L}{\partial O}$ has a size of $(n_h - k_h + p_h + s_h)/s_h \times (n_w - k_w + p_w + s_w)/s_w$, F has the same size, i.e., $k_h \times k_w$, and we will obtain $\frac{\partial L}{\partial I}$ with a size of $n_h \times n_w$, which is the same as that of I. Therefore, from a perspective of sizes, the above processes need to be "reversible." That is, the gradients returned in the backward pass should have the same size as the arrays where the gradients correspond to (or are applied to for update). For this purpose, a general rule needs to be maintained: the same padding and stride values need to be used in both passes. Usually, we do not change padding and stride values in different backpropagation iterations, so these values should stay the same in the training processes. Even in testing, the training network will still stick to the same paddings and strides if being applied to data of the same size.

The use of paddings will not change the operations introduced in the previous subsection too much. We just want to make sure the same padding(s) is adopted and also pay attention to the sizes of different arrays according to the equations given above. The treatment of strides requires more care. This is because, in the forward pass, the input array, I, is "downsampled" when a stride greater than 1 is used. That is, the output array O is smaller than that obtained with a stride of 1. Due to this reason, we will need to upsample the gradients with respect to O, i.e., $\frac{\partial L}{\partial O}$, before using it for calculating $\frac{\partial L}{\partial F}$ and $\frac{\partial L}{\partial I}$. This upsampling in the backward pass needs to be done with the same stride values as adopted in the forward pass for the same layer. For example, for a stride of s, the upsampling needs to add $s - 1$ 0s between the neighboring elements in the array to be upsampled.

8.2.2 ReLU

In traditional ANNs, most elements are differentiable. That is why, in the ANN chapter, continuous functions can be obtained for the derivatives (gradients) in backpropagation. By contrast, CNN has added special elements, such as ReLU, which is not differentiable at some points, and pooling, which changes the sizes of the data. Special treatment needed for these two CNN elements will be introduced in this and the following subsections.

The mathematical function of ReLU can be written as

$$\text{ReLU}(x) = \begin{cases} x, & x > 0 \\ 0, & x \leqslant 0 \end{cases} \tag{8.10}$$

or equivalently as

$$f(x) = \max(0, x) \tag{8.11}$$

This function is not differentiable at $x = 0$. To deal with this issue, we can set the derivative at $x = 0$ as 0 or 1. The choice will not affect the gradient to be used for updating the network. For example, the update on the weight in a typical example is $w = w + \alpha * (\hat{y} - y) \cdot \delta_{\text{ReLU}} \cdot \frac{\partial(w \cdot x - \theta)}{\partial w}$, where $\frac{\partial(w \cdot x - \theta)}{\partial w} = x$. As can be seen, at $x = 0$, the increment (second part on the right-hand side of the equation) equals 0 no matter the gradient of ReLU (δ_{ReLU}) is set as 0 or 1.

Other activation functions such as Linear, Linear threshold, Maxout, and Softmax will not be discussed here. More information can be found in more specific deep learning literature.

8.2.3 Pooling

As mentioned above, pooling changes the size of the feature map. This prevents the gradients from being passed from the elements after pooling to their corresponding elements in the feature map before pooling. A general solution is to make sure the gradient of every element (or pixel) can be passed backward to the elements that generated this element in the previous forward pass. In this way, the total gradients will stay the same, which is needed to avoid gradient explosion or vanishing. Different pooling methods, e.g., average pooling and max pooling, have different treatments.

For the average pooling, as illustrated in Fig. 8.8, a straightforward way is to evenly divide the gradient of an element for all the elements that are pooled to generate that element. For example, the yellow element in the feature map after pooling is generated by the four elements in the feature map before pooling in the

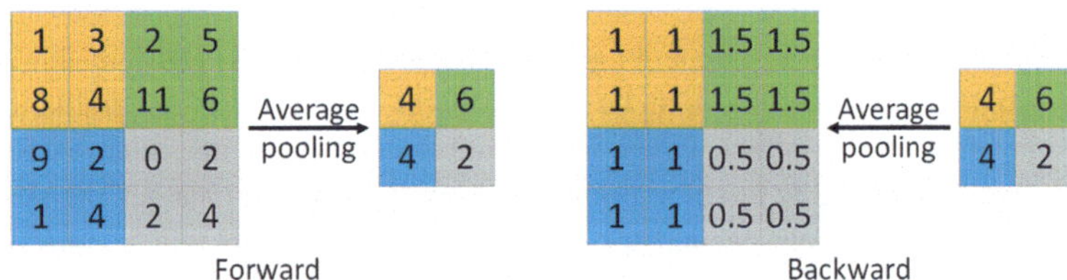

Fig. 8.8 Forward and backward passes in backpropagation of average pooling layer

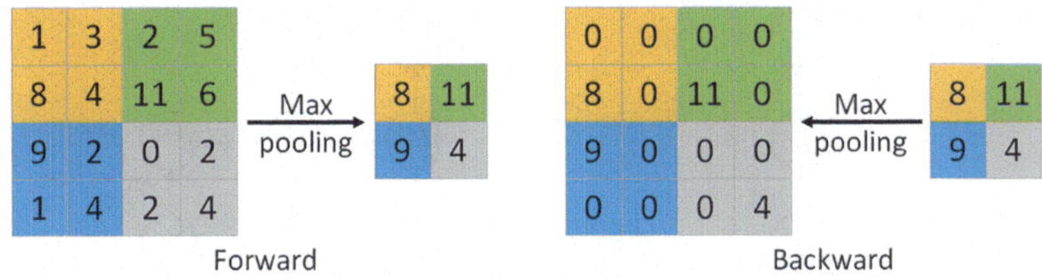

Fig. 8.9 Forward and backward passes in backpropagation of max pooling layer

forward process. Therefore, in the backward process, the element 4 is divided by 4. The obtained quotient, 1, is then assigned to those four elements.

For the max pooling, a common way is to give the gradient to the element with the maximum value in the feature map before pooling. As shown in Fig. 8.9, the element with a gradient of 8 (orange) passes this gradient to the lower left element in the orange area, which has the maximum value in the forward pass. Therefore, in the implementation, the location of the element that has the highest value in each small region for pooling needs to be recorded in the forward pass, so that the gradient can be passed to it in the following backward pass.

It is worthwhile to mention that, in the above figures, the same feature map after pooling is used for the two passes (i.e., forward and backward) to help the readers better understand the operations. In reality, the feature map after pooling and the gradients that need to be passed backward are totally different, though they are the same size.

8.3 Recurrent Neural Network

RNN has many different variations with distinct structures. Shown in Fig. 8.10 is one of the most common architectures. On the left is the schematic of a unit. The one in the middle shows a more detailed flow of the data through the unit. The subplot on the right illustrates how the unit unfolds over time.

The symbols in Fig. 8.10 have the following meanings:

x^t is the output at step t in the time series.
h^t is the state of the hidden layer at t, which is determined by both x^t and h^{t-1} according to the above network architecture.
o^t is the output at t, which is determined by the hidden state at the current step t.
L^t is the loss function at t.
y^t is the actual output (or label) at t.
$\tilde{y}^t$ is the predicted output at t.
σ_1 and σ_2 are the hidden layer activation function and output activation function, respectively.
U, W, and V are the arrays that include the network weights to be predicted.

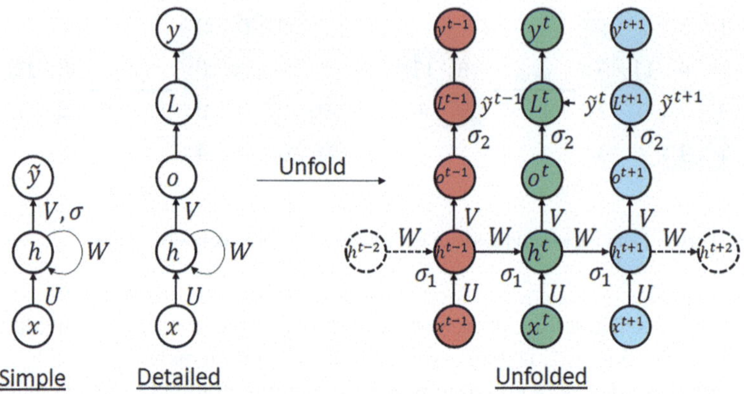

Fig. 8.10 Structure and workflow of RNN

These weights, i.e., U, W, and V, are shared through time. That is, though there are "multiple units" as the RNN is unfolded over time, there is essentially one unit. Thus, the backpropagation of RNN updates the weights of the same unit.

The above symbols were given without discussing the shape of the input x, which is handled as one sample here. In many cases, x has multiple attributes and thus needs to be formulated using a 1D array. In the following deductions for backpropagation, we will represent x as a column array with J elements (for the J attributes). In this way, the derived equations can be easily used for coding. The output y (actual) and $\tilde{y}$ (predicted) are 1D arrays with a shape of $K \times 1$. In implementation, we will still need to determine the shape of $\vec{h}$. Let us assume $\vec{h}$ has a shape of $N \times 1$. The shapes of other weight arrays can be determined by those of $\vec{x}$, $\vec{y}$, and $\vec{h}$. The data flow, especially the change of the shapes of arrays for data and weights, can be tracked in the following deductions.

8.3.1 Forward Pass

Let us first work on the forward pass. The following three operations will turn the input $\vec{x}_{J\times 1}$ into $\tilde{y}$:

$$\vec{h}^t_{N\times 1} = \sigma_1\left(\vec{z}^t_{N\times 1}\right) = \sigma_1\left(\bar{U}_{N\times J}\cdot x^t_{J\times 1} + \bar{W}_{N\times N}\cdot \vec{h}^{t-1}_{N\times 1} + \vec{b}_{N\times 1}\right) \tag{8.12}$$

$$\vec{o}^t_{K\times 1} = \bar{V}_{K\times N}\cdot \vec{h}^t_{N\times 1} + \vec{c}_{K\times 1} \tag{8.13}$$

$$\vec{\tilde{y}}^t_{K\times 1} = \sigma_2\left(\vec{o}^t_{K\times 1}\right) = \sigma_2\left(\bar{V}_{K\times N}\cdot \vec{h}^t_{N\times 1} + \vec{c}_{K\times 1}\right) \tag{8.14}$$

where σ_1 and σ_2 are the two activation functions used for the hidden layer and output layer, respectively.

The loss function at time step ℓ^t measures the distance between $\bar{\vec{y}}^t$ and $\vec{y}^t$. For example, the log likelihood function is usually used for classification tasks, and the following mean square root error function is used for regression tasks:

$$\ell = \sum_{t=1}^{\tau} \ell^t = \sum_{t=1}^{\tau} \frac{1}{2} \left(\bar{\vec{y}}^t_{K\times 1} - \vec{y}^t_{K\times 1} \right)^2 \tag{8.15}$$

8.3.2 Backward Pass

RNN's backward pass in which the network weights are updated via methods such as a gradient descent method is termed backpropagation through time (BPTT). The deduction of backpropagation rules, i.e., equations for the backpropagation of the gradients from the difference between the predicted and actual outputs, depends on the loss function and activation functions.

In the following, we will derive the general backpropagation equations. In this deduction, we will use the mean square root error for loss, tanh for σ_1, and Sigmoid for σ_2 to show more detailed equations that can be used for implementations. Accordingly, the three functions have the following derivatives, which are needed in the backpropagation:

$$\frac{\partial \ell^t}{\partial \bar{\vec{y}}^t} = \frac{\partial \left[\frac{1}{2} (\bar{\vec{y}}^t - \vec{y}^t)^2 \right]}{\partial \bar{\vec{y}}^t} = \bar{\vec{y}}^t - \vec{y}^t \tag{8.16}$$

$$\sigma_2' = \frac{\partial \sigma_2}{\partial \bar{\vec{y}}} = \bar{\vec{y}}^t \odot (1 - \bar{\vec{y}}^t) \tag{8.17}$$

$$\sigma_1' = \frac{\partial \sigma_1}{\partial \vec{h}} = 1 - (\vec{h}^t)^2 \tag{8.18}$$

The gradient of $\bar{V}_{K\times N}$ and $\vec{c}_{K\times 1}$ can be simply calculated as follows:

$$\frac{\partial \ell}{\partial \vec{c}}_{K\times 1} = \sum_{t=1}^{\tau} \frac{\partial \ell^t}{\partial \vec{c}} = \sum_{t=1}^{\tau} [(\bar{\vec{y}}^t - \vec{y}^t) \odot \bar{\vec{y}}^t \odot (1 - \bar{\vec{y}}^t)] \tag{8.19}$$

$$\frac{\partial \ell}{\partial \bar{V}}_{K\times N} = \sum_{t=1}^{\tau} \frac{\partial \ell^t}{\partial \bar{V}} = \sum_{t=1}^{\tau} [(\bar{\vec{y}}^t - \vec{y}^t) \odot \bar{\vec{y}}^t \odot (1 - \bar{\vec{y}}^t)]_{K\times 1} \cdot (h^t)^T_{1\times N} \tag{8.20}$$

The calculation of the gradients for $\bar{W}$, $\bar{U}$, and $\vec{b}$ is more complicated. As illustrated in the unfolded RNN, the gradient at a time step t comes from two parts: the loss caused by $\tilde{\vec{y}}^t - \vec{y}^t$ at the current step and that from the next time step via the hidden layer connection ($\vec{h}^t$ and $\vec{h}^{t+1}$). Thus, the gradient calculation will need to start from the last time step τ and then move backward (time) step by step.

$\bar{W}$, $\bar{U}$, and $\vec{b}$ are contained in the function for $\vec{h}^t$. Thus, the calculation of the gradients for these parameters will also involve $\frac{\partial \ell}{\partial \vec{h}^t}$ when applying the chain rule. For example, for $\bar{W}$, we have $\frac{\partial \ell}{\partial \bar{W}}_{N\times N} = \frac{\partial \ell}{\partial \vec{h}}_{N\times 1} \cdot \frac{\partial \vec{h}}{\partial \bar{W}}_{1\times N}$. To simplify the deduction, we can define the following intermediate variable:

$$\vec{\delta}^t_{N\times 1} = \frac{\partial \ell}{\partial \vec{h}^t_{N\times 1}} \tag{8.21}$$

Then, we can expand $\vec{\delta}^t$ to obtain the recursive relationship between steps:

$$\begin{aligned}\vec{\delta}^t_{N\times 1} =& \frac{\partial \vec{o}^t}{\partial \vec{h}^t}_{N\times K} \cdot \frac{\partial \ell}{\partial \vec{o}^t}_{K\times 1} + \frac{\partial \vec{h}^{t+1}}{\partial \vec{h}^t}_{N\times N} \cdot \frac{\partial \ell}{\partial \vec{h}^{t+1}}_{N\times 1} \\ =& \bar{V}^T_{N\times K} \cdot (\tilde{\vec{y}}^t - \vec{y}^t)_{K\times 1} + \bar{W}^T_{N\times N} \cdot \text{diag}\left[1 - (\vec{h}^{t+1})^2_{N\times 1}\right]_{N\times N} \cdot \vec{\delta}^{t+1}_{N\times 1}\end{aligned} \tag{8.22}$$

where the function diag($\vec{h}$) converts a vector $\vec{h}$ with a shape of $(N,)$ or $(N, 1)$ to a 2D array with a shape of (N, N) whose diagonal is $\vec{h}$.

$\vec{\delta}^{t+1}$ has a special case when $t = \tau$. At this last step, there is no gradient coming from the next step. Thus, it can be calculated directly:

$$\vec{\delta}^\tau_{N\times 1} = \left(\frac{\partial \vec{o}^\tau}{\partial \vec{h}^\tau}\right)_{N\times K} \cdot \frac{\partial \ell}{\partial O^T}_{K\times 1} = V^T_{N\times K} \cdot (\tilde{y}^\tau - y^\tau)_{K\times 1} \tag{8.23}$$

With $\vec{\delta}^t$, the gradients of $\bar{W}$, $\bar{U}$, and $\vec{b}$ can be calculated as

$$\frac{\partial \ell}{\partial \bar{W}}_{N\times N} = \sum_{t=1}^{\tau} \text{diag}[1 - (\vec{h}^t)^2]_{N\times N} \cdot \vec{\delta}^t_{N\times 1} \cdot (\vec{h}^{t-1})^T_{1\times N} \tag{8.24}$$

$$\frac{\partial \ell}{\partial \bar{U}}_{N\times J} = \sum_{t=1}^{\tau} \text{diag}[1 - (\vec{h}^t)^2]_{N\times N} \cdot \vec{\delta}^t_{N\times 1} \cdot (\vec{x}^t)^T_{1\times J} \tag{8.25}$$

$$\frac{\partial \ell}{\partial \vec{b}}_{N\times 1} = \sum_{t=1}^{\tau} \text{diag}[1 - (\vec{h}^t)^2]_{N\times N} \cdot \vec{\delta}^t_{N\times 1} \tag{8.26}$$

When implementing the above backward pass, we can start from the last time step. First, we calculate $\frac{\partial \ell}{\partial \vec{c}}$, $\frac{\partial \ell}{\partial \bar{V}}$, and $\vec{\delta}^t$. Then, we can compute $\frac{\partial \ell}{\partial \bar{W}}$, $\frac{\partial \ell}{\partial \bar{U}}$, and $\frac{\partial \ell}{\partial \vec{b}}$.

All the model parameters, i.e., $\vec{c}$, $\bar{V}$, $\bar{W}$, $\bar{U}$, and $\vec{b}$, which can be represented using a general parameter array, i.e., θ, can be updated using the gradient descent:

$$\theta = \theta - \alpha \cdot \frac{\partial \ell}{\partial \theta} \tag{8.27}$$

where α is the learning rate.

8.4 Practical Deep Learning Skills

8.4.1 Initialization

Overview

The training of ANNs including deep NNs is an optimization process via the decent of gradients. That is, we move downhill in a "hilly area" as a graphical representation of the loss function. In order to start the optimization, we will need to have a starting point, which may significantly determine the way and difficulty level of finding an optimum. This starting point is the initial values of the NN weights, which are determined by the initialization.

A very intuitive initialization strategy is to set all the model weights to 0. This is actually adopted for many optimization problems for simplicity. However, it will lead to problems in the training of ANNs. For example, if we use an activation function $\sigma(0) = 0$, then all the outputs in the forward pass and weight gradients in the backward pass will always be 0. As a result, the NN will not learn. If an activation function $\sigma(0) \neq 0$, then the weights tend to change "together," which will limit the power of backpropagation for searching the entire space for the optima.

Random initialization can solve this problem by assigning random numbers to different weights. Usually, these random numbers are generated according to some distribution functions, e.g., uniform distribution and normal distribution, which will make the initial weight values follow these distributions. Weights initialized with random distributions can still generate major issues if not appropriately set up.

If the weights are too small, then the output of the neurons will also be small and/or distributed in a very small range. This will likely cause vanishing gradient issues, especially when the number of layers increases in the NN. Also, for activation functions like Sigmoid, which is linear around 0, output values around this value will cause the activation function to lose its nonlinear capability.

If the weights are too large, then activation functions like Sigmoid tend to squeeze the values to the ends for large numbers, e.g., 0 and 1 for the output. This will lead to the saturation issue, which will cause vanishing gradients. As for activation functions that do not squash the input, e.g., ReLU, the output will be further enlarged by weights with large values in each layer. This will cause exploding gradient issues.

Therefore, the weights should be initialized in a way that they will not cause gradient issues and will stay in a reason range during backpropagation. Also, based on the above description, we can also see that the initialization also needs to be performed with a consideration of the adopted distribution(s) and activation function(s). In the following, we will first take a quick look at the normal distribution and the uniform distribution. Then see how common deep learning initialization methods can be proposed for these two distributions.

If we use a univariate normal distribution, we write it as $X \sim \mathcal{N}(\mu, \sigma^2)$, in which X is the random variable, which could correspond to the attribute in the data. When there are multiple attributes, then we will have multivariate normal distributions. Let us use a univariate normal distribution as an example:

$$f(x) = \frac{1}{\sqrt{2\pi}\sigma} \exp\left(-\frac{(x-\mu)^2}{2\sigma^2}\right) \tag{8.28}$$

If we use the uniform distribution, we can write it as $X \sim U(a, b)$. Accordingly, we have to follow the distribution equation for a single variate uniform distribution:

$$f(x) = \begin{cases} \frac{1}{b-a}, & a < x < b \\ 0, & \text{else} \end{cases} \tag{8.29}$$

Frequently, we generate random weights around 0, for example, using a normal distribution with a mean of 0. Accordingly, we have $\mu = 0$ for the normal distribution. If we take the same strategy for the uniform distribution, then the above equation can be rewritten as $X \sim U(-r, r)$:

$$f(x) = \begin{cases} \frac{1}{2r}, & -r < x < r \\ 0, & \text{else} \end{cases} \tag{8.30}$$

Therefore, the parameters that determine the scatter, such as the boundaries of the uniform distribution, i.e., r, and the variance of the normal distribution, i.e., μ, will primarily determine the characteristics of the random weights. Due to this fact, the primary goal for selecting appropriate initialization methods for deep learning is to find out how to set up these two parameters.

Xavier Initialization

A better understanding of what happens to the data via weights is needed before we can reasonably initialize the weights. So, let us first take a look at the forward pass. In particular, we can check the data operations and the resultant data changes at a typical layer, which can be summarized as the following two equations:

$$y^l = \sum_{i=1}^{n^l} w_i^l \cdot x_i^l \tag{8.31}$$

and

$$x^{l+1} = f(y^l) \tag{8.32}$$

in which x_i^l is the ith input (or ith element in the input array) in the lth layer, which has n^l elements, y^l is the output after the weight operations, and x^{l+1} is the input of the next (($l + 1$)th layer). x^{l+1} is also the output of the lth layer, which can be obtained by process y^l using the activation function.

In line with the above discussions, we will need to make sure the data will not be enlarged or reduced to avoid the saturation of the activation function. Meanwhile, we will also need to ensure y^l is at the linear range of the distribution function so that $x^{l+1} = f(y^l) \approx y^l$. Assuming the input of the current layer, i.e., x^l, has a mean of 0, then the mean and variance of the next layer, i.e., x^{l+1}, can be formulated as

Mean:

$$\mathbb{E}[x^{l+1}] = \mathbb{E}\left[\sum_{i+1}^{n^l} w_i^l \cdot x_i^l\right] = \sum_{i+1}^{n^l} \mathbb{E}\left[w_i^l\right] \cdot \mathbb{E}\left[x_i^l\right] = 0 \tag{8.33}$$

Variance:

$$\mathrm{Var}[x^{l+1}] = \mathrm{Var}\left[\sum_{i+1}^{n^l} w_i^l \cdot x_i^l\right] = n^l \mathrm{Var}\left[w_i^l\right] \cdot \mathrm{Var}\left[x_i^l\right] = 0 \tag{8.34}$$

The above two equations clearly show how the data changes as it moves through the operations in one layer. The mean of the data does not change, so we do not need to take any action. However, the variance of the data is enlarged by $n^l \cdot \mathrm{Var}[w_i^l]$ times. To make sure the variance does not change, we need to set

$$\mathrm{Var}\left[w_i^l\right] = \frac{1}{n^l} \tag{8.35}$$

Therefore, the variance of the distribution adopted for initializing the weights in the current layer should be $\frac{1}{n^l}$, in which n^l is the number of neurons in this layer.

The deduction is carried in the forward pass. To ensure that the data's variance does not change too much in both the forward and backward passes, we can use the following equation:

$$\mathrm{Var}\left[w_i^l\right] = \frac{2}{n^l + n^{l+1}} \tag{8.36}$$

Therefore, for the normal distribution, Xavier initialization adopts the following distribution: $\mathcal{N}(0, \sqrt{\frac{2}{n^l+n^{l+1}}})$.

For the uniform distribution, the key parameter r can be derived as follows:

$$\text{Var}[x \in U[-r, r]] = \frac{(r-(-r))^2}{12} = \text{Var}\left[w_i^l\right] = \frac{2}{n^l + n^{l+1}} \tag{8.37}$$

So we have

$$r = \sqrt{\frac{6}{n^l + n^{l+1}}} \tag{8.38}$$

He Initialization

He initialization is also called Kaiming initialization or MSRA initialization. It is proposed for the ReLU activation. When a layer adopts the ReLU activation function, it is observed that half of the neurons output 0. Accordingly, the variance of the distribution under this condition is approximately half of the variance under the condition of the logistic activation function. If we only consider the forward pass, the variance of w_i^l in the ideal condition is

$$n^l \cdot \text{Var}\left[w_i^l\right] = \frac{1}{2} \tag{8.39}$$

Thus we get

$$\text{Var}\left[w_i^l\right] = \frac{2}{n^l} \tag{8.40}$$

If we use the normal distribution, we should have $\mathcal{N}(0, \sqrt{\frac{2}{n^l}})$. If we use the uniform distribution between $[-r, r]$, we should have $r = \sqrt{\frac{6}{n^l}}$.

LeCun Initialization

The LeCun initialization is suitable for convolution layers. If we use the normal distribution, we should have $\mathcal{N}(0, \sqrt{\frac{1}{n^l}})$. If we use the uniform distribution between $[-r, r]$, we should have $r = \sqrt{\frac{3}{n^l}}$. What is tricky is the determination of n^l and n^{l+1}. We can use the following equations for the convolution layers:

$$n^l = \text{Number of feature maps} \times \text{Receptive field area} \tag{8.41}$$

$$n^{l+1} = \text{Number of feature maps} \times \text{Receptive field area/Max pool area} \tag{8.42}$$

where "receptive field area" is the calculated as kernel_height × kernel_width.

Batch Normalization

In general, batch normalization is not counted as an initialization technique. It can be viewed as an adjustment of the weights during backpropagation. If we view each backpropagation iteration as a brand new process, then batch normalization can be viewed as initialization as well. Batch normalization can also help us address gradient issues.

Batch normalization is proposed considering that the distribution of the data can change as it moves through different layers. For example, when using the Sigmoid function for activation, data tends to shift towards the lower and upper limits (i.e., 0 and 1). This will lead to vanishing gradients and slow down the convergence. Batch normalization can adjust the data distribution back to the standard or original condition, e.g., $\mathcal{N}(0, 1)$. This will ensure the data is distributed in the highly sensitive area of the activation function, which will enlarge the gradients to expedite the convergence.

8.4.2 *Optimization Methods*

Like other machine learning algorithms, deep learning aims to find a model that can best describe the knowledge behind the data. However, deep learning models, i.e., deep neural networks, are much more complicated than models established using other machine learning algorithms. Due to this fact, the search for the optimal model will require an iterative optimization process to minimize the loss function. A method or an algorithm that can dictate this search process is called a solver or an optimizer. Solvers are an essential part of deep learning because we usually need to specify which solver to use. Also, hyperparameters need to be determined to fix the solver. Common solvers used in machine learning include SGD, Momentum, NAG, Adagrad, Adadelta, RMSProp, AdaMax, and Nadam.

SGD

In a broad sense, SGD has three types: batch gradient descent, stochastic gradient descent, and mini-batch gradient descent. In many cases, SGD is thought to be mini-batch gradient descent. The essence of all of the three types is gradient descent as follows:

$$w_{t+1} = w_t + \Delta w_t \tag{8.43}$$

This gradient is adopted to update the model parameters as

$$\Delta w_t = -\eta \cdot g_t \tag{8.44}$$

where g_t is the gradient, w is the model parameter to be updated, the subscripts t and $t+1$ are the current and following steps, respectively, η is the learning rate, and f is the objective/loss function.

The gradient is calculated as

$$g_t = \nabla_w f(w_t) \tag{8.45}$$

Batch gradient descent uses all the samples, i.e., the whole training set as the batch, to calculate the gradient and then use this gradient to update the model parameters. If the loss function is a convex function, then we can expect to find the global optimal (minimal) value, i.e., the global minimal loss function value and the corresponding model parameters, if the learning rate is small enough. However, the loss function of deep learning problems is, in general, very complicated and usually is not convex. Thus, the identification of the global minimum is usually infeasible. Notwithstanding, a local minimal value can still ensure a decent model. Batch gradient descent can guarantee a local optimal value and can be relatively easy to implement. However, the method poses a high demand for memory as all the data needs to be processed together. Due to this reason, the gradient computation will take a long time.

Stochastic gradient descent takes a way that is completely opposite to batch gradient descent. Stochastic gradient descent calculates the gradient and updates the model after every sample is processed. Thus, this essentially uses the gradient determined by one sample, or viewed as the "slope" of a small area in the objective function, to determine which direction to go for optimizing the loss function. Because of the differences between samples—different samples may suggest much different or even opposite directions—sometimes it will take much more effort to find the best optimization direction. Also, due to the same reason, stochastic gradient descent is less likely to be trapped at a local optimal point compared with batch gradient descent. Meanwhile, stochastic gradient descent cannot utilize array operations to improve computational efficiency.

Mini-batch gradient descent is proposed to reach a compromise between batch gradient descent and stochastic gradient descent. It uses a subset of the training data as a mini-batch to calculate the gradient and update the model parameters. This method is adopted by many deep learning packages as the default SGD or even the default solver.

The selection of the learning rate in SGD is very difficult. In many cases, this can only be done via experience and/or trial and error. Please note that the difficulty of training deep neural networks is not only caused by the local minimum. The existence of saddle points can also trap the solver. A saddle point has zero

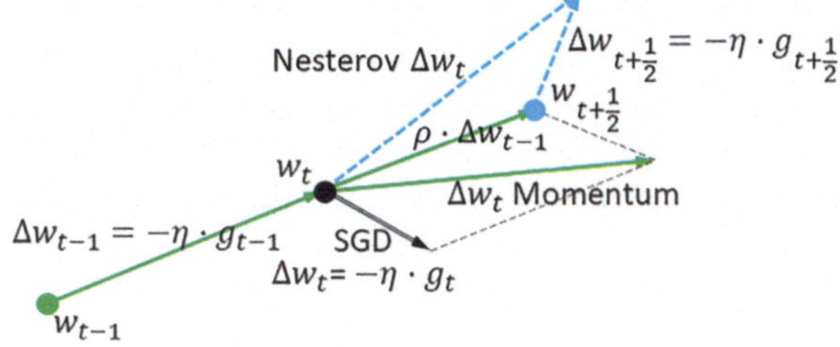

Fig. 8.11 Gradient descent method with manual learning rate

slopes (derivatives) in orthogonal directions. Thus, the solver may think the global optimum has been reached as zero gradients are found in both directions.

Momentum

As shown in Fig. 8.11, SGD only relies on the gradient of the current batch. Therefore, in the early stage of training, the update can be unstable, while in the late stage, the solver does not have enough power to jump off local optima or saddle points. The momentum method incorporates the gradient of the previous update step in addition to the use of the gradient of the current step. In this way, the optimization direction of the previous update is included in the gradient of the current step as a "momentum."

The update is calculated as

$$\Delta w_t = \rho \cdot \Delta w_{t-1} - \eta \cdot g_t = \rho \cdot \nabla_w f(w_{t-1}) - \eta \cdot g_t \tag{8.46}$$

Then, the update on the model parameter is performed as

$$w_{t+1} = w_t + \Delta w_t = w_t + \rho \cdot \nabla_w f(w_{t-1}) - \eta \cdot g_t \tag{8.47}$$

The gradient can be large in the early stage, so ρ typically adopts a relatively small value like 0.5. As the gradient decreases in the later training stage, ρ can be increased to values like 0.9.

Nesterov

The Nesterov or Nesterov momentum can be viewed as a modification of the momentum method. Compared with momentum, Nesterov uses the gradient at a "future" point for update. For this purpose, we first move from the current point by the momentum to reach this future point at $w_t + \rho \cdot \Delta w_{t-1}$ (denoted as Step $t - \frac{1}{2}$). Next, we calculate the gradient at this intermediate point, $g_{t+\frac{1}{2}}$. The momentum and the gradient will be added up as Δ_t for update. Mathematically, this can be written as

$$\Delta w_t = \rho \cdot \Delta w_{t-1} - \eta \cdot g_{t+\frac{1}{2}} = \rho \cdot \Delta w_{t-1} - \eta \cdot \nabla f_w(w_t + \rho \cdot \Delta w_{t-1}) \quad (8.48)$$

The learning rate in the above gradient descent is critical to the performance of this method. However, these gradient descent methods feature the use of a learning rate that is manually adjusted. The adjustment of the learning rate can significantly determine the learning outcome.

AdaGrad

AdaGrad is proposed to automatically adjust the learning rate. The update on the model parameters is performed as follows:

$$\Delta w_t = -\frac{\eta}{\sqrt{\sum_{\tau=1}^{t} g_\tau^2 + \epsilon}} \cdot g_t \quad (8.49)$$

where η is the initial learning rate, and ϵ is a small number used to avoid a zero denominator.

As can be seen in the above equation, the learning rate will gradually decrease as the update distance ($\sum \Delta w_t$) increases.

AdaDelta

AdaDelta is proposed to address three issues with AdaGrad: (1) the learning rate can only monotonically decrease, which can lead to excessively small learning rates in late training stages, (2) the units on two sides of the above AdaGrad update equation are not consistent, and (3) the initial learning rate still needs to be set manually.

As for (1), AdaDelta only uses the expectation of g_t^2 instead of the sum.

$$\Delta w_t = -\frac{\eta}{\sqrt{\mathbb{E}[g^2]_t + \epsilon}} \cdot g_t \quad (8.50)$$

The expected value can be calculated using the recent values as

$$\mathbb{E}[g^2]_t = \rho \cdot E[g^2]_{t-1} + (1-\rho) \cdot g_t^2 \quad (8.51)$$

As for (2) and (3), we can resort to Newton's method and obtain the following update equation:

$$\Delta w_t = -\frac{\sqrt{\mathbb{E}[w^2]_{t-1}}}{\sqrt{\mathbb{E}[g^2]_t + \epsilon}} \cdot g_t \quad (8.52)$$

RMSprop

RMSprop is a variation or a special case of AdaDelta. It is between AdaGrad and AdaDelta. First, the ρ value is fixed at 0.5.

$$\mathbb{E}[g^2]_t = 0.5 \cdot E[g^2]_{t-1} + (1 - 0.5) \cdot g_t^2 \tag{8.53}$$

Second, it still uses an initial learning rate as AdaGrad and the root mean square (RMS): $RMS|g|_t = \sqrt{\mathbb{E}[g^2]_t + \epsilon}$. Then, the update equation is as follows:

$$\Delta w_t = -\frac{\eta}{\sqrt{\mathbb{E}[g^2]_t + \epsilon}} \cdot g_t \tag{8.54}$$

The use of the initial (global) learning rate renders RMSprop suitable for unstable targets. Thus, it can generate relatively good results for RNN.

Adam

Adaptive moment estimation (Adam) is another method that can compute adaptive learning rates. Adam is essentially RMSprop with momentum. It uses the method of moments including the first- and second-order moments for calculating the learning rate.

First, the first-order and second-order moment estimation for computing the gradient is as follows:

$$m_t = \mu \cdot m_{t-1} + (1 - \mu) \cdot g_t \tag{8.55}$$

$$n_t = \nu \cdot n_{t-1} + (1 - \nu) \cdot g_t^2 \tag{8.56}$$

The above moments are adjusted as follows:

$$\hat{m}_t = \frac{m_t}{1 - \mu^t} \tag{8.57}$$

$$\hat{n}_t = \frac{n_t}{1 - \nu^t} \tag{8.58}$$

Then, the learning rate can be calculated as

$$\Delta w_t = -\frac{\hat{m}_t}{\sqrt{\hat{n}_t} + \epsilon} \cdot \eta \tag{8.59}$$

Adam fixes the ranges of the learning rate due to the use of the above adjustments, leading to stable variations of the learning rate. It integrates AdaGrad's advantage in handling sparse gradients and RMSprop's advantage in dealing with stable targets. Besides, Adam has a relatively low demand for internal memory and can set adaptive learning rates for different parameters. It is applicable to most non-convex optimization problems and is suitable for large datasets and high-dimensional data.

Nadam

Nadam can be viewed as an Adam variation with Nesterov momentum.

8.4.3 Data Preprocessing and Augmentation

Data preprocessing is a general machine learning topic. It includes data cleaning (removal of missing, noisy, and abnormal data), data integration (combination of data from different sources), data transformation (generalization, normalization, aggregation, attribute selection), and data reduction (dimension reduction). In deep learning, data resizing, i.e., adjusting the image size, and data normalization, i.e., rescaling the data to a certain range to facilitate solution and data analysis, are the most common data preprocessing work that we need to deal with. These data preprocessing techniques can be conveniently performed with many software packages.

Data augmentation is another useful technique in deep learning. In fact, it is sometimes necessary, especially when the data amount or diversity is limited. The data augmentation techniques in computer vision, e.g., CNN, and natural language processing (NLP), e.g., RNN, can be much different.

In computer vision applications, we have the following common data augmentation techniques.

Position augmentation

- Center crop: crop the given image at the center. The crop size is a parameter given by the user.
- Random crop: crop the given image at a random location.
- Random vertical flip: vertically flip the given image according to a given probability.
- Random horizontal flip: horizontally flip the given image according to a given probability.
- Random rotation: rotate the image by some angle.
- Resize: resize the size of the input image to a given size.

Color augmentation

- Brightness: one way to augment is to change the brightness of the image. The resultant image becomes darker or lighter compared to the original one.

- Contrast: the contrast is defined as the degree of separation between the darkest and brightest areas of an image. The contrast of the image can also be changed.
- Saturation: saturation is the separation between the colors of an image.

Advanced models for data augmentation

- Adversarial training/adversarial machine learning. Adversarial attacks are imperceptible changes to images (pixel-level changes) that can completely change the model prediction. In order to handle this issue, in adversarial training, images are transformed till the deep learning model is deceived and the model fails to correctly analyze the data.
- Generative adversarial networks (GANs). GANs have been widely used to generate synthetic images in a target domain.
- Neural style transfer. A series of convolution layers is trained such that the images are deconstructed where content and style can be separated. After separation, the content from an image is composed with the style of another image to create an augmented style image. Thus, the content remains the same, but the style is changed.

Data augmentation in NLP is less popular than that in the computer vision domain. It is, in general, difficult to automate the process of augmenting text data due to the complexity of a natural language. Common methods for data augmentation in NLP include:

- Easy data augmentation (EDA) operations: synonym replacement, word insertion, word swap, and word deletion
- Back translation: retranslating text from the target language back to its original language
- Contextualized word embeddings

8.5 Practice: Build AlexNet Using Keras to Address MNIST Image Classification

>>> More and Up-to-Date Course Materials including Practices @ AI-engineer.org <<<

The following code implements a simple NN for solving image classification problems associated with the MNIST dataset.

1. Please read the code and annotate the code line by line.
2. Please search online and modify the code to implement AlexNet.

```
# Use the following line to avoid a possible error
import os
os.environ["KMP_DUPLICATE_LIB_OK"]="TRUE"

# Import Dataset
```

```
from tensorflow.keras.datasets import mnist
(train_images, train_labels), (test_images, test_labels) = mnist.
    load_data()

# Build Deep Neural Network
from tensorflow import keras
from tensorflow.keras import layers
model = keras.Sequential([
    layers.Dense(512, activation="relu"),
    layers.Dense(10, activation="softmax")
])

# Compile Model
model.compile(optimizer="rmsprop",
              loss="sparse_categorical_crossentropy",
              metrics=["accuracy"])

# Preprocess Data
train_images = train_images.reshape((60000, 28 * 28))
train_images = train_images.astype("float32") / 255
test_images = test_images.reshape((10000, 28 * 28))
test_images = test_images.astype("float32") / 255

# Train the Model
model.fit(train_images, train_labels, epochs=5, batch_size=128)

# Test the Model
test_loss, test_acc = model.evaluate(test_images, test_labels)
print(f"test_acc: {test_acc}")
```

Hints:

1. You will need to change the "# Compile Model" section in the script file.

2. AlexNet takes a 2D input, so you will possibly need to specify the shape of the input (image) like "model.add(Conv2D(filters=96, input_shape=(28,28,1), kernel_size=(11,11), strides=(4,4), padding='same'))".

3. Some lines in the "# Preprocess Data" section may also need to be changed so that we can deal with 2D image data (28,28) instead of 1D flattened image data (784,).

Chapter 9
Ensemble Learning

9.1 Overview

This chapter introduces a very useful machine learning technique, ensemble learning, which can turn multiple models into a more powerful model. We will first study the basics of ensemble learning, including its definition, basic questions, major categories of algorithms, history, and challenges to figure out how it works, why it works, and what to use to make it work better. Next, three major categories of ensemble learning algorithms, i.e., bagging, boosting, and stacking, will be explained. The basic ideas and representative algorithms of these three categories of ensemble learning will be discussed with strict mathematical formulations and pseudo-code for guiding their applications.

9.2 Basics of Ensemble Learning

9.2.1 Definition

Ensemble learning is a machine learning technique in which multiple models are strategically generated and combined to obtain an ensemble as a model with better performance than that of individual constituent models. As shown in Fig. 9.1, these constituent models, which are called base models or weak models, can be generated with a basic machine learning algorithm such as linear model, SVM, decision tree, KNN, and NN, or their combinations. The former is called homogeneous, while the latter is heterogeneous. Currently, homogeneous base learners are more frequently used. Among them, CART decision tree and neural network are the most frequently used algorithms for constructing homogeneous base learners.

Essentially, the goal of ensemble learning is to create a model whose prediction bias and variance are comparable to or better than what can be obtained with a single

Z. "L." Liu, *Artificial Intelligence for Engineers*,
https://doi.org/10.1007/978-3-031-75953-6_9

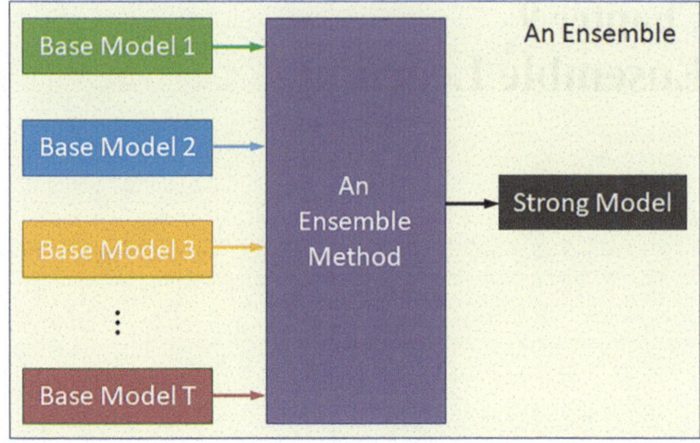

Fig. 9.1 Conceptual illustration of ensemble learning

base model. There are different strategies for creating an ensemble of base models, leading to different categories of ensemble learning algorithms. The idea behind ensemble learning can be described as "Union is Strength" or "Many hands provide great strength." Therefore, ensemble learning is also viewed as an optimization method that generates a strong learner from several weak learners.

The process of ensemble learning may contain extensive use of samplings, training of different models, selection and weighing of different models, and most importantly, combining the results from different base models to generate the final prediction for the whole ensemble. Common combination rules for regression tasks are primarily algebraic combiners, e.g., mean, sum, weighted sum, product, maximum, and median, and for classification tasks are primarily voting-based methods, e.g., majority vote and plurality voting. The simplest ensemble learning can be implemented by combining the predictions made by several base learners that are trained on the same training set, though the current mainstream ensemble learning does not refer to such a primitive way of ensembling.

In the remainder of this section, let us first take a quick look at some examples to understand "whether," "when," and "why" ensemble learning works. Then, the common categories of ensemble learning algorithms as well as concepts such as combination rules will be introduced. Finally, we will provide more details regarding the performance of ensemble learning by discussing bias and variance in association with different categories of methods.

9.2.2 Basic Questions

First, let us work on "whether" and "when" ensemble learning can generate a model that is better than the base models used for creating this model. Considering that we have not learned any ensemble learning strategies, let us start with a very simple example in which three cases use a basic combination strategy, plurality voting, as illustrated in Fig. 9.2. Plurality voting here means selecting the label that receives

	Sample 1	Sample 2	Sample 3
Model 1	-	+	+
Model 2	+	-	+
Model 3	+	+	-
Ensemble	+	+	+
True	+	+	+

(a)

	Sample 1	Sample 2	Sample 3
Model 1	-	+	+
Model 2	-	+	+
Model 3	-	+	+
Ensemble	-	+	+
True	-	+	+

(b)

	Sample 1	Sample 2	Sample 3
Model 1	-	-	+
Model 2	+	-	-
Model 3	-	+	-
Ensemble	-	-	-
True	+	+	+

(c)

Fig. 9.2 Examples of ensemble learning. (**a**) Ensemble improves performance. (**b**) Ensemble take no effect. (**c**) Ensemble reduces performance

the highest number of votes. As shown, in each case, the predictions of the three base models are combined in a simple way using plurality voting for a binary classification task.

A first look at the three examples can reveal one fact: a group of models, or the ensemble (model) represented by this group of models, does not necessarily exhibit a performance that is better than that of the base models. In fact, the three cases showed discrepant outcomes: (a) the ensemble improves the performance, (a) the ensemble does not take any effect, and (c) the ensemble reduces the performance. Thus, the result provides insights into the "whether" question: ensemble learning can generate better performance in some cases.

A closer look at the example can provide more hints for the answers to the "why" and when" questions. Case 1 shows "why" ensemble learning can improve the performance of base models: different models can complement each other. By contrast, Case 2 and Case 3 show "when" ensemble learning may not work. In Case 2, all the models provide the same predictions, and thus, predictions made by combining them make no improvements. That indicates that the base models need to be different from each other to some extent so that the ensemble can be meaningful. Case 3 implies that base models need to be good to some extent to ensure that the ensemble will lead to positive instead of negative changes in the performance. One extreme case is that an ensemble of models that provide random predictions cannot generate a better model, no matter how many such models are used or how these models are combined. In summary, the base models need to be good and diverse to some extent to ensure the ensemble can be effective.

Next, let us check a simple example that can help us understand why ensemble learning can work without touching complicated theories. As shown in Fig. 9.3, a binary classification problem can be solved with simple machine learning algorithms such as a linear model and an SVM. Such models can virtually generate a linear hyperplane, i.e., a straight line in the above 2D example, to separate the data into two groups. This can work well in linear problems.

However, this is not the case when a problem is nonlinear: an optimal boundary to separate the two categories is obviously not a straight line. As illustrated, three base models are generated using linear base models like SVM. Each of these base models can generate a straight line to separate the data. In this case, we can combine the predictions of these three models, i.e., the three straight lines, in some way to generate a nonlinear boundary for separating the two categories of data points.

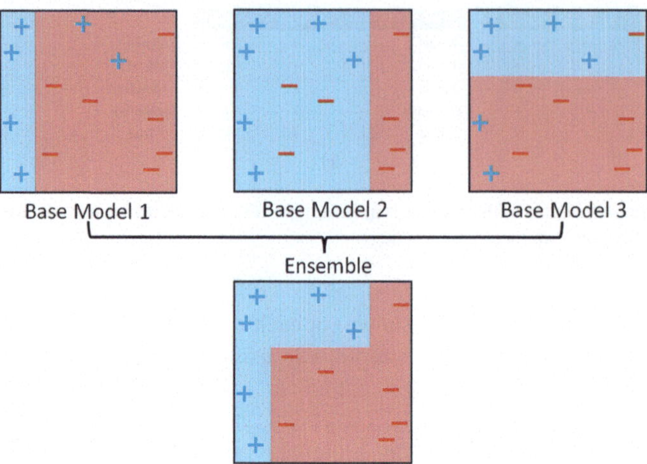

Fig. 9.3 Illustration of ensemble learning process

In this way, we obtain an ensemble that can better address this nonlinear binary classification problem by unifying the strengths of the three base models. This provides a very simple but intuitive demonstration to show why ensemble learning can work.

9.2.3 Categories of Ensemble Learning Methods

There are three common categories of ensemble (learning) methods: bagging (e.g., basic bagging), boosting (e.g., AdaBoost, Gradient Boosting), and stacking [98]. Because ensemble learning involves a process of making decisions with many decision-makers, these three categories are analogous to three concepts in politics. That is, bagging, boosting, and stacking are analogous to pluralist democracy, elitist democracy, and hierarchy, respectively.

As shown in Fig. 9.4, the idea behind bagging is "pluralism." For this purpose, many models can be trained parallelly (or simultaneously and independently), and these models generate the final prediction as the decision of the ensemble via some types of voting or averaging methods. We know that such a pluralistic democracy mechanism will help reduce the differences between prediction results as we attempt to find the average or majority. Therefore, bagging helps reduce the variance of the prediction. Bagging uses bootstrap sampling (sampling with displacement: placing drawn samples back for future sampling) to ensure diversity between base models.

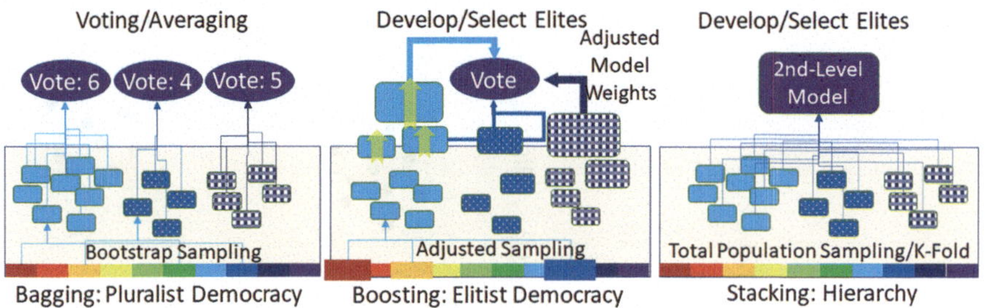

Fig. 9.4 Categories of ensemble learning algorithms

By contrast, the idea behind boosting is "elitism." For this purpose, the models are generated in a serial or sequential way so that late-generated models can improve the prediction accuracy based on the performance of early models in the series or sequence. As a result, we can generate "elites" or models with better performance. Such models are generated by focusing more on samples that are previously incorrectly predicted and are assigned greater weights in the combination of predictions. This effort at the search for better models essentially helps improve the bias of the prediction.

The third category of ensemble learning algorithms attempts to generate a better model from a group of base models by exploring better combination approaches. Intuitively, this can be done by simply combining the predictions of base models with simple combination rules. However, as can be seen, both bagging and stacking involve the use of such basic combination rules. Thus, simple combination via a basic combination rule is not frequently studied. In some literature, basic combination via a simple combination rule is introduced as a more basic category underneath or before bagging, boosting, and bagging or even merely as combination rules instead of a distinct category of ensemble learning algorithms. In fact, stacking as a more advanced combination approach is predominant in this third category of ensemble learning algorithms. This is why stacking is viewed as the third category in many places. Stacking differs from simple combination rules in that a machine learning algorithm is adopted to learn how to combine the results generated by the base learners for the final prediction. Therefore, this type of algorithm has a two-layer structure: base models and a combination model (or called combiner or final estimator).

9.2.4 Essence of Ensemble Learning

As mentioned above, bias and variance, which measure the performance of predictions, are important concepts that help us understand the goals and outcomes of ensemble learning methods [99]. Here, let us provide more information about them so that we can get a better understanding of the different categories of

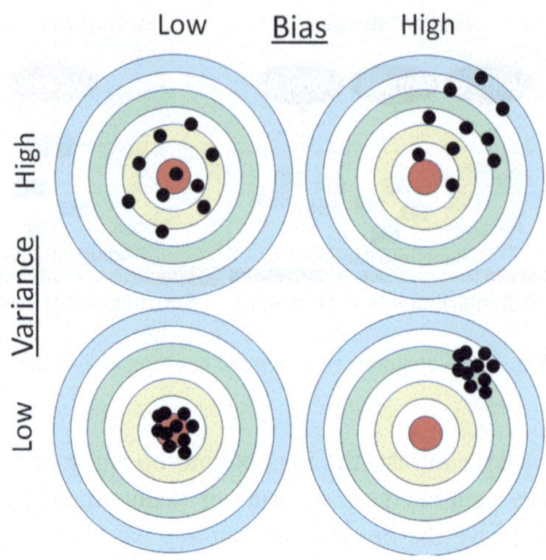

Fig. 9.5 Different conditions of bias and variance

algorithms. Figure 9.5 illustrates how bias and variance can reflect on the prediction results. In the context of ensemble learning, bias describes the differences between the prediction and the true label, while variance measures the difference between models, i.e., base models. A high bias implies that the model is not accurate, while a high variance may indicate overfitting.

The relationship between bias and variance in machine learning is interesting and essentially related to the understanding and implementation of ensemble learning. There is a tradeoff between these two parameters. In a simple way, as shown in Fig. 9.6, a more complicated model can provide better predictions in theory. Accordingly, the bias will be low. Due to the same reason, the model will turn out to be more sensitive to (new) data and thus cause a high variance, which may lead to poor performance in the testing data.

A good model should strike a balance between bias and variance. Ensemble learning is such an approach for striking the balance. Usually, bagging starts from the left of the optimum, so it usually has a high requirement on the bias. That is, the performance of the base models is expected to be relatively high. Boosting starts from the right of the optimum, so the base models can have relatively low performance, but the diversity between models needs to be relatively high to ensure a good ensemble learning outcome.

Therefore, in order to reduce the total error using ensemble learning, we will need to find ways to (1) reduce bias (performance of the base models), (2) improve variance (diversity of the base models), and (3) find better combination methods to generate a lower error with given bias and variance of the base models. Item 1 is associated with the selection and implementation of base models, so it is not in

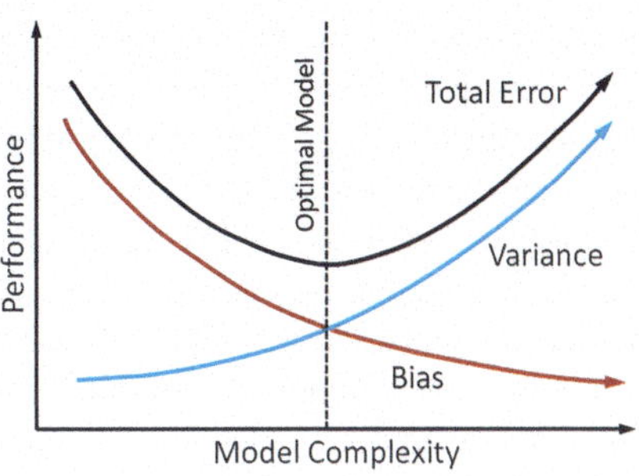

Fig. 9.6 Variation of bias and variance with model complexity

the context of ensemble learning. As for Item 2, disturbance to instances, attributes, output, and model parameters (e.g., hyperparameters) can enhance diversity. For example, the bootstrap sampling in bagging is a disturbance to instances, and the random selection of attributes in a random forest is a disturbance to attributes. Item 3 has a good example: stacking. The above understanding can be used as potential tricks for improving the performance of ensemble learning in more advanced applications of this technique.

9.2.5 History and Challenge

To obtain a complete understanding of ensemble learning, let us take a look at the history and challenges of ensemble learning. The concept of ensemble learning was first proposed by Dasarathy [100], who used linear models and KNN to form composite systems for classification. In the same year, a sample method that was later widely adopted in ensemble learning, i.e., bootstrap, was proposed by Efron [101]. Kearns [102] and Valiant (1989) proposed the concept of weak learning, which triggered discussions on whether a group of weak learners can create a strong learner [102, 103]. Schapire [104] provided a positive answer to the question and presented boosting [104]. A few years later, Schapire and Freund proposed the AdaBoost algorithm for boosting, which was awarded the famous Gödel Prize in the area of theoretical computer science [105]. Hensen (1990) proved the variance reduction characteristic of ensemble learning, which can be used to improve the performance of neural networks [106]. The concept of stacking was first proposed by Wolpert [107], which could help achieve performance better than that of individual learners [107].

In 1995, random forest as one of the most well-known ensemble learning (bagging) algorithms was proposed. In 1996, Breiman proposed bagging based on the understanding of the influence of disturbance on the structure of instances, which affects the variance and, consequently, the learning outcome [108]. Kalal (2010) showed the performance improvements of binary classifiers attributed to the treatment of the structure of instances, and bootstrapping was used for this purpose [109]. Deng et al. [110] suggested scalable stacking and learning for building deep architectures [110]. A more recent breakthrough is the release of XGBoost by Chen and Guestrin in 2016 [111], which was developed based on many existing techniques, such as GBDT (gradient boost and decision tree), regularization, shrinkage (change in learning rate), column subsampling, and parallelization, and was later widely adopted due to its outstanding performance.

Despite the advances, ensemble learning also faces challenges for further development. One major challenge is the time-demanding and computationally expensive process of training multiple (base) models. In addition to this general challenge, different categories of ensemble learning algorithms also have their own issues. As for bagging, the use of bootstrapping is based on many statistical hypotheses, whose validity affects the sampling accuracy and learning outcome. Boosting can be excessively sensitive to noise in labels, i.e., outlier, as it attempts to give incorrect predictions higher weights in the sequential learning process. Stacking is challenged by the determination of hyperparameters.

Ensemble learning can be considered in the following application scenarios:

- Selection of algorithm. It could be tricky to select an appropriate machine learning algorithm. Ensemble learning can help circumvent this issue as it can include many different types of algorithms. This helps save the effort of finding the best algorithm for the given data and reduces the risk of selecting an inappropriate model.
- Too much or little data. When a dataset is too large, we can split the dataset into subsets directly to train different sub-models; when the dataset is too small, bootstrap provides a good sampling tool for generating subsets from a small dataset.
- Complex problems. Many problems cannot be well addressed by a single machine learning model, such as classification problems with very complex, irregular (highly nonlinear) boundaries between different categories. Ensemble learning can address such problems by integrating the predictions of different sub-models.
- Data fusion. Many applications involve data from different sources. Such data from different sources can require the use of different models/algorithms for treatment. Ensemble learning can meet this need.
- Confidence estimate. Ensemble learning, by nature, can provide an estimate of confidence by investigating how decisions are made by different sub-models. For example, the situation in which most sub-models make the same prediction implies a high level of confidence. Though a high confidence level does not guarantee a good prediction, it was found that with proper training, a high confidence level usually implies a correct prediction and vice versa.

9.3 Bagging

Bagging attempts to obtain an ensemble learning model with high generalization. For this goal, the base learners need to be as mutually independent as possible. A straightforward way to ensure such independence is to divide the training data into mutually exclusive subsets. However, this may lead to insufficient samples for each subset, making the distributions of variables in the subsets different from those in the training data, especially when data is limited. An effective way to address this issue is bootstrap sampling. That is also where the name "bagging" comes: bootstrap aggregating.

Bagging is a very straightforward category of ensemble learning algorithms—it does not necessarily involve complicated theories to guide its implementation. As a result, no equations or deductions are needed in this section. Instead, in the following, extra information including bootstrap sampling, base learners, combination, and pseudo-code will be offered for the basic version of bagging or called bagging meta-estimator. Then, a popular and more advanced bagging algorithm, i.e., random forest, will be introduced. In particular, the changes that need to be made to move from basic bagging to random forest will be explained.

9.3.1 Basic Bagging

When bootstrap is used, we select a certain number of samples (between 1 and the total number of samples) to form a subset each time and put the selected samples back before generating the next subset. Though it appears to be very simple, this sampling method is very effective in striking a balance between achieving the independence of subsets and letting each subset represent the total dataset in terms of variable distributions (hints: each attribute represents a variable).

In some literature, each subset is also called a bootstrap sample. Therefore, a bootstrap sample is a small sample that is "bootstrapped" from a large sample, which corresponds to the original (training) dataset. Bootstrapping can be interpreted as a type of resampling where many subsets (or sub datasets) of the same size are repeatedly generated, with replacement, from a single original dataset (dataset).

One thing that needs to be mentioned is the out-of-bag (OOB) data. OOB refers to the samples that have not been selected after all the needed subsets are generated. We can use a very simple example to understand it. Let us assume we have a dataset consisting of I samples. We select 1 sample from it each time for I times to create I subsets. Then, we know that the percentage of samples that will not be selected in this process is $(1 - \frac{1}{I})^I$. The probability as I approaches infinite is $\lim_{I \to \infty}(1 - \frac{1}{I})^I = \frac{1}{e} \approx 0.368$. The real probability varies as the size of the subset and I change, but OOB exists due to such a probability. OOB is useful in several ways. First, it can be utilized to replace a separate testing dataset for error estimate, which is called OOB estimate, which is an unbiased estimate compared with the test set. In addition,

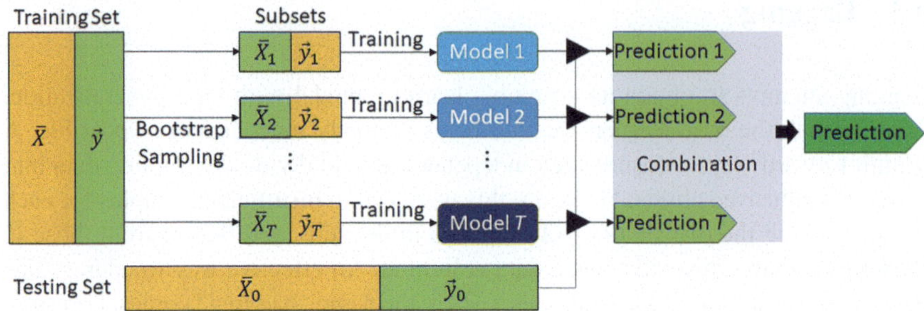

Fig. 9.7 Procedure of bagging

if decision trees are used as the base models, OOB can be employed to assist in pruning, estimating the posterior probability of different nodes and treatment of nodes with no samples. If neural networks are used as the base models, then OOB can be used for early stopping to diminish overfitting.

As shown in Fig. 9.7, with the T subsets, we can train T base models. These base models are then combined using one of the basic ensemble combination rules. That is, algebraic combiners such as mean, sum, weighted sum, product, maximum, and median are used for regression tasks, and voting-based methods such as majority vote and plurality voting are usually used for classification tasks.

The following is pseudo-code for implementing the above basic bagging.

Bagging:

Input: Training dataset $D = (x_1, y_1), (x_2, y_2), \ldots, (x_I, y_I)$; algorithms learning for base models $\mathfrak{L}$; number of base models T.
Initialize the base models: $h_t \in H \leftarrow \emptyset$
for $t = 1, 2, \ldots, T$ do[1]
  Generate a bootstrap sample (subset) from D: D_t
  Train each base model: $h_t \leftarrow \mathfrak{L}(D_t)$
  $H \leftarrow H \cup h_t$
end for
Obtain the final prediction by combining the predictions made by H.

An important fact needs to be mentioned in the implementation of bagging. As mentioned in the previous section, bagging primarily aims at the reduction of variance. Therefore, it works more effectively with algorithms that are more susceptible to disturbance in samples, such as decision trees without pruning and neural networks. This can be used to guide the selection of algorithms for base models.

In addition, as an effective ensemble learning method, the complexity of bagging is $T(O(h)+O(c))$, where $O(h)$ is the complexity of the base model and $O(s)$ is the

[1] Bagging is a parallel method, so this is not necessarily implemented as a loop. Instead, the generation of h_t is independent from each other and thus can be performed in a parallel way.

complexity of combination processes such as voting and averaging. In general, $O(c)$ is not high, while T is a constant; therefore, the complexity of training an ensemble with bagging has the same level of difficulty as that of training its base models. Finally, bagging has the advantage that it can be applied to multi-class classification without any modification; this is not the case in boosting algorithms, which will be introduced later.

9.3.2 Random Forest

Random forest is a special and improved version of the basic bagging algorithm. It is viewed as a special version of bagging because only decision trees will be used as the base models. That is also where "forest" comes from. It can be viewed as an improved version because, compared with the basic bagging, it introduces a random selection of attributes instead of using all the attributes every time, which causes disturbance to the attributes for improving diversity. This is where "random" comes from.

The basic premise of the algorithm is that building a small decision tree with a few features is a computationally cheap process. If we can build many small, weak decision trees in parallel, we can then combine the trees to form a single, strong learner by averaging or taking the majority vote. In practice, random forests are often found to be the most accurate learning algorithms to date. Pseudo-code for random forest is given below.

Random Forest:

Input: Training dataset $D = (x_1, y_1), (x_2, y_2), \ldots, (x_I, y_I)$; algorithms for base learning models $\mathfrak{L}$; number of base models T.

Initialize the base models: $H \leftarrow \emptyset$

for $t = 1, 2, \ldots, T$ do

  Generate a bootstrap sample (subset) from D: D_t

  Train each base model: $h_t \leftarrow \mathfrak{L}(D_t)$ via modified decision tree algorithm:

    At each node:

      $f \leftarrow$ a subset of F

      Split on the best feature in f

  $H \leftarrow H \cup h_t$

end for

Obtain the final prediction by combining the predictions made by H.

The algorithm works as follows. For every tree in the forest, a bootstrap sample is selected from D where D_t denotes the tth bootstrap. We then learn a decision tree using a modified decision tree learning algorithm. The modified decision tree works as follows: at each node of the tree, instead of examining all possible feature splits, we randomly select a subset of the features $f \subseteq F$, where F is the set of all the available features. The node then splits on the best feature in f rather than F. In practice, f is much, much smaller than F. The most computationally expensive link of decision tree learning is usually to decide which feature to split. By narrowing the set of features, we drastically speed up the learning of the tree.

9.4 Boosting

Boosting primarily focuses on reducing bias. Therefore, in theory, boosting can build a strong ensemble learner based on a group of weak learners with low biases. Thus, the essence of boosting can be understood as both the improvement and the optimized combination of multiple weak learners into a strong learner.

Boosting can be understood by comparing it with bagging. In bagging, an ensemble is created by making multiple different bootstrap samples (each ensemble sample is a subset) out of the same training dataset and obtaining a base model with each. The predictions made by all of the base models are combined to produce a prediction of the ensemble with an error lower than that of the base models. In this process, the disturbance to the samples was generated by bootstrap sampling. Bootstrap samples (subsets) are slightly different from each other, and consequently, base models trained with them are different.

By contrast, as shown in Fig. 9.8, boosting generates disturbance to the samples (original samples instead of bootstrap samples) by assigning different weights to the samples. The learning process is sequential: each step is taken to train one model. The weights are updated in each step so that incorrectly predicted samples in the previous step are given higher weights in the next step. In this way, the subsequent model works harder to correct errors made by the whole ensemble on the training dataset. In addition to the weights of samples, the weights of models for the later combination can also be adjusted based on their performance. All the samples will be combined using a combination rule. Therefore, the sequential training process in boosting can gradually improve the ensemble's performance.

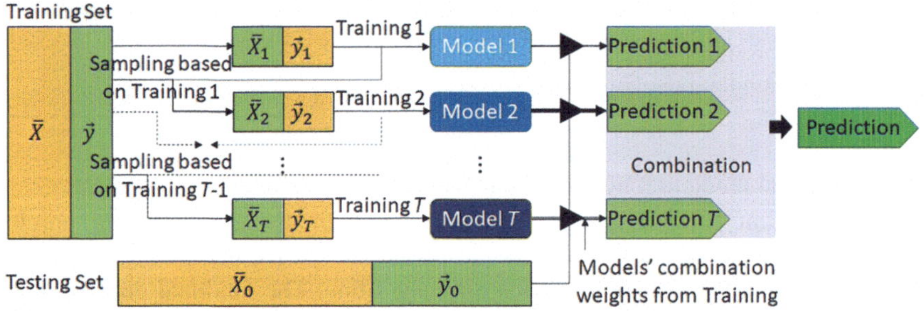

Fig. 9.8 Procedure of boosting

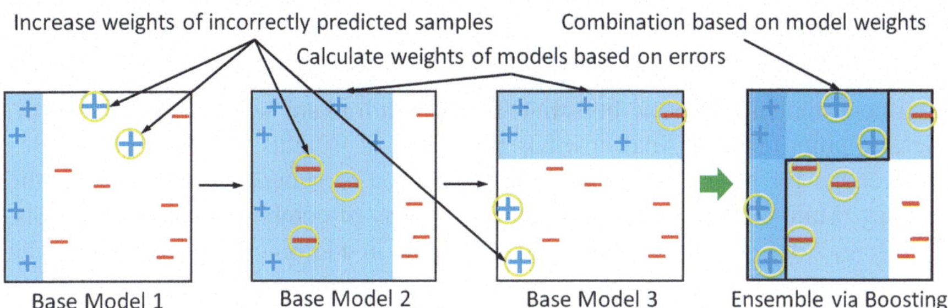

Fig. 9.9 Example of boosting

Figure 9.9 illustrates a very simple example of boosting. As can be seen, the incorrectly predicted samples, which are enlarged and circled, are given better considerations via greater weights for training the base model in the following step. The ensemble is a combination of all the base models according to their model weights, which are determined based on their performance.

Historically, boosting algorithms were challenging to implement. The situation changed with the advent of AdaBoost, which demonstrated how to implement boosting so that the technique could be used effectively. AdaBoost and modern Gradient Boosting work by sequentially adding models that correct the residual prediction errors of the existing models. As such, boosting methods are known to be effective, though constructing models with them can be slow, especially for large datasets. Gradient Boosting was merged with decision trees into gradient boosting decision trees (GBDT), and later improvements led to more powerful XGBoost, which became a preferred and often the top performing approach in machine learning competitions for classification and regression on tabular data. These advances have enabled boosting to become one of the most powerful learning ideas introduced in the last twenty years [112].

In fact, boosting algorithms could appear more complex than bagging and stacking algorithms, though the basic idea described above is similar and simple. As you may recall, bagging, as introduced in the previous section, does not even need an equation or a deduction to lay down the theory. By contrast, the theory for guiding boosting algorithms will require deductions and more equations. Thus, the implementation of boosting based on these derived equations appears slightly more complicated. Moreover, because of the sequential training process, boosting cannot be accelerated with parallel computing as bagging and thus requires more time.

In this section, AdaBoost and Gradient Boosting, as the two most representative boosting algorithms, will be introduced. We will first lay down the theories with necessary deductions to show how the major equations for implementations are obtained. Then, the implementation process will be detailed with pseudo-code for each method. After the introduction to the two methods, a few extra comments will be made on the comparison between boosting and bagging and on the application of boosting.

9.4.1 AdaBoost

AdaBoost implements basic boosting in a straightforward way [113]. It is worthwhile to point out that AdaBoost has two types of weights: weights for samples and weights for base models. The former is used to generate disturbance to the samples, while the latter is used to update the way of combining the base models. Therefore, AdaBoost updates the model weights in addition to the updates on the sample weights. The updates on these two types of weights usually need to be sought from the observations on the loss function. In the following, the loss function will be introduced first, and then, the theories for guiding the updates on the model weights and sample weights will be presented. Finally, pseudo-code will be provided to summarize the implementation.

The theory of AdaBoost can be derived in different ways. One of the easiest ways is based on the additive model, in which the prediction of the ensemble is calculated as a linear combination of the predictions made by all the base models:

$$H(x) = \sum_{t=1}^{T} \alpha_t h_t(x) \tag{9.1}$$

where $H(x)$ is (the prediction made by) the ensemble model and α_t is the weight of the model trained in step t: $h_t(x)$.

It is also very common that the above weighted sum is performed sequentially in a process in which we add base models one by one. Under this condition, the above equation can be reformulated as follows, for which more detail can be found in the following subsection for Gradient Boosting:

$$H_t(\bar{X}) = H_{t-1}(\bar{X}) + \alpha_t h_t(\bar{X}) \tag{9.2}$$

In AdaBoost, the weights of samples are described by the distribution of samples: $\mathcal{D}$. The loss function is usually constructed with this distribution and the exponential loss function:

$$\ell(H|\mathcal{D}) = \mathbb{E}_{\bar{X}\sim\mathcal{D}}[e^{-f(\bar{X})H(\bar{X})}] \tag{9.3}$$

where $f(\bar{X})$ is the true mapping (solution).

Thus, in all the T steps of training the ensemble for generating the T base models, $\mathcal{D}$ will be updated in each iteration. In the tth iteration step, the sample distribution $\mathcal{D}_t$ is calculated based on the ensemble obtained in the previous step using an equation to be introduced later. Then, samples will be drawn from the training set D using the distribution $\mathcal{D}_t$. Next, a new base model h_t will be trained with the data. The error of h_t will be used to update the weight of the base model α_t. The ensemble can be readily obtained with the ensemble from the previous step $H_{t-1}(\bar{X})$ and the base model h_t with its weight α_t from the current step.

In the following deduction of the theories for updating the two types of weights, let us use a simple binary classification problem. In this problem, let us assume f is the true solution: $f : \vec{y} = f(\bar{X})$ in which $y_i \in \{-1, +1\}$ for $\vec{y} = [y_1, y_2, \ldots, y_i, \ldots, y_I]^T$ and $\bar{X} = [\vec{x}_1^T, \vec{x}_2^T, \ldots, \vec{x}_I^T]^T$. Thus, the predictions made by the base models and the ensemble should also be either -1 or $+1$.

Loss Function

The gradient of the loss function for the above loss function can thus be written as

$$\begin{aligned}\frac{\partial \ell(H|\mathcal{D})}{\partial H(\bar{X})} &= e^{-f(\bar{X})H(\bar{X})} \cdot (-f(\bar{X})) \\ &= -e^{-H(\bar{X})} \cdot P(f(\bar{X}) = 1) + e^{H(\bar{X})} \cdot P(f(\bar{X}) = -1)\end{aligned} \tag{9.4}$$

The minimum of the loss function is achieved when the above gradient is zero, which points out the optimal ensemble:

$$H(\bar{X}) = \frac{1}{2} \ln \frac{P(f(\bar{X}) = 1)}{P(f(\bar{X}) = -1)} \tag{9.5}$$

The real prediction is made by the sign function of $H(\bar{X})$. The reason can be seen in the following equation:

$$\begin{aligned}\text{sign}(H(\bar{X})) =& \text{sign}\left(\frac{1}{2} \ln \frac{P(f(\bar{X}) = 1)}{P(f(\bar{X}) = -1)}\right) \\ =& \begin{cases} 1, & P(f(\bar{X}) = 1) \\ -1, & P(f(\bar{X}) = -1) \end{cases}\end{aligned} \tag{9.6}$$

Update on Model Weights

To derive the equation for the model weight update, we can first recall the error of the base model and then expand the loss function of the ensemble. The error of the base model h_t is

$$\epsilon_t = P_{\bar{X} \sim \mathcal{D}_t}(h_t(\bar{X}) \neq f(\bar{X})) \tag{9.7}$$

The contribution of the base model h_t to the total loss of the ensemble can be formulated as

$$\begin{aligned}\ell(\alpha_t h_t|\mathcal{D}_t) =&\mathbb{E}_{\bar{X}\sim\mathcal{D}_t}\left[e^{-f(\bar{X})\alpha_t h_t(\bar{X})}\right]\\ =&e^{-\alpha_t}P_{\bar{X}\sim\mathcal{D}_t}(f(\bar{X})=h_t(\bar{X}))+e^{\alpha_t}P_{\bar{X}\sim\mathcal{D}_t}(f(\bar{X})\neq h_t(\bar{X}))\\ =&e^{-\alpha_t}(1-\epsilon_t)+e^{\alpha_t}\epsilon_t\end{aligned} \tag{9.8}$$

The weights can be obtained by assessing the condition in which the derivative of the loss function with respect to α_t is zero:

$$\frac{\partial\ell(\alpha_t h_t|\mathcal{D}_t)}{\partial\alpha_t} = -e^{-\alpha_t}(1-\epsilon_t)+e^{\alpha_t}\epsilon_t \tag{9.9}$$

Then, we obtain

$$\alpha_t = \frac{1}{2}\ln\left(\frac{1-\epsilon_t}{\epsilon_t}\right) \tag{9.10}$$

Update on Sample Weights/Distribution

The loss function can be formulated based on the relationship between the ensemble and the current base model. In the following reformulation, Taylor's expansion of exponential function and the fact that $f^2(\bar{X}) = 1$ and $h_t^2(\bar{X}) = 1$ are utilized:

$$\begin{aligned}\ell(H_{t-1}+h_t|\mathcal{D}) =&\mathbb{E}_{\bar{X}\sim\mathcal{D}}\left[e^{-f(\bar{X})(H_{t-1}(\bar{X})+h_t(\bar{X}))}\right]\\ =&\mathbb{E}_{\bar{X}\sim\mathcal{D}}\left[e^{-f(\bar{X})H_{t-1}(\bar{X})}\cdot e^{-f(\bar{X})h_t(\bar{X})}\right]\\ =&\mathbb{E}_{\bar{X}\sim\mathcal{D}}\left[e^{-f(\bar{X})H_{t-1}(\bar{X})}\left(1-f(\bar{X})h_t(\bar{X})+\frac{f^2(\bar{X})h_t^2(\bar{X})}{2}\right)\right]\\ =&\mathbb{E}_{\bar{X}\sim\mathcal{D}}\left[e^{-f(\bar{X})H_{t-1}(\bar{X})}\left(1-f(\bar{X})h_t(\bar{X})+\frac{1}{2}\right)\right]\end{aligned} \tag{9.11}$$

Let us define a new distribution based on $\mathcal{D}$:

$$\mathcal{D}_t(\bar{X}) = \frac{\mathcal{D}(\bar{X})e^{-f(\bar{X})H_{t-1}(\bar{X})}}{\mathbb{E}_{\bar{X}\sim\mathcal{D}}\left[e^{-f(\bar{X})H_{t-1}\bar{X}}\right]} \tag{9.12}$$

The equation for updating $\mathcal{D}_t$ can be derived as follows:

$$
\begin{aligned}
\mathcal{D}_{t+1}(\bar{X}) =& \frac{\mathcal{D}(\bar{X})e^{-f(\bar{X})H_t(\bar{X})}}{\mathbb{E}_{\bar{X}\sim\mathcal{D}}\left[e^{-f(\bar{X})H_t(\bar{X})}\right]} \\
=& \frac{\mathcal{D}(\bar{X})e^{-f(\bar{X})H_{t-1}(\bar{X})} \cdot e^{-f(\bar{X})\alpha_t h_t(\bar{X})}}{\mathbb{E}_{\bar{X}\sim\mathcal{D}}\left[e^{-f(\bar{X})H_t(\bar{X})}\right]} \\
=& \mathcal{D}_t \cdot e^{-f(\bar{X})\alpha_t h_t(\bar{X})} \cdot \frac{\mathbb{E}_{\bar{X}\sim\mathcal{D}}\left[e^{-f(\bar{X})H_{t-1}(\bar{X})}\right]}{\mathbb{E}_{\bar{X}\sim\mathcal{D}}\left[e^{-f(\bar{X})H_t(\bar{X})}\right]} \\
=& \frac{\mathcal{D}_t}{Z_t} \cdot e^{-f(\bar{X})\alpha_t h_t(\bar{X})}
\end{aligned}
\tag{9.13}
$$

where Z_t is a parameter determined by the ensemble loss values in Step $t-1$ and Step t: $\frac{\mathbb{E}_{\bar{X}\sim\mathcal{D}}\left[e^{-f(\bar{X})H_t(\bar{X})}\right]}{\mathbb{E}_{\bar{X}\sim\mathcal{D}}\left[e^{-f(\bar{X})H_{t-1}(\bar{X})}\right]}$.

Pseudo-Code

Based on the theories from the above theories, pseudo-code for implementing AdaBoost for a binary classification problem is presented as follows.

AdaBoost:

Input: Training dataset $D = (x_1, y_1), (x_2, y_2), \ldots, (x_I, y_I)$; algorithms learning for base models $\mathfrak{L}$; number of base models T.

Initialize the sample distribution: $\mathcal{D}_1(\bar{X}) = \frac{1}{I}$

for $t = 1, 2, \ldots, T$ do

  Train a base model $h_t = \mathfrak{L}(D, \mathcal{D}_t)$

  Calculate the error: $\epsilon_t = P_{\bar{X}\sim\mathcal{D}_t}(h_t(\bar{X}) \neq f(\bar{X}))$

  if $\epsilon_t > 0.5$ then break

  $\alpha_t = \frac{1}{2}\ln(\frac{1-\epsilon_t}{\epsilon_t})$

  $\mathcal{D}_{t+1}(\bar{X}) = \frac{\mathcal{D}_t(\bar{X})}{Z_t} \times \begin{cases} \exp(-\alpha_t), & \text{if } h_t(\bar{x}) = f(\bar{x}) \\ \exp(\alpha_t), & \text{if } h_t(\bar{x}) \neq f(\bar{x}) \end{cases}$

end for

Output: Obtain the ensemble $H(\bar{X}) = \text{sign}(\sum_{t=1}^{T} \alpha_t h_t(\bar{X}))$

AdaBoost implements the basic boosting algorithm. It changes both the weights of the samples in the resampling process and the weights of the models in the combination. Moreover, it uses the exponential loss function, which makes this method sensitive to abnormal data points. Gradient Boosting, which is different from AdaBoost in many ways especially the introduction of gradient descent, will be introduced in the next subsection.

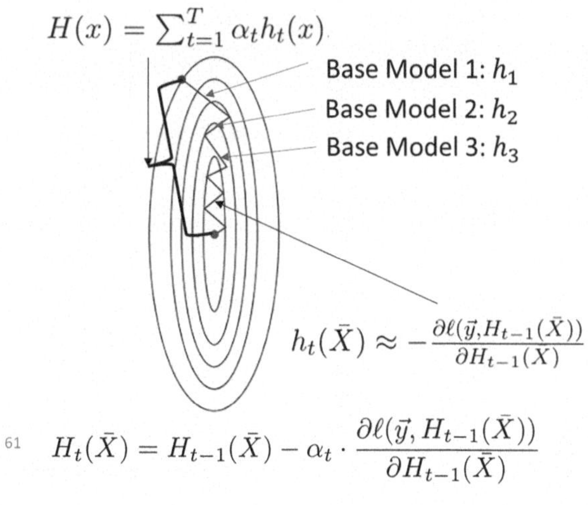

Fig. 9.10 Model update in gradient boosting

9.4.2 Gradient Boosting

Gradient Boosting integrates boosting and gradient descent [114]. As a typical ensemble learning algorithm, Gradient Boosting gradually develops new base models to improve the performance of the ensemble formed by the existing base models. As shown in Fig. 9.10, a new model is created to approximate the negative gradient of the loss function in the current step. Then, adding this base model to the current ensemble is equivalent to moving one step toward the minimum of the loss function. There is also a learning rate, which will be sought via the line search method.

To illustrate this integration, let us first recall the equation for gradient descent:

$$w = w - \alpha \cdot \frac{\partial \ell(w)}{\partial w} \tag{9.14}$$

where w can be a scalar or a tensor of any order as parameters for defining a model.

Next, let us rewrite the equation for calculating the ensemble, $H(x) = \sum_{t=1}^{T} \alpha_t h_t(x)$, into a form for iterative update. To simplify the deduction, we can assign the model weights as $\alpha_t = \frac{1}{T}$. Then, the equation for updating the ensemble becomes

$$\begin{aligned} H_t(\bar{X}) &= \frac{1}{t}\sum_{\tau=1}^{t} h_\tau(\bar{X}) = \frac{1}{t}\left[\sum_{\tau=1}^{t-1} h_\tau(\bar{X})\right] + \frac{1}{t}h_t(\bar{X}) = \frac{t-1}{t}H_{t-1}(\bar{X}) + \frac{1}{t}h_t(\bar{X}) \\ &= \left(1 - \frac{1}{t}\right) H_{t-1}(\bar{X}) + \frac{1}{t}h_t(\bar{X}) \end{aligned} \tag{9.15}$$

The above equation can be approximated as

$$H_t(\bar{X}) = H_{t-1}(\bar{X}) + \alpha_t h_t(\bar{X}) \tag{9.16}$$

The comparison of the above two equations can help us obtain the relationship between the new base model and the gradient: $h_t(\bar{X}) = -\frac{\partial \ell}{\partial w}$. Next, we know that the loss function is usually constructed with the label y_i and the output of the ensemble $H_t(\vec{x}_i)$:

$\ell(H) = \sum_{i=1}^{I} \ell(y_i, H_t(\vec{x}_i))$

Thus, the ensemble update equation constructed based on gradient descent is as

$$H_t(\bar{X}) = H_{t-1}(\bar{X}) - \alpha_t \cdot \frac{\partial \ell(\vec{y}, H_{t-1}(\bar{X}))}{\partial H_{t-1}(\bar{X})} \tag{9.17}$$

where the new base model in the current step for the update is $h_t(\bar{X}) \approx -\frac{\partial \ell(\vec{y}, H_{t-1}(\bar{X}))}{\partial H_{t-1}(\bar{X})}$. An approximation sign is used here because we are approaching the negative gradient with the base model, which is a training process.

Therefore, we know that the purpose of Gradient Boosting is essentially to minimize the following pseudo-residual (or called response in some literature), which stems from the difference between Eqs. 9.16 and 9.17:

$r = -\frac{\partial \ell(\vec{y}, H_{t-1}(\bar{X}))}{\partial H_{t-1}(\bar{X})} - h_t(\bar{X})$

Based on the above understanding, the process of training a gradient descent model can be implemented using the following pseudo-code.

Gradient Boosting:

Input: Training dataset $D = (x_1, y_1), (x_2, y_2), \ldots, (x_I, y_I)$; algorithm $\mathfrak{L}$ for learning base models h_t; number of base models T (also the number of steps for ensemble learning).

Obtain the initial ensemble: $H_0 = h_0 = \mathfrak{L}(D) = \arg\min_w \sum_{i=1}^{I} \ell(w)$

for $t = 1, 2, \ldots, T$ do

Calculate the negative gradient: $\vec{\delta} = -\frac{\partial \ell(\vec{y}, H_{t-1}(\bar{X}))}{\partial H_{t-1}(\bar{X})}$ or $\delta_i = -\frac{\partial \ell(y_i, H_{t-1}(\vec{x}_i))}{\partial H_{t-1}(\vec{x}_i)}, \quad i = 1, 2, \ldots, I$

Train the base model h_t by minimizing the residual: $w_t = \arg\min_w \sum_{i=1}^{I} r_t = \arg\min_w \sum_{i=1}^{I} [\delta_i - h_t(x_i; w)]^2$

Determine α_t using the line search to obtain the minimal ℓ:

$\alpha_t = \arg\min_\alpha \sum_{i=1}^{I} \ell(y_i, H_{t-1}(x_i) + \alpha \cdot h_t(x_i; w_t))$

Update the ensemble: $H_t(\bar{X}) = H_{t-1}(\bar{X}) + \alpha_t \cdot h_t(\bar{X}; w_t)$

end for

Output: $H_T(\bar{X})$

9.5 Stacking

Stacking is different from bagging and boosting. Instead of attempting to improve sampling and generating better base models, stacking focuses more on the improvements of combination via a second-level model. As illustrated in Fig. 9.11, we train the base models with the training dataset or its subsets. Then, the output of the base models, i.e., their predictions (or predicted labels), will be used as input for the second-level model (or called meta-model). The output of the meta-model will be the predictions of the ensemble.

There are many variations of stacking algorithms, depending on how to create the base models, especially sampling. In a basic stacking algorithm, the individual classification models are trained based on the complete training set. Accordingly, we use the total population sampling, and the models need to be heterogeneous to ensure different learning outcomes are obtained with the same training dataset. Another more popular variation is the stacking with k-fold cross-validation, which can help address the susceptibility to overfitting in the basic stacking. In this type of stacking, the training set is split into k folds, and in k successive rounds, $k-1$ folds are used to train the (first-level) base models, which can be homogeneous. The trained base models are then applied to the remaining 1 subset to generate predictions as input data for the second-level meta-model (or combiner). Bootstrap sampling can also be used. The procedure of implementing stacking using bootstrap is illustrated in the following pseudo-code.

Stacking:

Input: Training dataset $D = (x_1, y_1), (x_2, y_2), \ldots, (x_I, y_I)$; algorithms learning for base models $\mathfrak{L}$; number of base models T.

Initialize the base models: $H_1 \leftarrow \emptyset$

for $t = 1, 2, \ldots, T$ do

  Generate a bootstrap sample (subset) from D: D_t

  Train each base model: $h_t \leftarrow \mathfrak{L}(D_t)$

  $H_1 \leftarrow H_1 \cup h_t$

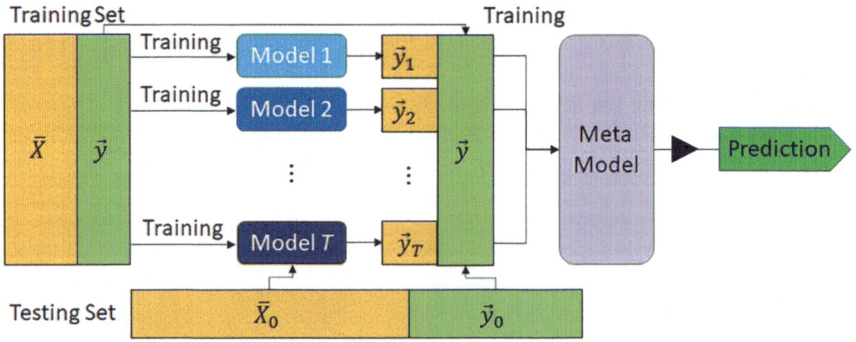

Fig. 9.11 Procedure of stacking

Integrate the output of the base learners as the dataset D' for training the meta model H_2

Train H_2 with dataset (D', y)

end for

Obtain the final prediction with a sequential implementation of H_1 and H_2.

9.6 Practice: Code and Evaluate Ensemble Learning Methods

>>>More and Up-to-Date Course Materials including Practices @ AI-engineer.org<<<

1. Run the following code to create a dataset using make_regression from Scikit-learn for a regression, split the data, and plot the training data.

```
# make_regression for generating a simple toy dataset to
    visualize regression algorithms
from sklearn.datasets import make_regression
import matplotlib.pyplot as plt
from sklearn.metrics import mean_squared_error

x,y=make_regression(n_samples=200, n_features=2, n_informative
    =2, n_targets=1,noise=10,random_state=1)
from sklearn.model_selection import train_test_split
x_train,x_test,y_train,y_test=train_test_split(x,y,
    random_state=10)
plt.figure(0)
plt.title("Training Data")
ax=plt.axes(projection='3d')
ax.scatter(x_train[:,0],x_train[:,1],y_train,color="k")
```

2. Develop computer code to implement bagging using decision trees from Scikit-learn (sklearn.tree.DecisionTreeRegressor) as the base learners (20 base models ($T = 20$); get 30 samples for each bootstrap sample; use the arithmetic mean for combination). Show the prediction errors (RMSE) made by the base models and the ensemble on the testing data in Step 1. Hints: (1) Bootstrap sampling can be easily done using code like

```
Index = np.random.randint(0, high=x_train.shape[0], size=s,
    dtype=int)
model = DecisionTreeRegressor()
model.fit(x_train[Index],y_train[Index])
```

(2) Base models can be saved to a Python list as

```
Models = []
Models.append(model)
```

3. Use bagging, boosting, and stacking algorithms from Scikit-learn (sklearn.ensemble.BaggingRegressor, sklearn.ensemble.GradientBoosting

Regressor, sklearn.ensemble.StackingRegressor) to solve the problem. Compare the results with those obtained using your own bagging code in Step 2.

4. XGBoost is one of the best ensemble learning techniques. Please install a new package, "xgboost," and use xgb.XGBRegressor to solve the regression problem. Compare the results with those obtained in Steps 2 and 3.
5. Optional: Print the testing samples with their given labels (2 attributes + 1 numeric label) and samples with their labels predicted by different ensemble models in Steps 2–4 in the 3D scatterplot (hints: using code that is similar to the plotting code given in Step 1).

Chapter 10
Clustering

10.1 Overview

This chapter introduces a major unsupervised learning topic: clustering. To ensure a good transition from supervised learning to unsupervised learning, we will first try to fill some possible knowledge gaps through a review of the basics of unsupervised learning. This includes changes from supervised to unsupervised learning, a framework of unsupervised learning, and an overview of clustering algorithms. Based on them, details will be provided for K-means, DBScan, GMM, and Agglomerative Hierarchical as the most representative algorithms for four main categories of clustering algorithms: centroid-based, density-based, distribution-based, and connectivity-based, respectively. Finally, the evaluation of clustering, including popular metrics and their use, will be discussed.

10.2 Basics of Unsupervised Learning

10.2.1 From Supervised Learning to Unsupervised Learning

So far, we have been discussing machine learning mostly in the context of supervised learning. This means we primarily focus on datasets consisting of input $\bar{X}$ and output $\vec{y}$, and accordingly, the goal of such a machine learning algorithm is to learn how to predict $\vec{y}$ based on $\bar{X}$. It is true that supervised learning traditionally predominates data science applications. However, other machine learning categories like unsupervised learning [115, 116] have been attracting more and more attention due to the explosive growth of data, in which labeled data is more frequently a small or even a rare slice.

Z. "L." Liu, *Artificial Intelligence for Engineers*,
https://doi.org/10.1007/978-3-031-75953-6_10

Starting from this chapter, we will explore unsupervised learning. Such machine learning algorithms do not process data with input-output pairs but, instead, only the inputs $\bar{X}$ or unlabeled samples. This fact would raise an obvious question: what do we need to predict if a label is not given as the output? In fact, the goal of unsupervised learning is not to find relationships between the unlabeled samples and labels as the two distinct parts of the data but, instead, to find relationships between different parts within the data. In other words, we need to discover some types of structures in the data. Different unsupervised learning algorithms work differently and discover very different kinds of structures.

In the following subsections, we will first work on the framework and classification of unsupervised learning. After outlining unsupervised learning in a generic manner, specific algorithms for one major type of unsupervised learning, i.e., clustering, will be introduced. Then, in the following chapters, we will use more unsupervised learning algorithms, i.e., dimension reduction and anomaly detection, to see how such algorithms fit into this framework.

10.2.2 Framework for Unsupervised Learning

The supervised learning algorithms introduced before have exhibited several key components: (1) a hypothesis (function), (2) a loss function, and (3) a method for minimizing the average loss function over the training data. Some of these components may change as we move to unsupervised learning. For example, how do we define a hypothesis function when there are no labels to predict? Likewise, how can we define a loss function when there are no true or predicted labels?

The hypothesis function in the unsupervised learning is defined as a mapping from input $\mathbb{R}^n$ back into this input space: $h_\theta : \mathbb{R}^n \rightarrow \mathbb{R}^n$. Thus, the goal of this hypothesis is to approximately reconstruct the input, i.e., $\vec{x}_i \approx h_\theta(\vec{x}_i)$ (or written as $\bar{X} \approx h_\theta(\bar{X})$ for all the data), just like $y_i \approx f_\theta(\vec{x}_i)$ in supervised learning. Under this condition, you may wonder why we would not propose an "easy" function $\vec{x}_i = \vec{x}_i$, which "reconstructs" the input perfectly yet makes little sense. The reason is that we need to additionally restrict our class of hypothesis functions or limit the hypothesis space so that the mapping $h_\theta(\vec{x}_i)$ is not a direct copy of $\vec{x}_i$ but a mapping that can help recover the data structure in a preferred way.

The loss function is always essential in unsupervised learning to specify the direction for the search for the best model. As indicated above, unsupervised learning needs to minimize the difference between the (output of) hypothesis function and the input itself. Accordingly, the loss function needs to present a quantitative way to measure this difference. That is, the loss function $\ell : \mathbb{R}^n \times \mathbb{R}^n \rightarrow \mathbb{R}_+$ is a measure of the difference between the predicted input with desired data structures and the original input. A common loss function is the least square function. Taking the ith sample as an example, we have its loss as

$$\ell(h_\theta(\vec{x}_i), \vec{x}_i) = \|h_\theta(\vec{x}_i) - \vec{x}_i\|_2. \tag{10.1}$$

Optimization as the process to minimize the above loss function is also essential in unsupervised learning. However, distinct from supervised learning, which employs methods like gradient descent to solve this optimization problem, the optimization in unsupervised learning may be more diverse. This is because the hypothesis functions in unsupervised learning often involve discrete or other terms that cannot be differentiated. This can be seen in algorithms like K-means for clustering. Moreover, the optimization method in some algorithms like the PCA for dimension reduction can provide an exact solution. Thus, the optimization may appear to be more of a "core" part in some unsupervised learning algorithms:

$$\min_{\theta} \frac{1}{I} \sum_{i=1}^{I} \ell(h_\theta(\vec{x}_i), \vec{x}_i) \tag{10.2}$$

10.2.3 Overview of Clustering

Clustering refers to a subcategory of unsupervised learning tasks, and the corresponding algorithms aim to find groupings in unlabeled data. These groupings are also frequently called "clusters." Every cluster is composed of a group of data points that are similar to each other. Thus, the data structures in clustering refer to the groupings of data points based on their similarities or connections. In the context of clustering, clustering sometimes refers to the result or process of a clustering task or called a cluster partition in some literature, while a cluster refers to a group of points in a clustering. Please be aware of the difference.

The way of performing a clustering task, e.g., selecting a suitable algorithm, is sometimes highly related to or even dependent on the nature of the data. On one hand, the selection of clustering algorithms requires knowledge about data. Meanwhile, we are usually expected to use clustering as a way of understanding the data. This leads to a "chicken or egg" paradox. In some cases, we may even know little about the data when selecting the algorithm.

Clustering algorithms can be classified based on how to find or assess the relationships between different data points. The following four common types of clustering models have been widely adopted [117].

Centroid-Based Clustering

In centroid-based clustering (or called centroid models in some literature), clusters are identified according to the proximity of the data points to the cluster centers or centroids. The centroid is searched to ensure that the data points are closest to the center. Data points are differentiated or clustered based on their relationships with multiple centroids in the data. K-means clustering and mean-shift clustering are the most popular algorithms in this category.

Density-Based Clustering
Density-based clustering algorithms search for areas with varying densities of data points in the data space. Those areas with a high concentration (or density) of data points surrounded by areas with low concentrations of data points are identified as groups. The clusters can be of any shape. DBScan and OPTICS are two of the most popular density-based clustering algorithms.

Distribution-Based Clustering
Distribution-based clustering groups data together based on the probabilities of how different data points belong to the same distribution. Usually, a distribution is assumed beforehand. Then, a center point is determined for each cluster. Next, the probability of a data point belonging to a cluster is determined by its distance to the cluster center point. This probability usually decreases as the distance to the center increases. Gaussian mixture theory (GMM) is the most common algorithm in this category.

Hierarchical (Connectivity) Clustering
Hierarchical, or connectivity-based clustering, usually involves top-bottom or bottom-up hierarchies to form clusters. This category of algorithms is suitable for hierarchical data and, due to the same reason, is more restrictive than the others. Despite the limitations, such algorithms may be also more efficient for specific kinds of data clusters. Agglomerative Hierarchical Algorithm is the most popular algorithm in this category.

In the following, we will provide details for K-means, DBScan, GMM, and Agglomerative Hierarchical as the most representative algorithms from the above four categories. In particular, more detail will be shared for K-means, which is a great example to illustrate the idea, process, and implementation of clustering, and for GMM, which is more math intensive and difficult to understand compared with other clustering algorithms.

10.3 K-Means Clustering

K-means algorithm aims to find K "centroids" (geometric centers), "centers," or "means" (i.e., $\mu_i \in \mathbb{R}^n$) for K clusters of points, so that the data points in each cluster are close to the corresponding center [118]. In this way, we can also associate every point with a specific (in fact, the closest) center, and this center can also serve as an indication of which cluster the point belongs to.

The implementation of K-means will involve an iterative process. This is because the centers that we specify at the beginning are not necessarily the "optimal" or "true" centers. Thus, after we cluster all the points according to these centers, we can calculate the new centers based on the clustered points, which may be different from the original ones. In this way, the locations of the centers can be updated. This process can be repeated until a criterion is met: further updates will not change the clustering.

In the following subsections, we will present a formal math description of K-means. Then, the implementation of this algorithm for a simple clustering task is given to illustrate the abovementioned iterative process. Next, two typical issues for K-means, i.e., the initialization of the centroids and the determination of the hyperparameters K, are discussed.

10.3.1 Math Framework of K-Means Algorithm

K-means is defined with a hypothesis function parameterized by $\bar{\mu}$, which are constructed with the coordinates of the centers:

$$\bar{\mu} = \{\vec{\mu}_1, \ldots, \vec{\mu}_k, \ldots, \vec{\mu}_k\} \tag{10.3}$$

where $\vec{\mu}_k \in \mathbb{R}^n$ is the vector of the coordinates of the kth center.

The hypothesis function, $h_{\bar{\mu}}(\vec{x})$, outputs the center that is the closest to the point $\vec{x}$. This can be mathematically formulated as

$$h_{\bar{\mu}}(\vec{x}) = \arg \min_{\vec{\mu} \in \{\vec{\mu}^{(1)}, \ldots, \vec{\mu}^{(k)}\}} \|\vec{\mu} - \vec{x}\|_2 \tag{10.4}$$

where the arg min operator returns the argument that minimizes the expression (as opposed to the min operator, which just returns the minimum value). That is, the expression outputs which center out of $\{ \vec{\mu}_1, \ldots, \vec{\mu}_k \}$ is closest to the sample represented by $\vec{x}$.

Then, for any point $\vec{x}$, the associated loss is the square of the distance between this point and the nearest center obtained using $h_{\bar{\mu}}(\vec{x})$:

$$\ell(h_{\bar{\mu}}(\vec{x}), \vec{x}) = \|h_{\bar{\mu}}(\vec{x}) - \vec{x}\|_2. \tag{10.5}$$

The above function is equivalent to the following formulation because this loss between a point and the nearest center is equal to the minimum of the losses between this point and all the centers:

$$\ell(h_{\bar{\mu}}(\vec{x}), \vec{x}) = \min_{\vec{\mu} \in \{\vec{\mu}_1, \ldots, \vec{\mu}_k\}} \|\vec{\mu} - \vec{x}\|_2. \tag{10.6}$$

Finally, clustering using K-means can be formulated as an optimization problem as Eq. 10.7:

$$\min_{\theta} \frac{1}{I} \sum_{i=1}^{I} \min_{\vec{\mu} \in \{\vec{\mu}_1, \ldots, \vec{\mu}_k\}} \|\vec{\mu} - \vec{x}_i\|_2. \tag{10.7}$$

As introduced, the solution to the above optimization problem involves an iterative process. The iterative process is implemented by assigning each point to its closest center and then updating each center to be the mean of these assigned points. This iterative process specific to K-means, as the standard method for training K-means models, is sometimes called Lloyd's algorithm or simply referred to as "K-means." Note that Lloyd's algorithm is an instance of the general expectation-maximization (or E-M) method for optimization, which is used by many machine learning algorithms. Besides, K-means, in a broad sense, can also be implemented using other methods, which will not be introduced here for simplicity.

There are many ways to initially assign cluster centers, but a common strategy is simply to choose k of the data points at random. Formally, the algorithm works as follows.

K-Means:
Input: Dataset $\vec{x}_i, i = 1, \ldots, I$
Initialize: $\mu_k = \text{RandomChoice}(\vec{x}_{(1:I)}),\ k = 1, \ldots, K$
Repeat until convergence:
  Assign point clusters: $y_i = \arg\min_k \|\vec{\mu}_k - \vec{x}_i\|_2^2$
  Compute new centers: $\vec{\mu}_k = \frac{\sum_{i=1}^{I} \vec{x}_i \cdot 1\{y_i=k\}}{\sum_{i=1}^{I} 1\{y_i=k\}}$.
where 1 is the indicator function, for which $1\{y_i = k\}$ yields 1 and $1\{y_i \neq k\}$ yields 0.

It can be proven that the above iterative process can guarantee that the loss function is gradually decreased. Moreover, because there are only a finite number of possible clusterings, the algorithm will converge after a finite number of steps. This process ends when we observe that the clusters stay unchanged from one iteration to the next.

10.3.2 Implementation of K-Means

In this subsection, we use a very simple clustering problem with three possible clusters of points to demonstrate the implementation of K-means. Let us first prepare the data. The three clusters of points were generated using normal distributions with three different centers.

```
# Prepare 3 clusters of data
I, J = (1000,2) # 1000 points, 2 attributes
np.random.seed(9527)
X = np.random.randn(I, J)
Mu0 = np.array([[0,-4], [-4,4], [2,2]]) # Centers of the 3
    clusters
X += Mu0[np.random.choice(np.arange(3), I),:] # Assign points to
    the 3 clusters
# plt.scatter(X[:,0], X[:,1]) # Plot original data
```

The following code gives out the implementation of the iterative optimization process and plots a screenshot for each step.

```
K = Mu0.shape[0] # Define the number of clusters to generate (e.g
    ., 3)
max_iter = 4 # Number of KNN iterations
colors = np.array(["C0", "C1", "C2"])
f,ax = plt.subplots(max_iter) # figsize=(6.0, 4.6*max_iter)
Mu = X[np.random.choice(X.shape[0],K),:]
for n in range(max_iter):
    D = np.empty((X.shape[0],K)) # 1000 points, each has 3
    distances to 3 centers
    for i in range(K): # Loop over all the centers (3 centers)
        D[:,i] = np.sum((X-Mu[i])**2,axis=1)
    y = np.argmin(D,axis=1)
    ax[n].scatter(X[:,0], X[:,1], edgecolors=colors[y],
    facecolors='none')
    ax[n].legend(['Iteration '+str(n+1)],frameon=False,
    scatterpoints=0)
    ax[n].scatter(Mu[:,0], Mu[:,1], c='k')
    Mu = np.array([np.mean(X[y==k],axis=0) for k in range(K)])
loss = np.linalg.norm(X - Mu[np.argmin(D,axis=1),:])**2/X.shape
    [0]
```

Figure 10.1 shows the first four iterations of the algorithm. As can be seen, the iterative process is already converged.

10.3.3 Initialization

K-mean is susceptible to the initialization: the initial locations of the centers. In the above example, if we adopt a different random initialization by adopting a different random seed as follows:

```
np.random.seed(1314)
```

then the first four iterations will be like Fig. 10.2.

As can be seen, we have not reached a converged solution. However, if we continue for more steps, we will still be able to converge to a solution. Such examples can help us reach a conclusion that inappropriate initialization for K-means can lead to local optima or/and slow convergence. Many methods are available to enable better initialization. One of the most popular K-means initialization methods is the K-means++ algorithm [119]. This algorithm employs an iterative sampling process based on probabilities calculated according to the distances to centers.

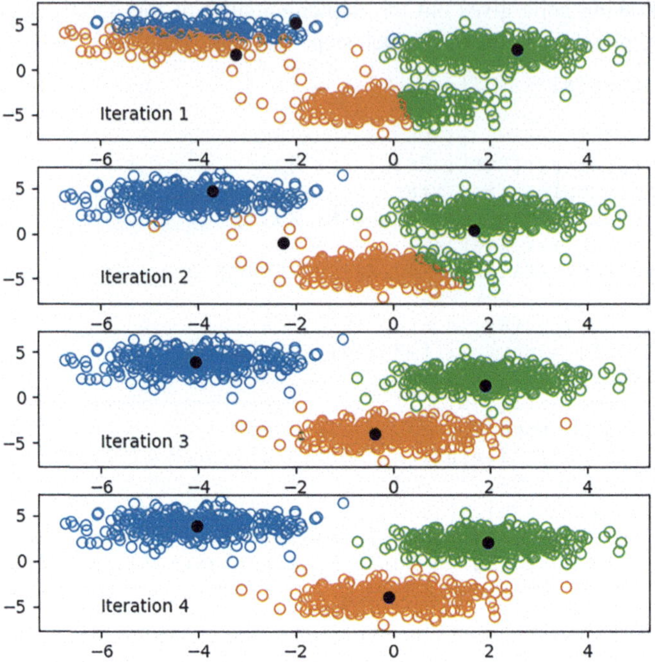

Fig. 10.1 Example of clustering process using K-means

K-Means++:
Input: Dataset $\vec{x}_i, i = 1, \ldots, I$
Output: $\vec{\mu}_k, k = 1, \ldots, K$
Initialize: $\mu_1 = \text{RandomChoice}(x_{(1:I)})$
For $k = 2, \ldots, K$
  Computing probabilities of selecting each point: $p_k = \frac{\min_{k'<k} \|\vec{\mu}_{k'} - \vec{x}_i\|_2}{\sum_{i'=1}^{I} \min_{k'<k} \|\vec{\mu}_{k'} - \vec{x}_{i'}\|_2}$.
  Select new centers based on probabilities: $\vec{\mu}_k = \text{RandomChoice}(x_{(1:I)}, p_{(1:I)})$.
where $\min_{k'<k} \|\vec{\mu}_{k'} - \vec{x}_i\|_2$ is the distance of point i to the nearest center and $\sum_{i'=1}^{I} \min_{k'<k} \|\vec{\mu}_{k'} - \vec{x}_{i'}\|_2$ is the total of the distances from all the points to their corresponding nearest centers.

10.3.4 Selection of K

K is the major hyperparameter for K-means. This selection of K value determines how many clusters will be obtained after the clustering is completed. In fact,

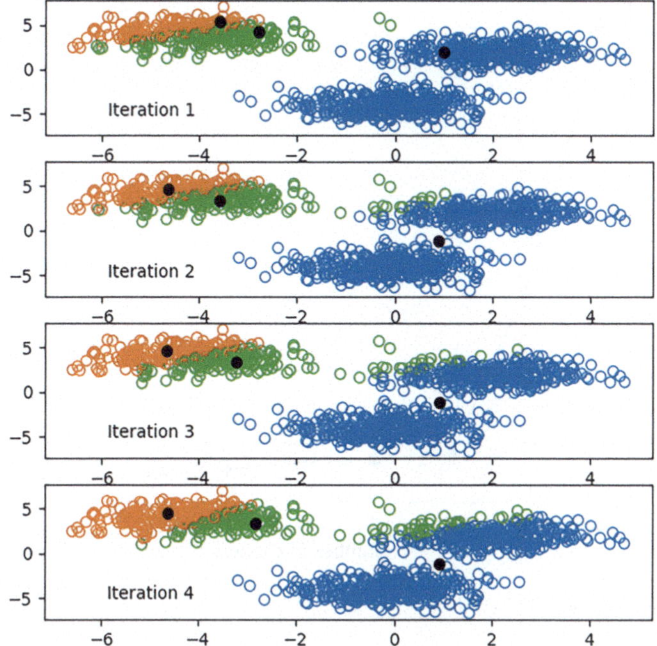

Fig. 10.2 Example of clustering process using K-means with a different initialization

the selection of K value is needed for a majority of clustering mechanisms. But unfortunately, the determination of this K is more of an imprecise art. In some cases, the K may be given because we have some knowledge about the data or a preference for the number of clusters. But in many other cases, we cannot get a quick overview of the data, e.g., in high-dimensional or high-volume data that is difficult to visualize, or a preference for the number of clusters is not given.

One strategy that can help yet still cannot guarantee any solution is to see if it is possible to find a suitable number of clusters by observing the corresponding loss values. Accordingly, we can plot the variation of the loss with respect to the number of clusters. As shown in Fig. 10.3, the loss drops quickly after three clusters. After that, adding more clusters does not help much. In this case, $K = 3$ appears to be a straightforward choice.

However, we need to point out that, though this practice can help in some contrived examples as above, it is generally hard to determine the optimal number of clusters.

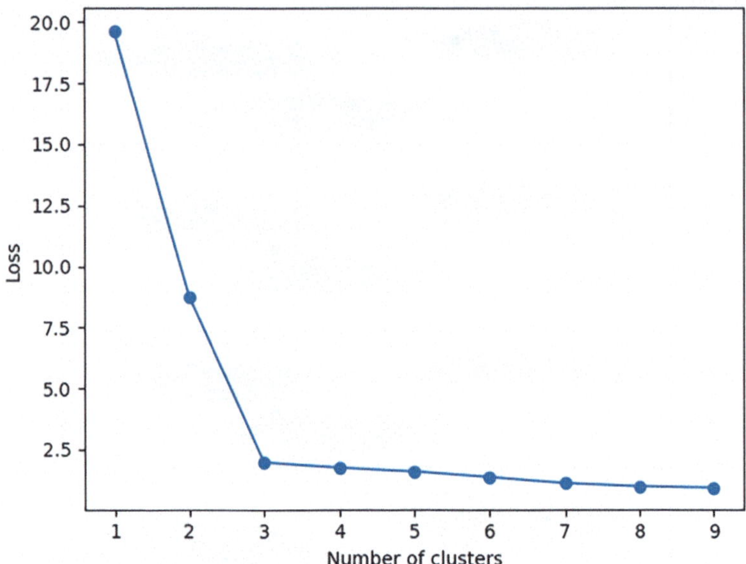

Fig. 10.3 Loss vs. number of clusters

10.3.5 Pros and Cons

Overall, K-means is a fast method because it does not involve demanding computations. However, this method may still suffer from issues that can be caused by its sensitivity to the initialization. That is, we may get different clusterings with different initial center locations.

10.4 Mean-Shift Clustering Algorithm

Another popular centroid-based clustering algorithm is mean shift. Mean shift features the use of sliding windows. This algorithm can help us find areas with a higher density or concentration of data points. Like K-means, this algorithm gradually updates the location of any center point with the mean of the points within the corresponding sliding window.

Figure 10.4 shows the process of using mean shift for clustering data points in a 2D space. As illustrated, we begin with an initial grid of seeds for search. Each seed point is associated with a circular sliding window that has its center at this point and a radius of r as the kernel. This algorithm is developed for a hill-climbing process to shift the kernel to an area with a higher density of data points in each step until convergence. In every step/iteration, the shift of the window toward a denser region is achieved by replacing the current center point with the mean of the points within

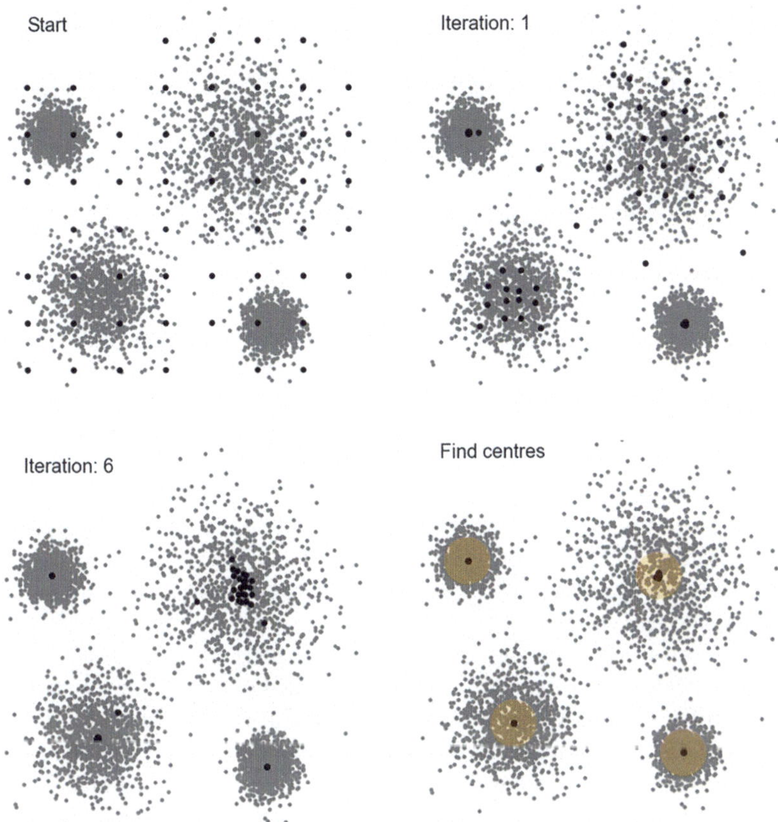

Fig. 10.4 Typical process of mean-shift clustering

the window. As a result, all the windows tend to gradually move toward the denser regions. The process is continued until every point is passed by at least one window. Meanwhile, some windows will overlap especially when some windows approach the same locations with a high density of points. Thus, we will need to count the number of times that every point is passed by different windows and determine the classification of every point by plurality in its window pass statistics.

A simple implementation of mean shift without introducing complex kernels can be guided using the following pseudo-code.

Mean Shift:

Input: Data set $\vec{x}_i, i = 1, \ldots, I$

Output: Classification of all points based on visit numbers

Repeat until all the points are visited:

Initialization: Randomly select K'[1] points, $\vec{\mu}_k, k = 1, \ldots, K'$, as the initial cluster centers.

for $k = 1, \ldots, K'$:

Repeat until the shift is small enough (converged):

Identify all the points in the ith circular window with a radius of r. Add 1 to the number of this point being visited by this window.

Obtain the vector between every point and the center. Add up vectors for all the points in the window to obtain the shift for updating this window: $\Delta\vec{u}_k = \frac{1}{I}\sum_{\vec{x}_i \in \bar{X}}(\vec{x}_i - \vec{\mu}_k)$

Move every window (parameterized by its center) according to its shift: $\vec{\mu}_k = \vec{\mu}_k + \Delta\vec{u}_k$

Combine windows when the distance between their centers is smaller than a threshold. Accordingly, the window visit statistics of all the relevant points should also be updated.

Classification: Assign every point to the cluster whose window visited this point the most within all the windows.

In more advanced variations of mean shift, kernels are introduced to better calculate the density distributions [120]. As illustrated in Fig. 10.5, every point can have a distribution, e.g., Gauss distribution, for the density distribution. Then, the total density distribution for one window is the sum of the density distributions for all the points within the window. Then, the shift vector is obtained as the direction for the fastest density increase, which can be obtained as the gradient of the total density function.

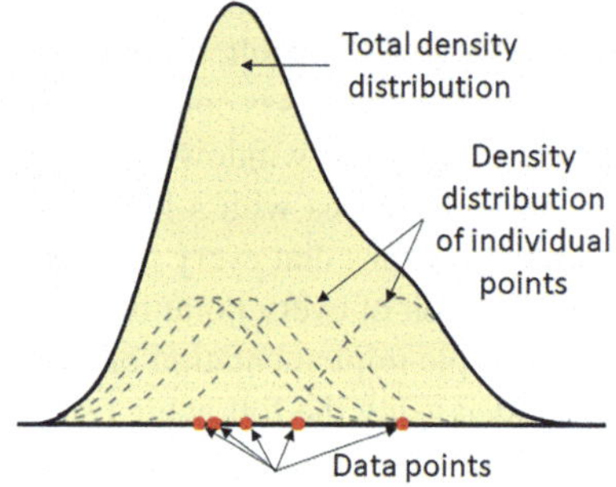

Fig. 10.5 Distribution determined by kernel functions

[1] $K' \geqslant K$ because sliding windows may merge later.

10.4.1 Pros and Cons

The mean-shift algorithm is advantageous in that it does not require us to specify the number of clusters as needed in K-means. Thus, it can help us more automatically and objectively understand the structure of the data via clustering. However, the selection of the window size poses some challenges and can be trivial in some cases.

10.5 Density-Based Spatial Clustering (DBScan)

DBScan is a density-based clustering algorithm [121]. Clustering methods such as K-means and hierarchical clustering are suitable for finding spherical-shaped or convex clusters, which are relatively well-separated and contain less noise and outliers. By contrast, DBScan can provide better performance for data containing clusters with arbitrary shape and noise.

DBScan is proposed based on the notion that clusters are regions with high densities of points. In addition, it introduces another key idea that the neighborhood of any point in a cluster defined by a given radius should contain at least a minimum number of points. To allow for these ideas, the DBScan algorithm defines two extra parameters: the threshold distance between points, ϵ, and the number of neighboring points, N_p.

ϵ or "eps" defines the neighborhood around a data point: two points are considered to be neighbors if their distance is smaller than or equal to ϵ. A small ϵ value will judge more data points to be outliers. On the contrary, large ϵ values will prompt the clusters to merge, and consequently, the majority of the data points will be in a few clusters.

N_p or "MinPts" is defined as the minimum number of neighboring points within a radius defined by ϵ. Large ϵ values should be considered for large datasets. As a rule of thumb, the minimum N_p can be related to the dimensions of the data J as $N_p >= J + 1$. In general, N_p should not be smaller than 3.

With the above extra concepts and parameters, we can identify three types of data points: core points, border points, and noise/outliers. A core point has more than N_p points within ϵ. A border point has fewer than N_p within ϵ, but it is in the neighborhood of a core point. A noise/outlier is neither a core point nor a border point.

In addition, points a and b are said to be density connected if there exists point c, which has a sufficient number of points in its neighbor ϵ, and both points a and b are within ϵ. This is a chaining process. So, if b is a neighbor of c, c is a neighbor of d, d is a neighbor of e, and e is a neighbor of a, then we can infer that b is a neighbor of a.

DBScan can be implemented in the following steps:

1. Identify all the neighbor points within ϵ, and mark the core points.
2. Create a new cluster for every core point that has not been assigned to a cluster.

3. For every core point, find all of its density-connected points and assign them to the same cluster as the core point.
4. Iterate through the remaining unvisited points in the dataset.
5. Those points that do not belong to any cluster are noise/outliers.

10.5.1 Pros and Cons

As introduced, DBScan can handle arbitrarily shaped and sized clusters. Also, like mean shift, DBScan does not require a preset number of clusters. One extra advantage is that this algorithm can be used to identify outliers as noise, which would be clustered as a regular data point in other clustering algorithms. However, it is noted that DBScan may not be very effective in the case of clusters with varying densities. Also, the setting of ϵ and N_p can be tricky especially when there is a change in the density levels. In addition, the determination of ϵ in high-dimensional data can be challenging.

10.6 Gaussian Mixture Models (GMM)

The GMMs are more flexible than the K-means clustering [122]. However, any GMM needs to start with an assumption that the data points are Gaussian distributed. Every cluster has a Gaussian distribution. Thus, two parameters are needed to describe the shape of each cluster, the mean and the standard deviation. Taking GMM for a 2D problem, for example, the distribution of points in any cluster can take an elliptical shape represented by a 2D Gaussian distribution.

We use an optimization algorithm known as expectation maximization (EM) to find out the parameters of the Gaussian for each cluster. In a typical clustering task, we need to cluster I points. Each point, or sample, has J features. We can imagine that the I points are distributed in a J-dimensional space. We assume the points can be clustered into several groups. In GMM, each group of points is assumed to follow a Gaussian distribution, $\mathcal{N}$. The goal of GMM is to search for the Gaussian distributions that best describe the distributions of points. Instead of a deterministic label stating that one point belongs to a specific group in KNN and K-means, any point in GMM has different probabilities of belonging to different groups. Therefore, the input $\bar{X}$ has a shape of $I \times J$. The probability that a point x_i is generated is

$$p_i = \sum_{k=1}^{K} [\pi_k \mathcal{N}_k(x_i|\mu, \Sigma)] \tag{10.8}$$

where π_k is the weight of the kth normal distribution (or group). To obtain the Gaussian distributions, we need to determine both the parameters for $\mathcal{N}$ and the weights π, which are related. For such a problem, the EM iterations are usually adopted. Accordingly, we first calculate some parameters such as those for $\mathcal{N}$ (i.e., μ and Σ) and then, based on the results, calculate the weights. The calculated weights can be used to update $\mathcal{N}$. This process will be continued until the improvement is less than a small tolerance.

The detailed EM procedure for implementing the GMM is as follows:

1. Assume the data can be clustered into K groups, e.g., 3. Generate μ and σ for all the groups. When there is more than one variable (dimensions), we use the covariance matrix Σ (e.g., in NumPy, we use np.cov(Sample, rowvar=False))[2] instead of standard deviation σ (square root of variance). Meanwhile, μ will become an array with J elements for each normal distribution (group).
2. Calculate the probabilities of each sample in different groups. Thus, there are $\mathcal{N}_1(\mu_1, \Sigma_1), \ldots, \mathcal{N}_K(\mu_K, \Sigma_K)$. For any sample x_i, we will have $\mathcal{N}_1(x_i|\mu_1, \Sigma_1), \ldots, \mathcal{N}_K(x_i|\mu_K, \Sigma_K)$. Calculate $\mathcal{N}_k(x_i|\mu_k, \Sigma_k)$ using the following equation:

$$\mathcal{N}_k(x_i|\mu, \Sigma) = \sqrt{\frac{1}{(2\pi)^J \det(\Sigma)}} \exp\left(-\frac{1}{2}(x_i - \mu)^T \Sigma^{-1}(x_i - \mu)\right) \tag{10.9}$$

   Therefore, there will be $I \times K$ probabilities in total.
3. Calculate the contribution of each $\mathcal{N}_k$ in the generation of any data point x_i:

$$\gamma_{i,k} = \frac{\pi_k \mathcal{N}_k(x_i|\mu, \Sigma)}{\sum_{k=1}^{K} [\pi_k \mathcal{N}_k(x_i|\mu, \Sigma)]} \tag{10.10}$$

   where $\gamma_{i,k}$ can also be viewed as the weight of the point in the Gaussian distribution $\mathcal{N}_k$.
4. Update the Gaussian distributions, that is, obtain the new μ and Σ for each Gaussian distribution. This is similar to updating the centroids of the groups in the K-means; the difference is that we use a (multivariate) distribution characterized by μ and Σ instead of the coordinates of the centroids and circles marking the group boundary. The new μ and Σ will be obtained with the weights of the points obtained in Step 3. In NumPy, this can be done as

```
Mu[k] = np.average(X,axis=0,weights=R[:,k])
E[k] = np.cov(X, rowvar=False,aweights=R[:,k])
```

   where $R[\,,k]$) has the same number of elements as X.

[2] rowrar = False specifies a row is a sample instead of a variable. Thus it generates a $J \times J$ covariance matrix.

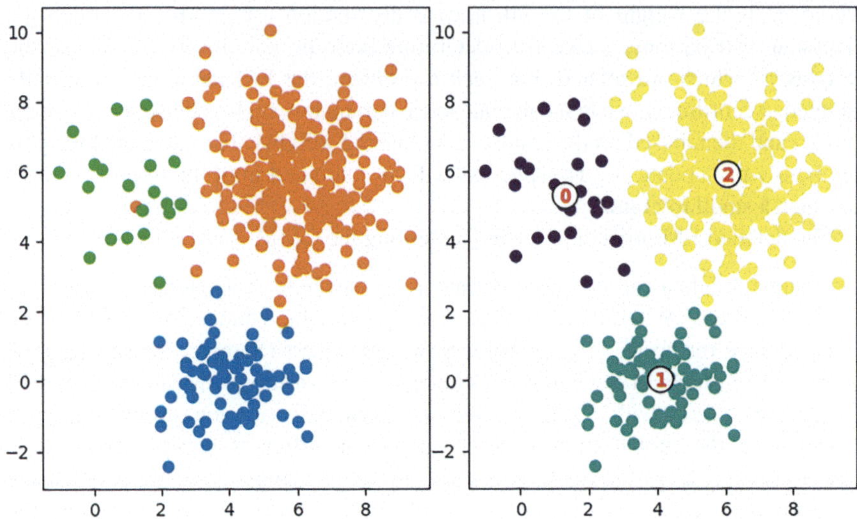

Fig. 10.6 Example of GMM: data generated with three normal distributions and their GMM clustering

5. Next, we will update the weights of the Gaussian distributions as

$$\pi_k = \frac{1}{I}\sum_{i=1}^{I}\gamma_{i,k} \tag{10.11}$$

6. Repeat Steps 2–5 until the difference between μ and/or Σ values in two consecutive steps, e.g., the root of the sum of square differences, is less than a tolerance, e.g., 1×10^{-3}. This generated the results in Fig. 10.6.

10.6.1 Pros and Cons

GMM maintains a higher level of flexibility in cluster covariance compared to the K-means clustering in light of the use of standard deviation. Also, GMM does not assume spherical clusters as those algorithms using Euclidean distances. The introduction of probabilities enables us to tell the probabilities that one point belongs to different clusters in addition to the clustering result. However, we need to be aware that GMM assumes Gaussian distributions, needs sufficient data for each cluster, and requires us to specify the number of clusters. In addition, this algorithm may be sensitive to outliers and initialization conditions and require more computing time.

10.7 Hierarchical Agglomerative Clustering (HAC)

There are two types of hierarchical clustering algorithms: top-down (divisive) and bottom-up (agglomerative) [123]. Figure 10.7 illustrates the divisive and agglomerative hierarchical clustering processes using a simple example with six data points.

Here, we will focus more on the bottom-up one. The bottom-up type of hierarchical clustering treats each data point as an individual cluster in the initial stage. It merges pairs of clusters until you have a single group containing all data points. Hence, it is also known as Hierarchical Agglomerative Clustering (HAC). In Fig. 10.7, the upper part shows the calculation of the linkage as quantified by the Euclidean distance in the agglomerative order. The data points are grouped into two clusters with a threshold. Therefore, we can use the following process for performing the HAC:

1. Start by marking each data point as a separate cluster.
2. Measure the distance between the two groups using a selected distance metric like the Euclidean distance. In a typical linkage method, e.g., the average linkage method, the distance between two clusters is the average distance between the data points in the two clusters.
3. Merge two clusters with the smallest average linkage into one as one iteration.
4. Repeat the above step until a threshold is reached.

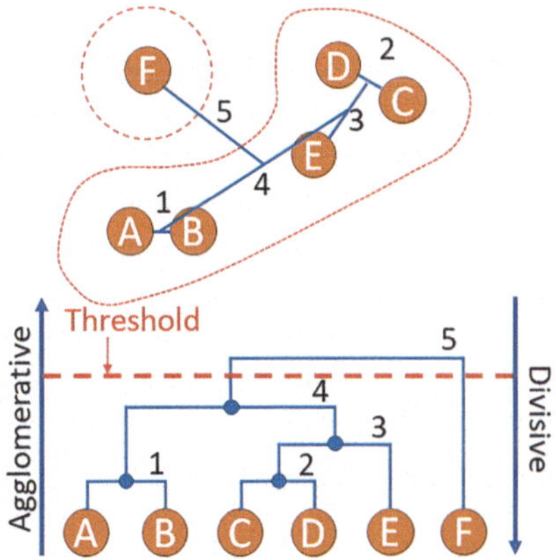

Fig. 10.7 Illustration of hierarchical clustering

10.7.1 Pros and Cons

HAC does not need to specify the number of clusters. In general, this algorithm is not sensitive to the selection of the distance metric. However, the number of clusters may be difficult to determine. Also, this algorithm is not as computationally efficient as other algorithms like K-means.

10.8 Evaluation of Clustering

10.8.1 Overview of Evaluation Metrics

The performance evaluation of clustering algorithms can be as significant and difficult as the clustering process. Many metrics can be employed for this purpose. There are two types of metrics: one is for "internal" evaluation, in which the clustering is evaluated as a single quality score, and the other is for "external" evaluation, in which the clustering is compared to an existing "ground truth" classification. We need to know some facts about metrics for both types of evaluation.

Internal evaluation is unique in that such metrics can also be used as the objective function of clustering. When using an internal evaluation metric as the objective function, we essentially cluster the data to achieve the highest metric value. Therefore, the meaning of using the same metric for evaluating the performance of the clustering task may be limited or questionable.

External evaluation suffers from a dilemma. That is, if we have "ground truth" labels, then we would not need to cluster the data again. In fact, such labels are usually not available in practice. Even if they are available, we also need to be aware that such labels only present one possible partitioning of the data. This does not necessarily exclude the possibility that there exists a different yet better clustering.

Therefore, no metric is perfect for evaluating the quality of a clustering. The selection needs to be done considering the data and purpose. Also, human experts' observations, which could be subjective, are still desired in many cases. Despite the needed precautions, evaluations with appropriate metrics can still present very useful evaluations of clustering results for both developing clustering algorithms and evaluating clustering tasks.

The evaluation metrics of clustering are distinct from the evaluation metrics for other AI topics and appear to be an integral part of clustering knowledge. Considering this fact, some common metrics with needed examples and code are provided here for a complete introduction. Extra information can be found in the appendices of this book for the evaluation metrics of AI applications.

10.8.2 Internal Evaluation

Internal evaluation employs internal information of the clustering process to evaluate the goodness of a clustering structure. Thus, the use of internal evaluation metrics does not need to involve the comparison to a reference model such as that represented by the true labels. A general idea of such metrics is to give better scores to algorithms that generate clusters with high similarity within a cluster and low similarity between clusters. Common internal evaluation metrics include the silhouette coefficient, Davies-Bouldin index, Dunn index, Calinski-Harabasz, etc.

As mentioned, internal evaluation needs to be used with two possible concerns in mind. First, if an internal metric is used as part of the objective function, then the use of the same or a related metric for performance evaluation is biased. Second, an internal metric presents a quantitative measure of how one algorithm is better than the other. However, it does not indicate that one algorithm produces more valid results than another.

Silhouette Coefficient

The silhouette coefficient is defined by the ratio between the average distance to elements in the same cluster (intracluster) and the average distance to elements in other clusters (intercluster). A silhouette coefficient can be obtained for each point, while an overall silhouette coefficient can be computed by averaging. A high silhouette value for a data point indicates that this point has been well clustered, where a point with a low value may be an outlier. This index can be used with K-means to determine the optimal number of clusters.

The silhouette coefficient for a sample i can be calculated as

$$S(i) = \frac{b(i) - a(i)}{\max\{a(i), b(i)\}} \tag{10.12}$$

where a is the intracluster distance, which is the average of the distances to all the other points in the same cluster, and b is the mean nearest-cluster distance, which is the smallest of the average distances to the other clusters. The above equation is equivalent to

$$S(i) = \begin{cases} 1 - \frac{a(i)}{b(i)} & \text{if } a(i) < b(i) \\ 0 & \text{if } a(i) = b(i) \\ \frac{b(i)}{a(i)} - 1 & \text{if } a(i) > b(i) \end{cases} \tag{10.13}$$

We can find from the above equation that $-1 \leqslant S(i) \leqslant 1$. The silhouette coefficient of a clustering of I data points can be calculated as

$$S = \frac{1}{I}\sum_{i=1}^{I} S(i) \tag{10.14}$$

A clustering with a silhouette coefficient greater than 0.5 is usually deemed acceptable, while a value lower than 0.2 may indicate the data does not contain an obvious cluster structure.

The silhouette coefficient can be calculated with functions from many data analysis packages. For example, sklearn.metrics.silhouette_score from Scikit-learn can be used as shown in the following code for evaluating the Iris dataset clustered by K-means:

```
from sklearn.metrics import silhouette_score
from sklearn.cluster import KMeans
dataset = datasets.load_iris()
X = dataset.data
y = dataset.target
kmeans_model = KMeans(n_clusters = 3, random_state = 1).fit(X)
labels = kmeans_model.labels_
silhouette_score(X, labels, metric = 'euclidean')
```

The output of the above code is as follows:

```
0.5528190123564091
```

Davies-Bouldin Index

The main idea of the Davies-Bouldin (DB) index is to calculate the similarity as a measure of the distance between every cluster and the cluster that is the most similar to it and then use the average of these similarity values for all the clusters to assess the clustering result. This process can be seen in Fig. 10.8.

The Davies-Bouldin index can be calculated using the following formula:

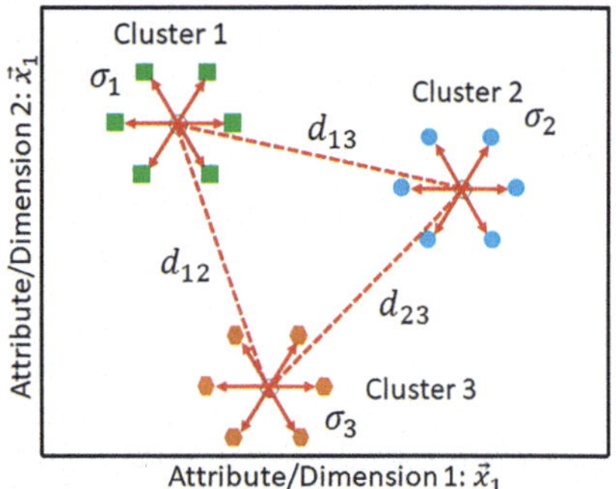

Fig. 10.8 Illustration of DB index using a simple example

$$\mathrm{DB} = \frac{1}{K}\sum_{i=1}^{K}\max_{j\neq i}\left(\frac{\sigma_i + \sigma_j}{d(c_i, c_j)}\right) \tag{10.15}$$

where K is the number of clusters, σ_i is the average distance of all the points in cluster i to its centroid c_i, and $d(c_i, c_j)$ (alternatively written as d_{ij} or $d(i, j)$) is the distance between the two centroids c_i and c_j for cluster i and j, respectively.

As the equation shows, the range of DB index is $[0, +\infty]$. A greater DB index requires the distance between clusters (denominator) is small, while the distances between points within the same cluster (numerator) are large, which corresponds to a poor clustering result. Therefore, greater DB index values imply worse clustering results and vice versa.

Dunn Index

The Dunn index can be used to identify dense and well-separated clusters. The metric is defined as the ratio between the minimal intercluster distance and the maximal intracluster distance. The Dunn index can be mathematically formulated as follows, which outputs one value for a clustering:

$$D = \frac{\min_{1\leq i<j\leq n} d(i, j)}{\max_{1\leq k\leq n} d'(k)} \tag{10.16}$$

where $d(i, j)$ is the distance between clusters i and j and $d'(k)$ measures the intracluster distance of cluster k, which is the maximum distance between any pair of points within cluster k. The intercluster distance $d(i, j)$ is usually defined by the distance between the centroids of the clusters. However, the intracluster distance $d'(k)$ may be defined in different ways, among which the maximal distance between any pair of elements in cluster k is the most common way.

A high Dunn index expects a high intracluster similarity and a low intercluster similarity, leading to a better clustering. Thus, higher Dunn index values are more desired.

10.8.3 External Evaluation

External evaluation uses data that is not used for clustering, such as true labels and external benchmarks, to evaluate the clustering results. Such metrics are proposed to measure the similarity between the clustering and external reference data.

Some aspects of external evaluation also deserve close attention. First, as mentioned, the meaning of clustering when the labels are given is questionable. When such a practice is needed, we will need to determine what information will be excluded for clustering but later used for the external evaluation. Second, the anomalies or incorrect labels need to be considered when clean data and true labels are not available. Also, the labels only present a way of classifying or labeling the data and thus do not include all the ways of interpreting the data. Thus, the external evaluation with external information, such as given labels, may not be suitable to tell whether the clustering algorithm successfully or well identifies good clustering structures that are not included in the external information.

Many external evaluation metrics are available such as the Rand index, Adjusted Rand Index, Normalized Mutual Information (NMI), Adjusted Mutual Information (AMI), Fowlkes-Mallows index, and the contingency matrix. Some other metrics like F-measure, Jaccard index, and Dice index are also frequently used. Information will be presented about the Rand index, Adjusted Rand Index, Fowlkes-Mallows index, and the contingency matrix.

Rand Index

The Rand index (RI) measures the similarity between a clustering (as the clusters obtained by a clustering algorithm) and the benchmark classification. A mathematical formulation of the Rand index is as follows:

$$RI = \frac{TP + TN}{TP + FP + FN + TN} \tag{10.17}$$

where TP is the number of true positives, TN is the number of true negatives, FP is the number of false positives, and FN is the number of false negatives. In a clustering task, TP is the number of pairs of points that are clustered together in both the clustering and the truth partition, and FP is the number of pairs of points that are clustered together in the predicted partition but not in the truth partition. FP and FN are defined using the point pairs in a similar way. Then, we should have $TP + TN + FP + FN = \binom{I}{2}$, in which I is the number of points in the dataset.

The RI value ranges within [0, 1]. Higher RI values indicate a higher similarity with the reference system or clustering result. $RI = 0$ indicates two clusterings (the considered clustering and the reference one) have no overlaps, while $RI = 1$ appears when two clusterings are identical.

One issue with the Rand index is that false positives and false negatives are equally weighted, which may be undesired in some clustering applications. The adjusted Rand index is proposed to address this issue. Other metrics, such as the F-measure, can also help address this concern.

Adjusted Rand Index

Adjusted Rand Index (ARI) is proposed to improve the original or raw Rand index. The adjustment changes the range of the index values from [0, 1] to [−1,1]. The RI score is "adjusted for chance" into ARI as follows:

$$RI_{\text{Adjusted}} = (RI - RI_{\text{Expected}})/(\max(RI) - RI_{\text{Expected}}) \tag{10.18}$$

The following presents three examples of using the ARI function from Scikit-learn. The use of this equation requires two parameters, i.e., labels_true, which are ground-truth class labels, and labels_pred, which are clusters labels to evaluate.

Example

```
from sklearn.metrics.cluster import adjusted_rand_score
labels_true = [1, 0, 1, 2, 1, 3]
labels_pred = [0, 1, 1, 2, 3, 2]
print(adjusted_rand_score(labels_true , labels_pred))
labels_true1 = [1, 0, 1, 2, 1, 3]
labels_pred1 = [1, 1, 1, 2, 1, 3]
print(adjusted_rand_score(labels_true1, labels_pred1))
labels_true2 = [1, 1, 1, 2, 1, 3]
labels_pred2 = [1, 1, 1, 2, 1, 3]
print(adjusted_rand_score(labels_true2, labels_pred2))
```

Output

```
-0.19047619047619047
0.5454545454545455
1.0
```

As can be seen, perfect labeling or 100% similarity between a clustering and the clustering represented by the true labels has an ARI of 1, whereas ARI values of 0 or negative indicate poor clustering results.

Normalized Mutual Information (NMI)

Mutual Information-based scores are a category of metrics that evaluate the similarity between a clustering and a reference via quantifying the mutual independence between the variables in these two systems. This metric originates from the mutual information in the information theory, which uses the information entropy. Typical metrics in this category include the Normalized Mutual Information (NMI) and the Adjusted Mutual Information (AMI). The detailed mathematical formulations are not presented here yet can be found in the appendices. Instead, a simple use of these two metrics via a function from Scikit-learn is provided in the following:

```
from sklearn.metrics.cluster import normalized_mutual_info_score,
    adjusted_mutual_info_score
labels_true = [1, 0, 1, 2, 1, 3]
labels_pred = [1, 1, 1, 2, 1, 3]
print('The NMI is: ', normalized_mutual_info_score(labels_true,
   labels_pred))
print('The AMI is: ', adjusted_mutual_info_score(labels_true,
   labels_pred))
```

The output of the above code is

```
The NMI is:  0.822328362394686
The AMI is:  0.6160962485695716
```

Fowlkes-Mallows Index

The Fowlkes-Mallows index (FM or FMI) also measures the similarity between two clusterings. Mathematically, FM can be defined as follows:

$$FM = \sqrt{\frac{TP}{TP+FP} \cdot \frac{TP}{TP+FN}} \tag{10.19}$$

where TP, FP, and FN have the same meanings for clustering tasks as explained in the introduction to RI.

As can be seen from the above equation, mathematically, FM is the geometric mean of the precision P and recall R. Due to this reason, this metric is also known as the G-measure, while the F-measure is their harmonic mean. FM can also be easily implemented using functions from existing packages. The following presents a simple example:

```
from sklearn.metrics import fowlkes_mallows_score
labels_true = [1, 0, 1, 2, 1, 3]
labels_pred = [1, 1, 1, 2, 1, 3]
fowlkes_mallows_score(labels_true , labels_pred)
```

Output

```
0.7071067811865476
```

Higher values of the Fowlkes-Mallows index imply that the clusters and the benchmark classifications are more similar.

Contingency Matrix

The contingency matrix is another simple but powerful tool for assessing the performance of a clustering algorithm when the ground truth is known. This metric presents a table to report the intersection cardinality for every trusted pair of [true, predicted]. It compares the different classes/labels in the predicted (clustering) and the true label (reference system). The number in a cell or element in this matrix shows the number of samples that have the corresponding predicted and true labels. Therefore, the confusion matrix for classification problems is a square contingency matrix. The following code shows how to obtain this metric using a function from Scikit-learn easily.

Example

```
from sklearn.metrics.cluster import contingency_matrix
labels_true = [1, 0, 1, 2, 1, 2]
labels_pred = [1, 0, 3, 2, 1, 3]
contingency_matrix(labels_true , labels_pred)
```

Output

```
array([[1, 0, 0, 0],
       [0, 2, 0, 1],
       [0, 0, 1, 1]])
```

In the above example, the obtained contingency matrix has three rows for the three labels in the true label, i.e., 0,1, and 2, and four columns for the four labels in the predicted labels, i.e., 0,1,2,3. Thus, the element value in the second row and second column, i.e., 2, tells that there are two samples that have a predicted label "1" (2nd true label) and a true label of "1" (2nd predicted label). Likewise, the element value in the second row and fourth column, i.e., 1, tells that there is one sample that has a predicted label of "1" (2nd true label) and a true label of "3" (4th predicted label).

10.9 Practice: Test and Modify Clustering Code for Problem-Solving

>>> More and up-to-date course materials including practices @ AI-engineer.org <<<

1. Please run and understand the code for K-means in the book.
2. Please run and understand the following code for GMM. A separate data file titled "clustering_data.txt" is needed as the input for the execution.

```
import numpy as np
X_train = np.loadtxt('./clustering_data.txt')
X_1 = X_train[0:75]
X_2 = X_train[75:75+250]
X_3 = X_train[75+250:75+250+20]

# Cluster the data
import numpy as np

n_clusters = 3 # K in the literature
e = 1e-3
centroids = np.linspace(np.min(X_train,axis=0),np.max(X_train,
    axis=0),n_clusters+2)# np.empty((n_clusters,X_train.shape
    [1]))
centroids = centroids[1:-1]

Mu = centroids
Sigma = np.ones((n_clusters,1))@(np.std(X_train, axis=0)/
    n_clusters).reshape(1,X_train.shape[1]) # '/n_clusters' is
     a guess
E = np.tensordot(np.ones(n_clusters),np.cov(X_train,rowvar=
    False),axes=0)
Pi = np.ones(n_clusters)*1/n_clusters
```

```
labels = np.empty((X_train.shape[0]))
R = np.empty((X_train.shape[0],n_clusters))

def normal(X,mu,E):
    N = np.empty((X.shape[0],mu.shape[0]))
    if X.ndim == 1:
        sigma = np.std(X)
        for i in range(X.shape[0]):
            N[i] = 1/(sigma**2*(2*np.pi))**0.5 * np.exp
    (-0.5*((X-mu)/sigma)**2)
    else:
        for i in range(X.shape[0]):
            for j in range(mu.shape[0]):
                N[i,j] = 1/(np.linalg.det(E[j])*(2*np.pi)**X.
    shape[1])**0.5 * np.exp( -0.5*(X[i]-mu[j]).T @ np.linalg.
    inv(E[j]) @ (X[i]-mu[j]) )
    return N

while True:  #EM Algorithm
#for nn in range(100):
    # Label all the points (obtain R): Expectation (E) Step
    N = normal(X_train,Mu,E)
    R = N*Pi/( np.sum(N*Pi,axis=1).reshape((N.shape[0],1)) @
    np.ones((1,N.shape[1])))
    centroids_old = Mu.copy()

    # Update centroids (update normal distributions (clusters)
    )
    for k in range(n_clusters):
        Mu[k] = np.average(X_train,axis=0,weights=R[:,k])
        E[k] = np.cov(X_train, rowvar=False,aweights=R[:,k])
        Pi[k] = np.mean(R[:,k])
    centroids = Mu


    labels = np.argmax(R,axis=1)
    error = np.sum((centroids-centroids_old)**2) ** 0.5
    print(error,labels.sum())

    if(error < e):
        break
    if(error < e):
        break

# Plot the data
import matplotlib.pyplot as plt
fig, ax = plt.subplots(1, 2, figsize=(10, 10))
ax[0].scatter(X_1[:, 0], X_1[:, 1])
ax[0].scatter(X_2[:, 0], X_2[:, 1])
ax[0].scatter(X_3[:, 0], X_3[:, 1])
ax[0].set_aspect('equal')
ax[1].scatter(X_train[:, 0], X_train[:, 1], c=labels)
ax[1].scatter(centroids[:, 0], centroids[:, 1], marker='o',
```

```
                c="white", alpha=1, s=200, edgecolor='k')
for i, c in enumerate(centroids):
    ax[1].scatter(c[0], c[1], marker='$%d$' % i, s=50, alpha
    =1, edgecolor='r')
ax[1].set_aspect('equal')
plt.tight_layout()
```

3. Switch the datasets in Step 1 and Step 2 and run both K-means and GMM code to solve the new problems.
4. Add one more cluster of data to the dataset in Step 1 (to obtain 4 clusters). Then, cluster the new dataset with both K-means code in Step 1 and GMM code in Step 2.

Chapter 11
Dimension Reduction

11.1 Overview

This chapter discussed dimensionality reduction. First, we will try to understand the concepts and possible uses of this unsupervised learning technique. Next, the classification of dimensionality reduction methods and popular algorithms will be reviewed. Then, two categories of dimensionality reduction methods, i.e., feature selection and feature extraction, will be introduced. In particular, two feature extraction algorithms, i.e., principal component analysis and linear discriminant analysis, will be explained in detail, considering their significance and predominance in dimensionality reduction studies. For each of them, we will try to understand the idea first, then formulate the idea using mathematical equations, and finally show how to implement the method with the equations.

11.2 Basics of Dimension Reduction

11.2.1 Concepts and Needs

Dimensionality reduction, or called dimension reduction, is the transformation of data from a high-dimensional space, e.g., with a large number of features (also counted as random variables or attributes), into a relatively low-dimensional space, e.g., with fewer features than the original data while retaining essential properties of the original data. Dimension reduction could be very useful or/and needed for a variety of reasons as follows:

- The size of data can be significantly reduced as the dimensionality decreases. Thus, dimensionality reduction can help compress data. Such compression can help us save effort in handling data such as storage space and transmitting time.

Z. "L." Liu, *Artificial Intelligence for Engineers*,
https://doi.org/10.1007/978-3-031-75953-6_11

- Low-dimensional data can, in general, reduce the time for computing and training.
- Many algorithms do not show good performance when the data dimensionality is too high, and dimension reduction can improve their performance by transforming the data to a low-dimensional space.
- Dimension reduction can help address the curse of dimensionality: various problems would arise when analyzing and organizing data in high-dimensional spaces yet not occur in low-dimensional settings. A common theme of these problems is that, when the dimensionality increases, the volume of the space increases so fast that the available data becomes sparse.
- Dimension reduction can help reduce the redundancy in data, such as features that are dependent on others. For example, if a feature is a linear combination of the other feature values, this feature can be removed without hurting the value of the data. One application is denoising in signal processing.
- Dimension reduction can facilitate data visualization, which can become difficult as data dimensionality becomes high. For example, it is relatively easy to plot 2D or 3D data for visualization, but this job becomes much more difficult as the dimensionality gets above 3.

Therefore, dimensionality reduction helps remove the redundant or less significant variables. In this way, it can help us find the major variables or the functions of such variables (e.g., a linear combination of the variables) that make a major contribution to the task of interest like classification. In practice, dimensionality reduction is used a lot in fields that deal with large numbers of observations (number of instances) and/or large numbers of variables (size/dimension of individual instance), such as signal processing, speech recognition, bioinformatics, and geoinformatics.

11.2.2 Popular Methods and Classification

Many methods have been proposed for reducing the dimensionality of data [124]. These methods can be classified according to different criteria. The most common classification is to divide the methods into two major categories: feature selection and feature projection. Feature selection seeks to find a subset of the input variables (or features). Therefore, we select features directly according to some criteria, such as the filter strategy (e.g., information gain) and wrapper strategy (e.g., search guided by accuracy), and drop less desired features to reduce the data dimensionality. By contrast, feature extraction tends to project data in a high-dimensional space to a space of fewer dimensions, in which the combinations of more desired features are extracted. Accordingly, instead of selecting features directly, we try to extract those feature functions (e.g., combinations) that better reflect the data characteristics (e.g., unsupervised learning) or better serve the data analysis tasks (e.g., supervised learning) by finding a low-dimensional space to map the data to. Feature extraction can be further divided into two subcategories based on whether the projection is made to a lower-dimensional space directly or to a manifold in a lower-dimensional space.

Feature selection methods [125] primarily refer to those data reduction techniques studied in statistics for variable selection and are now mostly discussed in high-dimensional regression analysis. Classical methods include the missing value ratio, low variance filter, high correlation filter, random forest, backward feature extraction, and forward feature selection. Feature projection methods are more popular in the dimensionality reduction literature and are even used to represent feature projection in a narrow sense [126]. This category includes the most popular dimensionality reduction methods such as principal component(s) analysis (PCA) [127], linear discriminant analysis (LDA) [128], independent component analysis (ICA) [129], Isomap [130], and multifactor dimensionality reduction (MDR) [131].

In addition to the above classification, dimensionality reduction methods can also be classified based on whether they process labeled or unlabeled data. A typical example is that PCA was developed for unlabeled data. Thus, PCA is mostly used for clustering in unsupervised learning. On the contrary, LDA was proposed for processing labeled data. For example, it is usually used to condense data with labels for classification tasks. Another way of classifying dimensionality reduction methods depends on whether they are linear or nonlinear in nature. For example, PCA, LDA, and ICA are linear while locally linear embedding (LLE) [132], Isomap, MDR, and kernel PCA are nonlinear. Table 11.1 lists popular dimension reduction methods, which are categorized according to multiple criteria.

It is important to point out that dimensionality reduction is different from other machine learning topics in that it contains significantly different methods for the same purpose rather than one method and its variants. Therefore, popular dimensionality reduction methods seem to have little common theoretical basis and may be much different from each other. This is unlike other machine learning topics, such as Bayesian classifiers, in which most methods are built on Bayes' theorem. Due to this reason, this chapter for dimensionality reduction is organized in a way slightly different from the other chapters. We will first briefly introduce the common feature selection methods, which are relatively straightforward and easy to understand. Then, two of the most popular/classic dimensionality reduction methods, i.e., PCA and LDA, will be introduced to cover both their theories and implementations.

11.3 Common Feature Selection Methods

The following are common feature selection methods.

Missing Value Ratio

The missing value ratio is a criterion that we can intuitively select for removing attributes. Missing attribute values for some samples are very common in real-world data. In some cases, we choose to fill the missing value, which may be represented

Table 11.1 Popular dimension reduction methods and their classifications

Methods	Selection/extraction	Linear/nonlinear	Supervised/unsupervised
Missing value ratio	Selection	Linear	Both
Low variance filter	Selection	Linear	Both
High correlation filter	Selection	Linear	Both
Random forest	Selection	Linear	Both
Backward feature extraction	Selection	Linear	Both
Forward feature selection	Selection	Linear	Both
Principal component analysis (PCA)	Extraction	Linear	Unsupervised
Independent component analysis (ICA)	Extraction	Linear	Unsupervised
Singular value decomposition (SVD)	Extraction	Linear	Unsupervised
Latent semantic analysis (LSA)	Extraction	Linear	Unsupervised
Linear discriminant analysis (LDA)	Extraction	Linear	Supervised
Factor analysis (FA)	Extraction	Linear	Supervised
Canonical correlation analysis (CCA)	Extraction	Linear	Supervised
Partial least squares (PLS)	Selection	Linear	Supervised
Locally linear embedding (LLE)	Extraction	Nonlinear	Unsupervised
Laplacian Eigenmaps (LE)	Extraction	Nonlinear	Unsupervised
Metric multidimensional scaling (MDS)	Extraction	Nonlinear	Unsupervised
Isomap[a]	Extraction	Nonlinear	Unsupervised
t-distributed stochastic neighbor embedding (t-SNE)	Extraction	Nonlinear	Unsupervised
Uniform manifold approximation and Projection (UMAP)	Extraction	Nonlinear	Both
Multifactor dimensionality reduction (MDR)	Extraction	Nonlinear	Unsupervised
Kernel PCA	Extraction	Nonlinear	Unsupervised
Autoencoder	Extraction	Nonlinear	Unsupervised

[a] Extend MDS by incorporating the geodesic distances imposed by a weighted graph

by “nan” or “null,” with values such as ”0.” However, when too many samples have missing values for a specific attribute, we may consider deleting that attribute. The deletion of an attribute is equivalent to removing a variable containing little information or eliminating an insignificant direction/dimension.

For this purpose, we need to define a threshold value to determine whether an attribute is insignificant enough to remove. A threshold or critical missing value ratio is proposed for this purpose. When the missing value ratio is higher than the threshold, we delete the corresponding attribute or dimension. Thus, a lower threshold indicates a more aggressive dimension reduction strategy.

Low Variance Filter
The low variance filter uses the variance of the values of an attribute to determine whether to remove the corresponding dimension. In general, we assume data with a lower variance has less information. That is, data points that cluster very close to each other contain little information. Based on this understanding, we can calculate the variance of the values of one attribute and then delete the attributes with the lowest variance values or values lower than a prescribed threshold. It is noted that variance is related to data ranges. Therefore, normalization is usually needed so that variance values of different attributes can be compared.

High Correlation Filter
The high correlation filter utilizes the correlation between different variables for dimension selection. If two variables corresponding to two attributes exhibit high correlation, which can be calculated using different correlation equations, then, we believe they exhibit similar trends or provide similar information. Thus, the coexistence of these highly correlated variables will reduce the performance of some models constructed with the data.

In general, we should keep input variables that are highly correlated with the target/label. However, for two variables in the input, we should consider removing one of them if their correlation is higher than a value of 0.5 or 0.6.

Random Forest
Random forest can be employed for feature selection with some extra modifications [133]. This is because parameters such as the Gini index and out-of-bag error rates can be related to the significance of the attribute. Take the Gini index as an example. The significance of an attribute at a node at which the attribute is used for splitting data equals the change of the Gini index during data splitting. Because the random forest algorithm grows multiple trees with data generated via bagging sampling, we need to sum up the Gini index values for an attribute that may appear multiple times in the same trees and/or across different trees. Then, the total Gini index values of different attributes will be compared as their importance measures to determine which attributes are more significant. Those attributes with higher total Gini index values should be selected or retained. Fortunately, such values usually can be directly obtained from machine learning packages such as Scikit-learn.

Backward Feature Elimination
In backward feature elimination, we first train a model with data containing all the attributes. Then, we exclude one attribute of the data each time to train new models. If we have J attributes, we would get J new models. We can select the attribute corresponding to the model whose performance is closest to the model trained with all the attributes. This attribute is believed to be the one with the lowest influence on the model performance. After deleting this attribute, the above process can be repeated to remove another attribute. The dimension can be reduced in this way.

Forward Feature Selection

Forward feature selection employs a process that is opposite to that of backward feature elimination. That is, we first train J models with data that only contains one of the J attributes. The attribute corresponding to the model with the best performance is selected as the optimal attribute. This process is repeated so that we will have more and more attributes until the performance of the model cannot be further improved.

11.4 Feature Extraction Method 1: Principal Component Analysis

As introduced in the previous section, PCA is one of the most important dimensionality reduction methods. In fact, it is one of the most popular dimensionality reduction algorithms. Despite the fragmented nature of dimensionality reduction knowledge, PCA is viewed as a must-know by most people. It has wide applications in data compression, redundancy removal, and noise reduction. In the following, the concept and basic idea of PCA will be introduced first. Then, the theoretical basis, especially the deduction of major equations for PCA's implementations, will be introduced. Next, the procedure and guidelines for implementing PCA will be presented. Lastly, kernel PCA as a way to extend PCA from a linear dimensionality reduction method into nonlinear ones will be discussed.

11.4.1 Concept and Main Idea

From the name of PCA, i.e., primary component(s) analysis, we can guess that the major idea of PCA is to find out the major components of data and represent the data only with these major components for the purpose of dimensionality reduction. In a more strict mathematical description, we attempt to implement dimensionality reduction for data with J dimensions. Because the data is unlabeled, each data point or called instance will have J features (or attributes). Please be aware that the data will have multiple data points, say I, which is the number of instances. This number of data points cannot be confused with the dimensionality of the data, though both can determine the total size of the data. Our goal of dimensionality reduction is to reduce the dimension J rather than the number of instances I. Specificially, our goal is to reduce the dimensionality of the data, J, into a small one, J' ($J' < J$). This dimensionality reduction is conducted in the hope that the data after dimensionality reduction can still represent the original data. That is, we hope the loss of essential information can be minimized while reducing the data dimensions.

Next, let us use a simple example to illustrate the above concepts and understand what it means by reducing dimensionality and finding major components. In the example shown in Fig. 11.1, we hope to reduce the dimensionality of the data represented by the points. The data is plotted in a two-dimensional space, i.e., have

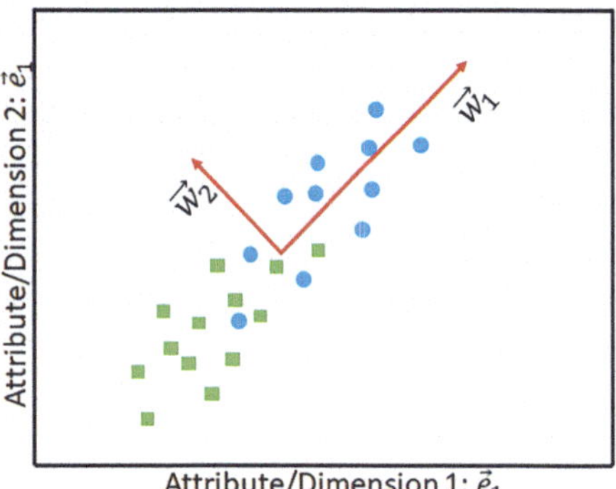

Fig. 11.1 Main idea of dimensionality reduction

two axes, and accordingly, each data point has two elements (coordinates $\vec{e}_1$ and $\vec{e}_2$). Obviously, we can and only can reduce the data dimensionality from 2 to 1. Thus, the purpose of PCA, in this case, is to find a direction or an axis to which the data can be projected. Therefore, a key to PCA is how to find this direction/axis as the new dimension, which can also be understood as the principal component in this PCA example.

Instead of revealing which direction is the best directly, let us see two candidate directions, i.e., $\vec{w}_1$ and $\vec{w}_2$, which represent the directions of the two straight lines pass through the center the data points. A comparison of the two directions could give us some hints for identifying better directions for projection. Many of us, in fact, can point out that $\vec{w}_1$ is a better direction: when data is projected onto this direction or axis, more information of the original data can be retained. But why?

There are two explanations. The first explanation is that the data points are closer to this axis. In a more strict description, we can state that the total distance between all the points and this axis is smaller. The second explanation is that the projected points on $\vec{w}_1$ are more separated from each other. That is, the variance of the projected data is higher, which can facilitate the differentiation between the projected data. In fact, these two explanations can be generalized as the criteria for dimensionality reduction with PCA. But, in a more general case, we attempt to find J' ($1 \leqslant J' < J$) directions (axes) for projection so that the criteria can be met in the best way. In the later deduction, we will also show that the two criteria are equivalent: we can use either of them to find the same best directions for projections (or understood as the best components).

11.4.2 Theoretical Basis

Deduction Based on Minimum Distance

As mentioned above, there are two criteria for selecting the best projection direction(s) or the principal components. Let us first see how to establish the theoretical basis based on the criterion of minimum distance. In order to obtain a general deduction, we will use generalized data and space. The 2D data and space in Fig. 11.1 may still be used to assist us in understanding the process. However, please be aware that the distance and the axis/plane to which the data is projected will be generalized into the Euclidean distance and hyperplane, which are not easy to visualize as those lines and planes in 1D and 2D spaces.

We will start with data consisting of I data points (instance), and each data point has J dimensions (attributes). Accordingly, we can use a matrix $\bar{X}$ as follows to represent the data:

$$\bar{X} = \left[\vec{x}_1^T, \vec{x}_2^T, \ldots, \vec{x}_I^T\right]^T = \begin{bmatrix} x_{11} & x_{12} & \cdots x_{1J} \\ \vdots & \ddots & \vdots \\ x_{I1} & x_{I2} & \cdots x_{IJ} \end{bmatrix}_{I \times J} \tag{11.1}$$

where a data point $\vec{x}_i$ has J attributes and is formulated using a column array as $\left[x_{i1}, x_{i2}, \ldots, x_{iJ}\right]^T$.

It should be noted that PCA takes zero-centered data as the input. In general, data like that in Fig. 11.1 is not zero centered. If that is the case, we will need to take the following action for zero centering:

$$\vec{x}_i \leftarrow \vec{x}_i - \frac{1}{I}\sum_{i=1}^{I} \vec{x}_i \tag{11.2}$$

After this preprocessing operation, the result is that the origin of the coordinate system will be moved to the center of the data. In Fig. 11.1, the origin is moved to the data center marked by the intersection of $\vec{w}_1$ and $\vec{w}_2$. The following deduction will be performed with the zero-centered data. For such data, the following condition is met:

$$\sum_{i=1}^{I} \vec{x}_i = 0 \tag{11.3}$$

To generate the low-dimensional space, we first rotate the current coordinate system to obtain a new one $\bar{W}_0 = \left[\vec{w}_1, \vec{w}_2, \ldots, \vec{w}_J\right]_{(J\times J)}$, in which $\vec{w}_j = [w_{1j}, w_{2j}, \ldots, w_{Jj}]^T$. Each vector $\vec{w}_j$ represents an axis, which can also be understood as a direction or a dimension. $\vec{w}_j$ has J components, which are the projections of the unit vector along this axis onto the J axes of the original coordinate system. This new coordinate system also has J dimensions, which are represented by an orthonormal basis consisting of J unit vectors that are orthogonal to each other. Therefore, we should have the following relationships for any pair of vectors from this basis: $\vec{w}_j^T \cdot \vec{w}_j = \|\vec{w}_j\| = 1$ and $\vec{w}_j^T \cdot \vec{w}_{j'} = 0$ when $j \neq j'$.

However, our purpose is not to obtain this new coordinate system. Instead, we aim to find the best projection directions in a lower-dimensional space. That is, we attempt to reduce the dimensionality from J to J' ($J \leqslant J'$), so we will need to drop several axes in the above new coordinate system so that we will have $\bar{W} = \left[\vec{w}_1, \vec{w}_2, \ldots, \vec{w}_{J'}\right]_{J \times J'}$. For this projection array $\bar{W}$, we have the following relationships because it consists of an orthonormal basis: $\bar{W}^T_{(J' \times J)} \cdot \bar{W}_{(J \times J')} = \bar{I}_{(J' \times J')}$ and $\bar{W}_{(J \times J')} \cdot \bar{W}^T_{(J' \times J)} = \bar{I}_{(J \times J)}$, in which $\bar{I}$ is an identity matrix.

This reduced coordinate system with J' axes represents a new lower-dimensional space. Alternatively, we can understand it as a subspace formed by the reduced axes, to which the data in the high-dimensional space is projected. We know a projection can be obtained as the dot product of the tensor (data as an array) to be projected and the tensor representing the direction. Therefore, we can obtain the data in the new low-dimensional space as

$$\bar{Z}_{(I \times J')} = [\vec{z}_1^T, \vec{z}_2^T, \ldots, \vec{z}_I^T]^T = \bar{X}_{(I \times J)} \cdot \bar{W}_{(J \times J')} \tag{11.4}$$

where the projected data point $\vec{z}_i$ is essentially represented by the coordinates of this data point in the new low-dimensional space. The array for the new system $\bar{W}$ has the dimensions (or called size in the general context of arrays) of $J \times J'$. Thus, for each data point, we have

$$\vec{z}_{i(J' \times 1)} = \bar{W}^T_{(J' \times J)} \cdot \vec{x}_{i(J \times 1)} \tag{11.5}$$

If we project the data in the low-dimensional space back to the original high-dimensional space, we can take the following operation:

$$\hat{\bar{X}}_{(I \times J)} = \bar{Z}_{(I \times J')} \cdot \bar{W}^T_{(J' \times J)} \tag{11.6}$$

Or, we can write this equation in terms of individual data points as follows:

$$\hat{\vec{x}}_{i(J \times 1)} = \bar{W}_{(J \times J')} \cdot \vec{z}_{i(J' \times 1)} \tag{11.7}$$

where $\hat{\bar{X}}_{(I \times J)}$ are the coordinates of the projected data in the original high-dimensional space. For example, $\vec{x}_i$ is the coordinates of data point i in the original high-dimensional space, $\vec{z}_i$ is the projection of this data point in the low-dimensional space (coordinates of the projected point in the new coordinate system), and $\hat{\vec{x}}_i$ is the coordinates of this projected point in the original high-dimensional space. Then, the distance between the original data point and the "hyperplane" is the distance between the coordinates of the original point and its projected point in the same (high-dimensional) space: $\|\vec{x}_i - \hat{\vec{x}}_i\|_2$. According to the minimum distance criterion, we will need to find a "hyperplane" (coordinate systems of low-dimensional space $\bar{W}$), for which the total distance between all the points and the "hyperplane" is the smallest. That is, we will need to minimize the following equation:

$$\sum_{i=1}^{I} \|\vec{x}_i - \hat{\vec{x}}_i\|_2 \tag{11.8}$$

where $\|\|_2$ is the $\ell - 2$ norm (or called Euclidean norm), which is defined as $\|\vec{e}_i\|_2 = \sqrt{e_{i1}^2 + e_{i2}^2 + \cdots + e_{iI}^2}$.

Next, we can reformulate this equation based on the above relationships and $\bar{W}^T \cdot \bar{W} = I_{(J \times J')}$ as follows:

$$\begin{aligned}
\sum_{i=1}^{I} \|\vec{x}_i - \hat{\vec{x}}_i\|_2 &= \sum_{i=1}^{I} \|\vec{x}_i - \bar{W} \cdot \vec{z}_i\|_2 \\
&= \sum_{i=1}^{I} \vec{x}_i^T \cdot \vec{x}_i - 2\sum_{i=1}^{I} (\bar{W} \cdot \vec{z}_i)^T \cdot \vec{x}_i + \sum_{i=1}^{I} (\bar{W} \cdot \vec{z}_i)^T \cdot (\bar{W} \cdot \vec{z}_i) \\
&= \sum_{i=1}^{I} \vec{x}_i^T \cdot \vec{x}_i - 2\sum_{i=1}^{I} \vec{z}_i^T \cdot (\bar{W}^T \cdot \vec{x}_i) + \sum_{i=1}^{I} \vec{z}_i^T \cdot \vec{z}_i \\
&= \sum_{i=1}^{I} \vec{x}_i^T \cdot \vec{x}_i - \sum_{i=1}^{I} \vec{z}_i^T \cdot \vec{z}_i \\
&= \sum_{i=1}^{I} \sum_{j=1}^{J} x_{ji} x_{ij} - \sum_{i=1}^{I} \sum_{j=1}^{J'} z_{ji} z_{ij} = \bar{X}^T \cdot\cdot \bar{X} - \bar{Z}^T \cdot\cdot \bar{Z}
\end{aligned} \tag{11.9}$$

where $\bar{X}^T \cdot\cdot \bar{X}$ is a double contraction operation, which will "consolidate" two axes/orders corresponding to the two variables involved in the operation. For example, $\bar{X}^T$ and $\bar{X}$ are both second-order tensors or can be viewed as second-order arrays, and their double contraction result will be a scalar. To further the deduction, we will recall the one property of the trace of the product of two matrices: $tr(\bar{A} \cdot \bar{B}) = \sum_{j=1}^{J} (\bar{A} \cdot \bar{B})_{jj} = \sum_{j=1}^{J} (\sum_{i=1}^{I} A_{ij} B_{ji}) = \bar{A} \cdot\cdot \bar{B}$. Hence, the equation becomes

$$\begin{aligned}
\sum_{i=1}^{I} \|\vec{x}_i - \hat{\vec{x}}_i\|_2 &= \sum_{i=1}^{I} \vec{x}_i^T \cdot \vec{x}_i - \sum_{i=1}^{I} \vec{z}_i^T \cdot \vec{z}_i \\
&= \bar{X}^T \cdot\cdot \bar{X} - \bar{Z}^T \cdot\cdot \bar{Z} \\
&= tr(\bar{X}^T \cdot \bar{X}) - tr(\bar{Z}^T \cdot \bar{Z}) \\
&= tr(\bar{X} \cdot \bar{X}^T) - tr(\bar{Z} \cdot \bar{Z}^T)
\end{aligned} \tag{11.10}$$

The first term, $\sum_{i=1}^{I} \vec{x}_i^T \cdot \vec{x}_i = tr(\bar{X} \cdot \bar{X}^T)$, is a constant depending solely on the original data, so it is not related to projection. Therefore, only the second term controls the projection. Therefore, the PCA problem can be mathematically formulated as

$$\arg\min_{\vec{w}} - tr(\bar{Z}^T \cdot \bar{Z}) \quad s.t. \quad \bar{W}^T \cdot \bar{W} = I_{(J \times J')} \tag{11.11}$$

The loss function $-tr(\bar{Z}^T \cdot \bar{Z})$ to be minimized can be further reformulated as

$$- tr(\bar{Z}^T \cdot \bar{Z}) = -tr((\bar{X} \cdot \bar{W})^T \cdot (\bar{X} \cdot \bar{W})) = -tr(\bar{W}^T \cdot (\bar{X}^T \cdot \bar{X}) \cdot \bar{W}) \tag{11.12}$$

Using the above equation and the method of Lagrangian multipliers, we can convert the above constrained optimization problem into an unconstrained optimization problem with the following objective function (or loss function in this case):

$$J(\bar{W}) = -tr(\bar{W}^T \cdot (\bar{X}^T \cdot \bar{X}) \cdot \bar{W} + \lambda(\bar{W}^T \cdot \bar{W} - \bar{I})) \tag{11.13}$$

The minimum value of the loss function is achieved when the derivation of the above function with respect to $\bar{W}$ is zero: $-(\bar{X}^T \cdot \bar{X}) \cdot \bar{W} + \lambda \bar{W} = 0$. Rearranging this equation, we obtain the following solution:

$$(\bar{X}^T \cdot \bar{X})_{(J \times J)} \bar{W}_{(J \times J')} = \lambda \bar{W}_{(J \times J')} \tag{11.14}$$

The above equation corresponds to a typical eigenvalue problem. That is, $\bar{W}$ is the matrix consisting of the eigenvectors of $\bar{X}^T \cdot \bar{X}$, while λ is a diagonal matrix whose diagonal elements are the eigenvalues of $\bar{X}^T \cdot \bar{X}$ and the other elements are 0. This provides a way of finding the best projection (array) in PCA. If we can obtain the projection array using the above equation, then we can use $\bar{Z} = \bar{X} \cdot \bar{W}$ to easily obtain the data in the low-dimensional space.

Deduction Based on Maximum Variance

The theoretical basis of PCA can also be established based on the criterion of the maximum variance, leading to the same equations for PCA implementation.

Let us start from the same zero-centered data $\bar{X}$ and try to project to a new low-dimensional space via a projection array $\bar{W}_{(J \times J')}$. Thus, for any data point $\vec{x}_i$, its projection in the new space (or coordinate system $\bar{W}$) is $\bar{W}^T \cdot \vec{x}_i$. Then, the covariance matrix of the projected data $\bar{X} \cdot \bar{W}$ will be $(\bar{X} \cdot \bar{W} - \bar{0})^T \cdot (\bar{X} \cdot \bar{W} - \bar{0}) = \bar{W}^T \cdot (\bar{X}^T \cdot \bar{X}) \cdot \bar{W} = \bar{Z}^T \cdot \bar{Z}$. $\vec{0}$ is the matrix of the mean values for variance calculation, whose elements are 0 because the data is zero centered. The sum of the variance is the sum of the diagonals of this matrix: $tr(\bar{W}^T \cdot (\bar{X}^T \cdot \bar{X}) \cdot \bar{W})$. Then, the PCA problem becomes the following constrained optimization problem:

$$\arg\max_{\bar{W}} \quad tr(\bar{W}^T \cdot (\bar{X}^T \cdot \bar{X}) \cdot \bar{W}) \quad s.t. \quad \bar{W}^T \cdot \bar{W} = \bar{I}_{(J' \times J')} \tag{11.15}$$

The above equation aiming to maximize the variance can be reformulated as a minimization problem:

$$\underset{\bar{W}}{\arg\min} \quad -tr(\bar{W}^T \cdot (\bar{X}^T \cdot \bar{X}) \cdot \bar{W}) \quad s.t. \quad \bar{W}^T \cdot \bar{W} = \bar{I}_{(J' \times J')} \tag{11.16}$$

Similarly, we can convert the above constrained optimization problem into an unconstrained optimization problem with the following objective function using the method of Lagrangian multipliers:

$$J(\bar{W}) = -tr(\bar{W}^T \cdot (\bar{X}^T \cdot \bar{X}) \cdot \bar{W} + \lambda(\bar{W}^T \cdot \bar{W} - \bar{I}_{(J' \times J')})) \tag{11.17}$$

Then, we can get the following solution to the optimization problem:

$$(\bar{X}^T \cdot \bar{X})_{(J \times J)} \cdot \bar{W}_{(J \times J')} = \lambda \bar{W}_{(J \times J')} \tag{11.18}$$

This result is the same as what was deduced based on the minimum distance criterion.

11.4.3 Implementation

From the above introduction to PCA's theoretical basis, we can see that the major step in implementing PCA is to find eigenvalues and eigenvectors of the covariance matrix, $X \cdot X^T$. To reduce the dimensionality, we adopt the major eigenvalues (with bigger values) and their corresponding eigenvectors. Then, use the projection made of the selected eigenvectors to project the data from the high-dimensional space to the low-dimensional one. The detailed implementation procedure is outlined using the following pseudo-code.

PCA:

Input: a set of data with I samples and J attributes: $\bar{X}_{(I \times J)} = [\vec{x}_1^T, \vec{x}_2^T, \ldots, \vec{x}_I^T]^T$.

Output: a set of data with I samples and reduced dimensionality (J' attributes): $Z_{(I \times J')}$.

Perform zero centering for all the data samples: $\vec{x}_i \leftarrow \vec{x}_i - \frac{1}{I}\sum_{i=1}^{I} \vec{x}_i$

Calculate the covariance matrix $(\bar{X}^T \cdot \bar{X})_{(J \times J)}$

Obtain the eigenvalues and eigenvectors of $\bar{X}^T \cdot \bar{X}$

Select the J' highest eigenvalues and their corresponding eigenvectors. Form the projection array $\bar{W}$ with the selected eigenvectors. These eigenvectors need to be standardized (if not) before use.

Project the data to the lower space to reduce its dimensionality: $\bar{Z} = \bar{X} \cdot \bar{W}$ or $\vec{z}_i = W^T \cdot \vec{x}_i$

Sometimes, we do not specify a J' value. Instead, we give out a threshold value t, which is defined as follows. This offers another way to define how hard the dimensionality will be reduced:

$$\frac{\sum_{j=1}^{J'} \lambda_j}{\sum_{j=1}^{J} \lambda_j} \geqslant t \tag{11.19}$$

The following is a simple example of practicing PCA.

We have a dataset consisting of ten 2D data points: (2.5,2.4), (0.5,0.7), (2.2,2.9), (1.9,2.2), (3.1,3.0), (2.3, 2.7), (2, 1.6), (1, 1.1), (1.5, 1.6), (1.1, 0.9). The goal is to use PCA to reduce the dimensionality of the data to 1.

First, we carry out zero centering to subtract the mean of the two features, i.e., (1.81, 1.91), from all the data points. The zero-centered data is (0.69, 0.49), (−1.31, −1.21), (0.39, 0.99), (0.09, 0.29), (1.29, 1.09), (0.49, 0.79), (0.19, −0.31), (−0.81, −0.81), (−0.31, −0.31), (−0.71, −1.01).

Next, we calculate the covariance matrix, which is as follows:

$$(\bar{X}^T \cdot \bar{X}) = \begin{bmatrix} 5.549 & 5.539 \\ 5.539 & 6.449 \end{bmatrix} \tag{11.20}$$

The eigenvalues of the matrix are [11.55624941, 0.44175059], and the corresponding eigenvectors are $[-0.6778734, -0.73517866]^T$ and $[-0.73517866, 0.6778734]^T$. So, we selected $[-0.6778734, -0.73517866]^T$ to make $\bar{W}$. Then, we apply $\bar{Z} = \bar{X} \cdot \bar{W}$ and obtain the following ten data points in 1D: [−0.827970186, 1.77758033, −0.992197494, −0.274210416, −1.67580142, −0.912949103, 0.0991094375, 1.14457216, 0.438046137, 1.22382056].

11.5 Feature Extraction Method 2: Linear Discriminant Analysis

LDA is another classic linear dimensionality reduction method. It also appears to be a must-know in the area of dimensionality reduction and has been widely applied in computer vision especially object detection. Distinct from PCA, LDA is a supervised learning method, so it is used with labeled data. Therefore, it is necessary to see how LDA works so that we can get a decent understanding of dimensionality reduction.

11.5.1 Concept and Main Idea

Like PCA, LDA also attempts to project data in a high-dimensional space to a low-dimensional space. However, the criterion for selecting the direction of projection is different. In PCA, we try to let the projected data spread out as much as possible (e.g., the maximum variance of the projected data). This is because, in PCA, we have no other information about the data points, such as labels, so what we can do

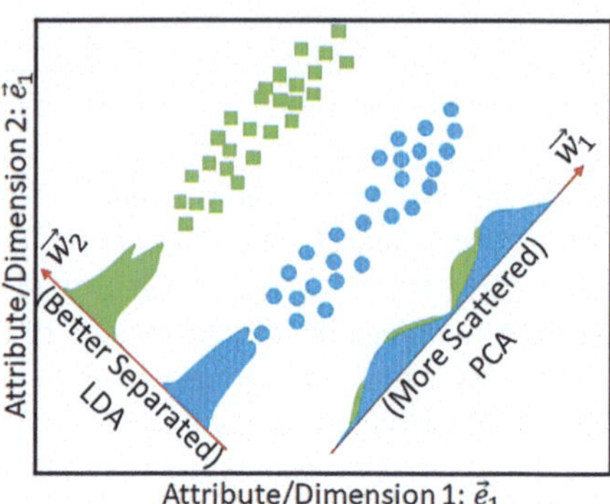

Fig. 11.2 Comparison and understanding of LDA and PCA

is to let the points be far away from each other so that can be easily differentiated. However, LDA deals with labeled data. Thus, the optimum way of projection is sought to meet the goal that the projected data points can reflect their classification the best. To illustrate this, let us check the example in Fig. 11.2.

When performing PCA, all the data points would appear to be the same because there are no labels. Therefore, the best way of projecting the data from this 2D space to a 1D one (i.e., a line), as shown in Fig. 11.2, is to project them onto a line parallel to the length direction ($\vec{w}_1$, southwest-northeast). But in LDA, we usually deal with labeled data. The projection direction along which the two categories of data can be best differentiated is the width direction ($\vec{w}_2$, northwest-southeast). In fact, we can make this selection intuitively without knowing too much about LDA. But, how to describe the rule mathematically?

If we take a look at the figure, we can see that the projected data in both PCA and LDA will exhibit their distribution characteristics. In PCA, because there is no label, so all the data will form one distribution function. That is why, in PCA deductions, we saw parameters such as the mean and variance of the projected data. In LDA, we also have these parameters. However, now we should have such parameters for the distribution functions for different categories. In the above example, the best way of projection for differentiating the data belonging to the two categories is to maximize the distance between the centers of the two categories while minimizing the variance of each distribution. The former ensures the two categories are far from each other, while the latter ensures that each distribution is narrow so that the overlap can be small. These two criteria need to be met or balanced to reach the best classification result. Therefore, in LDA, we utilize the information from the label to guide us in finding the best direction for dimensionality reduction.

In fact, we can extract the distribution information by performing LDA for dimensionality reduction, which can in return provide a tool for aiding in classifying other data. Despite the extra benefit, we will still primarily focus on the use of LDA for dimensionality reduction. In the following, we will show how to derive the major equations for guiding LDA in binary classification to lay down the theoretical basis. The use of LDA for multiple-class classification will be given out based on that.

Then, the procedure and other information for implementing LDA will be presented. Finally, the implementation of LDA for classification will be briefly described.

11.5.2 Theoretical Basis

Rayleigh Quotient and Generalized Rayleigh Quotient

Before we start the deduction, we should obtain some knowledge about the Rayleigh quotient and generalized Rayleigh quotient, which is needed for the deduction. First, let us take a look at the definition of the Rayleigh quotient [134], which is defined as a function $R(\bar{A}, \vec{x})$:

$$R(\bar{A}, \vec{x}) = \frac{\vec{x}^H \cdot \bar{A} \cdot \vec{x}}{\vec{x}^H \cdot \vec{x}} \tag{11.21}$$

where $\vec{x}$ is a nonzero vector, $\bar{A}$ is the Hermitian matrix which satisfies $\bar{A}^H = \bar{A}$, and $\vec{x}^H$ is the conjugate transpose of $\vec{x}$. The above is the general definition of complex numbers. When there are only real numbers, we have $\vec{x}^H = \vec{x}^T$ and $\bar{A}^H = \bar{A}^T$, which is what we will consider in machine learning as we most likely only deal with real numbers in this area. That is, if $\bar{A}$ in the above equation is real and meets the condition of $\bar{A}^T = \bar{A}$, then $\bar{A}$ is the Hermitian matrix.

One reason that we introduce the Rayleigh quotient here is that it has a very important property, which we will employ later. That is, the quotient R is between the minimum and maximum eigenvalues of $\bar{A}$:

$$\lambda_{\min} \leqslant R(\bar{A}, \vec{x}) = \frac{\vec{x}^H \cdot \bar{A} \cdot \vec{x}}{\vec{x}^H \cdot \vec{x}} \leqslant \lambda_{\max} \tag{11.22}$$

When $\vec{x}$ represents an orthonormal basis, that is, $\vec{x}^H \cdot \vec{x} = \vec{I}$, the Rayleigh quotient definition is simplified into $R(\bar{A}, \vec{x}) = \vec{x}^H \cdot \bar{A} \cdot \vec{x}$. Next, let us take a look at the generalized Rayleigh quotient, which we will see in the later deduction:

$$R(\bar{A}, \bar{B}, \vec{x}) = \frac{\vec{x}^H \cdot \bar{A} \cdot \vec{x}}{\vec{x}^H \cdot \bar{B} \cdot \vec{x}} \tag{11.23}$$

where $\bar{B}$ is also a Hermitian matrix. In order to get insights into the minimum and maximum of the quotient, we can reformulate the generalized Rayleigh quotient with $\vec{x} = \bar{B}^{-\frac{1}{2}} \cdot \vec{x'}$, leading to

$$R(\bar{A}, \bar{B}, \vec{x'}) = \frac{\vec{x'}^H \cdot \bar{B}^{-\frac{1}{2}} \cdot \bar{A} \cdot \bar{B}^{-\frac{1}{2}} \vec{x'}}{\vec{x'}^H \cdot \vec{x'}} \tag{11.24}$$

Based on the property of Rayleigh quotient, we can see that the quotient $R(\bar{A}, \bar{B}, \vec{x}')$ is bounded by the minimum and maximum eigenvalues of $\bar{B}^{-\frac{1}{2}} \cdot \bar{A} \cdot \bar{B}^{-\frac{1}{2}}$. In fact, the eigenvalues of $\bar{B}^{-\frac{1}{2}} \cdot \bar{A} \cdot \bar{B}^{-\frac{1}{2}}$ are identical to those of $\bar{B}^{-1}\bar{A}$. Therefore, the value of $R(\bar{A}, \bar{B}, \vec{x}')$ is greater than or equal to the smallest eigenvalue of $\bar{B}^{-1}\bar{A}$ and smaller than or equal to the biggest eigenvalue of $\bar{B}^{-1}\bar{A}$.

Binary Classification

In this subsection, we will see how LDA works with data labeled for binary classification. In particular, we will derive the equations needed for implementing LDA. This will give us a more in-depth understanding of LDA, which lays down the theoretical basis. In the following subsections, we will extend the basis so that data with more categories can also be considered.

The data that will be considered is the following dataset $D = \{(\vec{x}_1, y_1), (\vec{x}_2, y_2), \ldots, (\vec{x}_I, y_I)\}$, in which each instance has J features and is associated with a label $y_i \in \{0, 1\}$. Here, we define I_k as the number of samples in the kth category ($k \in \{0, 1\}$ for binary classification), and $\bar{X}_k$ is the matrix for all the instances in the kth category with a size of $I_k \times J$. $\vec{\mu}_k$ is the mean vector (as a column array with a size of $J \times 1$) of the kth category, and $\bar{\Sigma}_k$ is the covariance matrix of this category (as a 2D array with a size of $J \times J$).

The mean vector $\vec{\mu}_k$ can be obtained via its definition as

$$\vec{\mu}_k = \frac{1}{I_k} \sum_{\vec{x}_i \in \bar{X}_k} \vec{x}_i, \quad k \in \{0, 1\} \tag{11.25}$$

Thus, $\vec{\mu}_k$ has the same size as $\vec{x}_i$: $(J \times 1)$

The covariance matrix $\bar{\Sigma}_k$ can be obtained according to the definition of the covariance matrix as

$$\bar{\Sigma}_k = \sum_{\vec{x}_i \in \bar{X}_k} (\vec{x}_i - \vec{\mu}_k)_{(J \times 1)} \cdot (\vec{x}_i - \vec{\mu}_k)^T_{(1 \times J)} = (\bar{X}_k - \bar{I}_{(I_k \times 1)} \cdot \vec{\mu}_k^T)^T \cdot (\bar{X}_k - \bar{I}_{(I_k \times 1)} \cdot \vec{\mu}_k^T) \tag{11.26}$$

where $\bar{I}_{(I_k \times 1)}$ is a unit matrix with a shape of $(I_k \times 1)$, in which all the elements are 1, and $\bar{\Sigma}_k$ has a shape of $(J \times J)$.

There are two classes; thus, we can project the data onto a line. In fact, we will see in later deductions that we can only project it to a line, i.e., 1D space, as the maximum dimensionality of the low-dimensional space to be projected to in LDA is bounded by the number of categories, which we utilize to find the way of projection. We can use a vector $\vec{w}$ to represent a line. Then, for any instance $\vec{x}_i$, its projection on the line $\vec{w}$ is $\vec{w}^T \cdot \vec{x}_i$. Accordingly, projections of the two class centers, i.e., $\vec{\mu}_0$ and $\vec{\mu}_1$, are $\vec{w}^T \cdot \vec{\mu}_0$ and $\vec{w}^T \cdot \vec{\mu}_1$, respectively. Considering that LDA aims to maximize the distance between the centers, so we will need to find a way of projection whereby the distance between the projected centers, e.g., $\|\vec{w}^T \cdot \vec{\mu}_0 - \vec{w}^T \cdot \vec{\mu}_1\|_2$, will reach the maximum value. Likewise, the variance of the projected belonging to the two groups are $\vec{w}^T \cdot \bar{\Sigma}_0 \cdot \vec{w}$ and $\vec{w}^T \cdot \bar{\Sigma}_1 \cdot \vec{w}$. To make the projected points belonging to the same group be distributed together as much as possible, we will need to minimize the (total) variance of the projected data, i.e., $\vec{w}^T \cdot \bar{\Sigma}_0 \cdot \vec{w} + \vec{w}^T \cdot \bar{\Sigma}_1 \cdot \vec{w}$. Combining these two criteria, we will reach the following optimization problem:

$$\arg\max_{\vec{w}} J(\vec{w}) = \frac{\|\vec{w}^T \cdot \vec{\mu}_0 - \vec{w}^T \cdot \vec{\mu}_1\|_2}{\vec{w}^T \cdot \bar{\Sigma}_0 \cdot \vec{w} + \vec{w}^T \cdot \bar{\Sigma}_1 \cdot \vec{w}} = \frac{\vec{w}^T \cdot (\vec{\mu}_0 - \vec{\mu}_1) \cdot (\vec{\mu}_0 - \vec{\mu}_1)^T \cdot \vec{w}}{\vec{w}^T \cdot (\bar{\Sigma}_0 + \bar{\Sigma}_1) \cdot \vec{w}} \tag{11.27}$$

Then, we define new quantities to simplify the equation. The first quantity is the interclass variance matrix, $\bar{S}_b$, which quantifies the distance between classes and is thus related to the numerator of the above objective function:

$$S_b = (\vec{\mu}_0 - \vec{\mu}_1) \cdot (\vec{\mu}_0 - \vec{\mu}_1)^T \tag{11.28}$$

It is not difficult to find that $\bar{S}_b$ has a size of $(J \times J)$.

The second quantity is the intra-class variance matrix, $\bar{S}_w$, which defines the distance between different points within the same class in the low-dimensional space, i.e., the line in this binary classification case:

$$\begin{aligned}
\bar{S}_w =& \bar{\Sigma}_0 + \bar{\Sigma}_1 = \sum_{\vec{x}_i \in X_0} (\vec{x}_i - \vec{\mu}_0) \cdot (\vec{x}_i - \vec{\mu}_0)^T + \sum_{\vec{x}_i \in X_1} (\vec{x}_i - \vec{\mu}_1) \cdot (\vec{x}_i - \vec{\mu}_1)^T \\
=& (\bar{X}_0 - \bar{I}_{(I_0 \times 1)} \cdot \vec{\mu}_0^T)^{T} \cdot (\bar{X}_0 - \bar{I}_{(I_0 \times 1)} \cdot \vec{\mu}_0^T) \\
& + (\bar{X}_1 - \bar{I}_{(I_1 \times 1)} \cdot \vec{\mu}_1^T)^T \cdot (\bar{X}_1 - \bar{I}_{(I_1 \times 1)} \cdot \vec{\mu}_1^T)
\end{aligned} \tag{11.29}$$

We can also find that $\bar{S}_w$ has a size of $(J \times J)$. With these two quantifies, the optimization problem can be rewritten as

$$\arg\max_{\vec{w}} J(\vec{w}) = \frac{\vec{w}^T \cdot \bar{S}_b \cdot \vec{w}}{\vec{w}^T \cdot \bar{S}_w \cdot \vec{w}} \tag{11.30}$$

It is interesting to notice that the above objective function is a generalized Rayleigh quotient that we introduced earlier. Using the property of the generalized Rayleigh quotient, we know that the maximum value of $J(\vec{w})$ is the biggest eigenvalue of $\bar{S}_w^{-1} \cdot \bar{S}_b$, and $\vec{w}_{\max}$ is the maximum eigenvector, while the optimal solution is $\vec{w}^* = \bar{S}_w^{-\frac{1}{2}} \cdot \vec{w}_{\max}$.

However, in LDA for binary classification data, we usually do not use the above eigenvalue problem as what we do in PCA. Instead, an easier process is adopted based on the following relationship. That is, in binary classification problems, $S_b \cdot \vec{w}$ is parallel to $\vec{\mu}_0 - \vec{\mu}_1$, which gives out a shortcut for finding the optimal direction for projection. Thus, we can let $S_b \cdot \vec{w} = \lambda(\vec{\mu}_0 - \vec{\mu}_1)$. Then, substituting this equation into $(S_w^{-1} \cdot S_b) \cdot \vec{w} = \lambda \vec{w}$, we can obtain an equation for calculating $\vec{w}$: $\vec{w} = S_w^{-1} \cdot (\vec{\mu}_0 - \vec{\mu}_1)$.

Multiclass Classification

The data that will be considered is the following dataset $\bar{D} = \{(\vec{x}_1, y_i), (\vec{x}_2, y_2), \ldots, (\vec{x}_I, y_I)\}$, in which each instance has J features and is associated with a label $y_i \in \{C_1, C_2, \ldots, C_K\}$. We define I_k as the number of samples in the kth category ($k \in \{0, 1, \ldots, K\}$), and $\bar{X}_k$ is the matrix for all the instances in the kth category with a size of $I_k \times J$. $\vec{\mu}_k$ is the mean vector (as a column array with a size of $J \times 1$) of the kth category, and $\bar{\Sigma}_k$ is the covariance matrix of this category.

Now, we try to project this data with K categories and J features (dimensions) to a low-dimensional space, which is not necessarily a line anymore. To generalize the deduction, we can view this low-dimensional space as a "hyperplane." If the dimensionality of this "hyperplane" is J', then the orthonormal basis for defining the low-dimensional space, i.e., the way of projection, is $\bar{W}_{(J\times J')} = [\vec{w}_0, \vec{w}_1, \ldots, \vec{w}_{J'}]$. Then, the objective function of the optimization problem becomes

$$J(\bar{W}) = \frac{\bar{W}^T \cdot \bar{S}_b \cdot \bar{W}}{\bar{W}^T \cdot \bar{S}_w \cdot \bar{W}} \tag{11.31}$$

where $\bar{S}_b$ and $\bar{S}_w$ are defined using the following equations.

$$\bar{S}_b = \sum_{k=1}^{K} I_k(\vec{\mu}_k - \vec{\mu}) \cdot (\vec{\mu}_k - \vec{\mu})^T \tag{11.32}$$

where $\vec{\mu}$ is the mean of all the instances and can be calculated as

$$\vec{\mu} = \frac{1}{I} \sum_{\vec{x}_i \in X} \vec{x}_i \tag{11.33}$$

$$\begin{aligned} \bar{S}_w &= \sum_{k=1}^{K} \sum_{\vec{x}_i \in X_k} (\vec{x}_i - \vec{\mu}_k) \cdot (\vec{x}_i - \vec{\mu}_k)^T \\ &= \sum_{k=1}^{K} (\bar{X}_k - \bar{I}_{(I_k\times 1)} \cdot \vec{\mu}_k^T)^T \cdot (\bar{X}_k - \bar{I}_{(I_k\times 1)} \cdot \vec{\mu}_k^T) \end{aligned} \tag{11.34}$$

Unlike what we are faced with in binary classification, this objective function may not be a scalar, but a matrix with a shape of $J' \times J'$. Therefore, it cannot be optimized as a scalar function. To solve the problem, we can use some alternative objective functions. A common objective function for the optimization problem of LDA with multiclass classification data is

$$J(\bar{W}) = \frac{\prod_{diag} \bar{W}^T \cdot \bar{S}_b \cdot \bar{W}}{\prod_{diag} \bar{W}^T \cdot \bar{S}_w \cdot \bar{W}} \tag{11.35}$$

where $\prod_{diag} \bar{W}^T \cdot \bar{S}_b \cdot \bar{W}$ is the product of all the diagonal elements of the matrix $\bar{W}^T \cdot \bar{S}_b \cdot \bar{W}$, which has a shape of $J' \times J'$.

The above objective function can be further reformatted to be related to Rayleigh quotients as follows:

$$J(\bar{W}) = \frac{\prod_{diag} \bar{W}^T \cdot \bar{S}_b \cdot \bar{W}}{\prod_{diag} \bar{W}^T \cdot \bar{S}_w \cdot \bar{W}} = \frac{\prod_{j=1}^{J'} \vec{w}_j^T \cdot \bar{S}_b \cdot \vec{w}_j}{\prod_{j=1}^{J'} \vec{w}_j^T \cdot \bar{S}_w \cdot \vec{w}_j} = \prod_{j=1}^{J'} \frac{\vec{w}_j^T \cdot \bar{S}_b \cdot \vec{w}_j}{\vec{w}_j^T \cdot \bar{S}_w \cdot \vec{w}_j} \tag{11.36}$$

As can be seen, the objective function is the product of J' Rayleigh quotients, and $\bar{W}$ consists of the J' eigenvectors. This is because in LDA, the search for the best way of projection utilizes the information of classification. In detail, $\bar{S}_b$ is made of $\vec{\mu}_k - \vec{\mu}$, so the rank of this matrix is bounded by the number of $\vec{\mu}_k$. Meanwhile, once we know $K - 1$ out of all the K mean vectors (i.e., $\vec{\mu}_k, k \in \{1, \ldots, K\}$), the remaining one can be inferred because the average of all the mean vectors, i.e., $\vec{\mu}$, is present. Therefore, we will have $K - 1$ independent eigenvalues. This means, in general, we can use LDA to reduce the dimensionality to $K - 1$ or lower. The rigorous proof will need to employ the nullity theorem, which is not provided here.

11.5.3 Implementation

The implementation of LDA is also pretty straightforward. But we may need to pay attention to two points. First, LDA uses labeled data, which could cause differences from the procedure of PCA implementation. Second, LDA for binary classification data can be slightly different from data with more classes. This latter point is because binary classification data provides an easier way for implementation, though LDA for binary classification data can also be processed using the general procedure for multiple classification data. In the following, the procedure for general LDA implementation will be laid down. The simplification that can be made for binary data will be briefly mentioned afterward.

LDA:
Input: a set of labeled data with I samples and J attributes: $D = \{(\vec{x}_1, y_1), (\vec{x}_2, y_2), \ldots, (\vec{x}_I, y_I)\}$, which belong to K categories as $y_i \in \{C_1, C_2, \ldots, C_K\}$.
Output: a set of data with I samples and reduced dimensionality (J' attributes): D'.
Prepare the data for dimensionality reduction: $\bar{X}_{(I \times J)} = [\vec{x}_1^T, \vec{x}_2^T, \ldots, \vec{x}_I^T]^T$. As can be seen, labels are excluded and each instance is a vertical array.
Calculate the mean array $\vec{\mu}$ as $\vec{\mu} = \frac{1}{I} \sum_{\vec{x}_i \in X} \vec{x}_i$.

Calculate $\bar{S}_b$ and $\bar{S}_w$ with $\bar{S}_b = \sum_{k=1}^{K} I_k(\vec{\mu}_k - \vec{\mu}) \cdot (\vec{\mu}_k - \vec{\mu})^T$ and $\sum_{k=1}^{K}(\bar{X}_k - \bar{I}_{(I_k \times 1)} \cdot \vec{\mu}_k^T)^T \cdot (\bar{X}_k - \bar{I}_{(I_k \times 1)} \cdot \vec{\mu}_k^T)$.
Compute the J' eigenvalues and eigenvectors of $\bar{S}_w^{-1} \cdot \bar{S}_b$. Use the eigenvectors to form the projection array $\bar{W}$.
Project the data to the lower space to reduce its dimensionality: $\bar{Z} = \bar{X} \cdot \bar{W}$ or $\vec{z}_i = \bar{W}^T \cdot \vec{x}_i$
Prepare the output dataset $D' = \{(\vec{z}_1, y_1), (\vec{z}_2, y_2), \ldots, (\vec{z}_I, y_I)\}$

To compare the data before and after dimensionality reduction. The data in the low-dimensional space can be projected back to the high-dimensional space using the equation $\hat{\bar{X}} = \bar{W}^T \cdot \bar{Z}$ or $\vec{x}_i = \bar{W} \cdot \vec{z}_i$ (same as PCA). For example, a dataset D with three features can be plotted in a 3D space, e.g., as points in 3D. If we project the data to 2D space, then we will get points with two features (axes). In order to compare, we can use the above equation to convert the data back to 3D space. In a typical data visualization for comparison, we will see the data converted back, which are points in a 3D space (three feature values are different from those of the original points) and the projected points (with two features) distributed on a plane (2D).

In the above procedure, if we have binary classification data, we can either use the above process or replace the eigenvalue calculation step with $\vec{w} = \bar{S}_w^{-1} \cdot (\vec{\mu}_0 - \vec{\mu}_1)$ for finding the best projection direction.

11.6 Practice: Develop and Modify Code for PCA and LDA

>>> More and Up-to-Date Course Materials including Practices @ AI-engineer.org <<<

1. Develop PCA code to reduce the following 2D data to 1D.

```
# Create high-dimensional data as input for dimensionality
    reduction: X with dimension of I (instances) by J (
    features)
X_raw = np.array([[2.5,2.4], [0.5,0.7], [2.2,2.9], [1.9,2.2],
    [3.1,3.0],\
                [2.3, 2.7], [2, 1.6], [1, 1.1], [1.5, 1.6],
    [1.1, 0.9]])

I,J = X_raw.shape
J_prime = 1 # Please specify J_prime: 1 <= J_prime < J

X = X_raw - np.mean(X_raw,axis=0).reshape(1,J) # X is zero-
    centered
```

Hints: You will need to at least finish the following steps: (attention: incomplete code)

```
Lambda,V = np.linalg.eig(...)
# Find the first few eigenvalues and eigenvectors
idx = Lambda.argsort()[::-1]
Lambda = Lambda[idx]
V = V[:,idx]

W = ...
Z = ...
X_hat = ...

# The following visualization is applicable to 2D data (
    original)
plt.plot(X[:,0],X[:,1],'b.')
plt.plot(X_hat[:,0],X_hat[:,1],'ro')
```

2. Please try to understand the following LDA code and use it to solve the dimension reduction problem using the Iris dataset from Scikit-learn. Iris has 150 samples in three classes, and each sample has four attributes. You can use the first 50 samples for X_0 (NumPy array for the class 1 dataset) and Samples 50-100 as X_1. For visualization, you can only use three attributes (e.g., X[:,0],X[:,1], and X[:,2]).

```
import numpy as np
from sklearn import datasets
import matplotlib.pyplot as plt

J = X.shape[1]
Mu_0 = np.mean(X_0,axis=0).reshape((J,1)) # Mu is J(4) by 1
Mu_1 = np.mean(X_1,axis=0).reshape((J,1))
S_b = (Mu_0-Mu_1)@(Mu_0-Mu_1).T # S_b is J by J

Epsilon_0 = (X_0-Mu_0.T).T@(X_0-Mu_0.T) # X_0-Mu_0.T uses
    broadcasting for I0 by 4 and 1 by 4
Epsilon_1 = (X_1-Mu_1.T).T@(X_1-Mu_1.T)
S_w = Epsilon_0+Epsilon_1

# Multiclass way
J_prime = 1
Lambda,V = np.linalg.eig(np.linalg.inv(S_w)@S_b)
# Find the first few eigenvalues and eigenvectors
idx = Lambda.argsort()[::-1]
Lambda = Lambda[idx]
V = V[:,idx]
W = V[:,:J_prime]

Z = X @ W # (I by J) @ (J by J') yields (J by J')
X_hat = Z @ W.T # (I by J') @ (J' by J) yields (I by J)

# The above data is 4D, so we can only select 3D the most for
    visualization as follows
ax = plt.axes(projection='3d')
ax.scatter3D(X[:,0],X[:,1],X[:,2],'b.')
ax.scatter3D(X_hat[:,0],X_hat[:,1],X_hat[:,2],'ro')
```

Chapter 12
Anomaly Detection

12.1 Overview

This chapter introduces anomaly detection—an unsupervised learning topic that has been gaining more and more attention. We will start the introduction with some basic knowledge about anomaly detection including its definition, classification, major concepts, and popular algorithms. Skipping the so-called rule-based methods, we will first present information about statistics-based methods to bridge possible gaps between the anomaly detection in machine learning and traditional anomaly detection practice. Then, more detailed information will be provided for machine learning-based anomaly detection. Considering anomaly detection is usually handled by modifying existing machine learning algorithms, the popular anomaly detection algorithms from supervised machine learning, unsupervised machine learning, and semisupervised machine learning will be discussed one by one. In the final, issues in the practice of anomaly detection will be summarized.

12.2 Basics of Anomaly Detection

Anomaly detection, also called novelty detection, outlier detection, forgery detection, or out-of-distribution detection in different areas, is intended to identify rare items, events, or observations that significantly differ from the majority of the data and do not conform to a well-defined notion of normal behavior. Anomaly detection has been applied to a variety of areas such as fraud detection, web hack detection, medical (disease) detection, sensor network anomaly detection, IoT bid data anomaly detection, log anomaly detection, and industrial hazard detection [135, 136].

Z. "L." Liu, *Artificial Intelligence for Engineers*,
https://doi.org/10.1007/978-3-031-75953-6_12

The data that can be processed for anomaly detection include continuous data (e.g., time series data), text data (logs), and multidimensional data (e.g., images). Data anomalies can appear as point anomalies (data points different from the others), group anomalies (single points in a group appear normal while the whole group behaves differently from the others as a group), and background anomalies (data only appears different on certain backgrounds).

In a broad sense, the available methods for anomaly detection can be roughly grouped into rule-based methods, statistics-based methods, and machine learning-based methods. Among them, anomaly detection methods based on machine learning algorithms are anomaly detection in a narrow sense in the state of the art. The machine learning-based methods can be further categorized into supervised, unsupervised, and semisupervised methods.

The key in rule-based methods is to obtain rules that are used to judge the anomaly. Such rules can be obtained automatically using algorithms or manually by experts. Then, such rules are used to assess the patterns or behaviors to identify whether they are normal. The advantage of such methods is that they can accurately and conveniently identify the anomalies that follow the anomaly rule. The disadvantages include difficulties in identifying rules and troubles. For example, such rules may be incomplete and need to be updated frequently. When the database of rules is large, the comparison may also be time-consuming.

The core of the statistics-based method is to assume the data follows a certain type of probabilistic distribution and then use data to conduct coefficient estimation. The most common methods in this category are 3σ, Boxplot, Grubbs, and Z-score. Such methods are more applicable to low-dimensional data and have good robustness. However, they may suffer from heavy reliance on the adopted statistical assumptions.

Machine learning is the major category of anomaly detection methods [137, 138]. In unsupervised learning, common methods can be divided into five groups: statistics based, distance based, density based, clustering based, and tree based. In semisupervised learning, labeled data are usually normal data points, and popular methods include one-class SVM, autoencoder, and GMM. In supervised meaning, we usually need to pay attention to data labeling and imbalanced data as possible issues, and such methods are suitable for considering data with new classes. Common methods in this category include linear models, SVM, and artificial neural networks. Table 12.1 gives a list of some common methods.

12.3 Statistics-Based Methods

12.3.1 3 Sigma

Three-sigma (or 3 σ) rule is a simple yet widely adopted method for anomaly detection. This method is established based on the assumption that the data follows a normal distribution and only contains random errors. As shown in Fig. 12.1, the

Table 12.1 Popular anomaly detection methods

Category	Algorithm	Criteria
Distribution	3sigma	$x > \mu + 3\sigma$ or $x < \mu - 3\sigma$
	Z-score	Z-score > 3
	Boxplot	$x > q3 + 1.5\text{IQR}$ or $x < q1 - 1.5\text{IQR}$
	Grubbs Testing	Z-score > Grubbs limit
Distance	KNN	KNN distance > threshold
Density	LOF	LOF > threshold
	COF	COF > threshold
	SOS	Anomaly probability > threshold
Clustering	DBSCAN	Cannot be clustered (label = −1)
Tree	iForest	Anomaly score > threshold
Dimension reduction based	PCA	Dimension bias > threshold
	Autoencoder	Error > threshold
Classification	One-class SVM	Points out of the hyperplane (label = −1)
Prediction	Moving average, RIMA	Errors + distribution-based methods

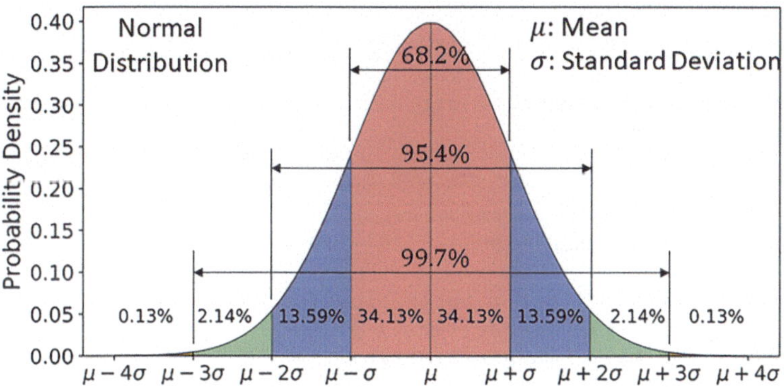

Fig. 12.1 Three sigma: 99.7% of the data are within three standard deviations of the mean

standard deviation of this normal distribution, σ, quantifies how the data points cluster around the mean. This rule determines that, when a data point is far from the mean, e.g., 3σ, this point is an anomaly.

According to the characteristics of normal distributions, we can obtain the following probabilities:

- The probability that a data point appears in the range of $\mu - \sigma$ and $\mu + \sigma$ is 0.6827.
- The probability that a data point appears in the range of $\mu - 2\sigma$ and $\mu + 2\sigma$ is 0.9545.

- The probability that a data point appears in the range of $\mu - 3\sigma$ and $\mu + 3\sigma$ is 0.9973.

Therefore, the probability of a data point falling out of the normal data range, i.e., between $\mu - 3\sigma$ and $\mu + 3\sigma$, is only 0.0027 (0.27%). Such a point is = far from the mean, where most data points cluster, so it is believed to be anomalous.

The following code presents a simple implementation of the three-sigma rule. The Python function gives out the lower and upper bounds for defining the range of normal data. Any data point that is smaller than the lower bound or greater than the higher bound is an anomaly.

```
def three_sigma(s):
    mu, std = np.mean(s), np.std(s)
    lower, upper = mu-3*std, mu+3*std
    return lower, upper
```

Three sigma and similar methods are limited to the assumptions of the distribution, e.g., normal distribution. Also, they require a large number of measurements to ensure a good estimate of σ to judge anomalies. The above function can be employed to process data in the following way as a simple implementation of the three-sigma method.

```
import numpy as np
import pandas as pd
import matplotlib.pyplot as plt
np.random.seed(42)
anomalies = []
normal = []
# Generate data for demonstration
data = np.random.randn(50000) * 20 + 20

lower_limit, upper_limit = three_sigma(data)
print("Lower Limit: ",lower_limit)
print("Upper Limit: ",upper_limit)
# Anomaly detection
for outlier in data:
    if outlier > upper_limit or outlier < lower_limit:
        anomalies.append(outlier)
    else:
        normal.append(outlier)
data_3sigma = pd.DataFrame(anomalies,columns=["Anomalies"]),pd.
    DataFrame(normal,columns=["Normal"])
```

The output of the above program is as follows. The original data and the data processed using the three-sigma method can be checked in arrays "data" and "data_3sigma," respectively.

```
Lower Limit: -40.01800995575444
Upper Limit: 80.00117401256483
```

12.3.2 Z-Score

Z-score is a common method for detecting anomalies in 1D data. Like three sigma, a normal distribution is also assumed for the data. The Z-score can be calculated as follows when the mean and standard deviation are known:

$$z_i = \frac{x_i - \mu}{\sigma} \tag{12.1}$$

where z_i and x_i are the Z-score and attribute value of the ith sample, respectively.

Z-score is a measure of the distance between a point and the mean. A large absolute value of the Z-score implies an anomaly. Z-score can be easily calculated using the following Python function:

```
def Z_score(s):
    Z_score = (s - np.mean(s)) / np.std(s)
    return Z_score
```

If a point is 3σ from the mean μ, then the Z-score is 3. Threshold Z-score values can be used as the criteria for identifying anomalies. For example, the selection of Z-score=3 as the threshold for identifying anomalies is equivalent to the three-sigma method.

12.3.3 Boxplot

The boxplot method employs the interquartile range (IQR) to find anomalies. The idea of boxplot is illustrated in the box plot (also called box or whisker plot) in Fig. 12.2. The boxplot presents a standardized way of displaying data distribution in terms of a five-number summary: "minimum," first quartile (Q1), median, third quartile (Q3), and "maximum." This plot can be used to detect outliers easily. In addition, this plot can tell us whether the data is symmetrical, how tightly your data is clustered, and whether/how the data is skewed.

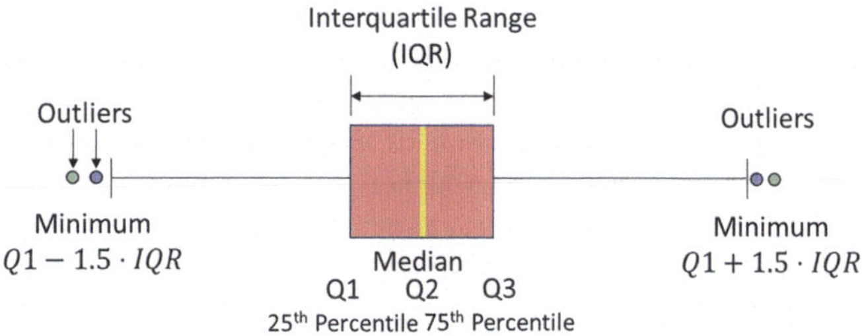

Fig. 12.2 Major points in boxplot

The key points in boxplot have the following meanings:

- Median (Q2/50th percentile): the middle value of the dataset.
- First quartile (Q1/25th percentile): the middle number between the smallest number and the median of the dataset.
- third quartile (Q3/75th Percentile): the middle value between the median and the greatest number of the dataset.
- Interquartile range (IQR): the distance between the 25th and 75th percentile, which tells how spread the middle values are.
- "Minimum": $Q1 - 1.5 * IQR$, which gives out the lower bound of normal values in boxplot.
- "Maximum": $Q3+1.5*IQR$, which gives out the higher bound of normal values in boxplot.

In boxplot, data points that are out of the range defined by the lower and upper bounds are determined as anomalies. Here, they are the anomalies. These can be defined using the following code.

```
import numpy as np
# Input data should be a NumPy array
def boxplot(data):
    q1, q3 = np.quantile(data, 0.25), np.quantile(data,0.75)
    IQR = q3 - q1
    lower, upper = q1 - 1.5*IQR, q3 + 1.5*IQR
    return lower, upper
```

Quantiles are those values that split sorted data or a probability distribution into equal parts. Quartiles are the three values that divide ordered datasets into four equal parts. Thus, quartiles are a special type of quantile.

In the boxplot method, anomalies are defined as the data points lower than $Q1 - 1.5 \times \text{IQR}$ or greater than $Q3 + 1.5 \times \text{IQR}$. The above function can be used in a way similar to the code given for three sigma for implementation.

12.3.4 Grubbs Hypothesis Test

Some statistical tests, such as Grubbs' test, Chi-square test, and Dixon's Q test, are designed to detect data outliers. A hypothesis test is a procedure for determining whether a hypothesis can be supported or rejected based on sample data. Such hypothesis tests involve the comparison of a test statistic from the data to an appropriate distribution to determine whether a given hypothesis is supported by the data. This subsection is dedicated to introducing Grubbs' test [139].

Grubbs' test is also known as the maximum normed residual test. It assumes that the dataset being tested for outliers is normally distributed. Two hypotheses, the null and alternative hypotheses, are used in Grubbs' test.

- H0: There are no outliers in the dataset.
- Ha: There is at least one outlier in the dataset.

The test statistic for a two-tailed test is as follows:

$$G = \frac{\max_{i=1,\ldots,I} |x_i - \mu|}{\sigma} \tag{12.2}$$

where x_i is ith ($i \in \{1, \ldots, I\}$) sample to be tested, μ is the mean of the dataset, and σ is the standard deviation.

The test statistic for a left-tailed test is as follows:

$$G = \frac{\mu - x_{\min}}{\sigma} \tag{12.3}$$

The test statistic for a right-tailed test is as follows:

$$G = \frac{x_{\max} - \mu}{\sigma} \tag{12.4}$$

The above statistic needs to be compared against the critical G value, which can be calculated as follows:

$$G_{\text{critical}} = \frac{I-1}{\sqrt{I}} \sqrt{\frac{t^2}{I-2+t^2}} \tag{12.5}$$

where I is the number of samples drawn from the population and t is the upper critical value of the Student's t-distribution, which has $I-2$ degrees of freedom. Thus, t in a two-tailed test is $t_{\alpha/(2I),I-2}$ and $t_{\alpha/(I),I-2}$ for a one-tailed test, which α is the significance level.

The outlier is determined by assessing the hypotheses as follows:

- $G_{\text{test}} < G_{\text{critical}}$: the data point is not an outlier and keeps the point in the dataset.
- $G_{\text{test}} > G_{\text{critical}}$ the data point is an outlier.

The Grubs hypothesis test for outlier detection can be implemented in the following procedure:

(1) Sort the samples from the lowest value to the highest.
(2) Calculate the mean and deviation.
(3) Calculate the statistic (G value) for the data point to be considered.
(4) Calculate the critical G value or the Grubbs threshold: G_{critical}.
(5) Compare G against G_{critical} to test the hypothesis to decide whether the point is an outlier.

In practice, the Grubbs threshold is obtained by two parameters: the significance level α (lower values mean more strict with the check) and the number of samples. The above procedure can be repeated to assess multiple data points.

The Grubbs hypothesis test for anomaly detection suffers from the following limitations:

- Can only be applied to 1D data.
- The data is assumed to follow a normal distribution or an approximate normal distribution.
- Cannot yield the normal range.
- Identify anomalies one at a time and thus may be time-consuming.

12.4 Supervised Learning Methods

12.4.1 Why Not Use Binary Classification for Anomaly Detection?

An intuition that will pop out when people start looking into anomaly detection, especially for people who are familiar with supervised learning but not anomaly detection, is "*why not use binary classification for anomaly detection in which anomalies and the other data are treated as two categories?*"

One major reason is that we identify anomalies based on whether a data point is similar to the others. Thus, the normal data points are usually similar to each other and can be treated as one category. However, anomalous data points can be much different from each other and the normal data points. Such points may be different from any data or data types that the model has been trained with. As a result, the model has limited or adequate "experience" and "capability" to treat such data.

Another reason is class or data imbalance. In many cases, we may have a very limited collection of anomalous data points. Some of the anomalies might not have appeared when an anomaly detection model was developed. For example, many credit fraud cases present conditions that can hardly be known before they really happen, and there may be no cases before we create an anomaly detection model.

Due to the above reasons, anomaly detection cannot be simply treated using a binary classification method, though possibilities are still there for applying such methods for special detection applications. As a result, anomaly detection is investigated as an independent machine learning area.

12.4.2 Modification of Supervised Classification Methods for Anomaly Detection

Common supervised classifiers can be considered for anomaly detection, depending on the characteristics of the data. In general, anomalous samples are much less than the normal samples. Thus, we should deal with data imbalance using techniques such as upsampling, downsampling, and threshold adjustment, which will be further discussed in the last section of this chapter. In addition, precision and recall instead of accuracy should be used in the model evaluation.

12.5 Unsupervised Machine Learning Methods

12.5.1 Overview

In most applications, we only have unlabeled data, and interestingly, unlabeled data contains information for effective anomaly detection. Thus, unsupervised learning methods have been the most widely used ones in anomaly detection.

12.5.2 Probabilistic Distribution Based: HBOS

Histogram-Based Outlier Score (HBOS) [140] is the most common probabilistic distribution-based anomaly detection method. Similar to the Naive Bayes method, this method assumes all the attributes are independent of each other. Then, a histogram for each feature can be generated. Next, we can obtain the probability of every sample as the product of the probabilities of its attributes from the corresponding histograms. The samples with probabilities that are much lower than the others are viewed as anomalies.

Thus, HBOS tends to find anomalies in datasets that rarely occur in the data. The attributes of such data points may be significantly different from those of the normal samples. It is noted that the assumption of independent features renders it a much faster method than multivariate approaches at the cost of compromised precision. Therefore, HBOS is suitable for applications involving great amounts of data. Due to the same reason, HBOS heavily relies on the assumption of independent features, which may not be valid in many cases. It is also worthwhile to point out that this method is not suitable for tasks with lots of anomalies.

12.5.3 Distance Based: KNN

Distance-based methods assume anomalies are far from the normal data points. K-nearest neighbors is the most popular method in this category.

A typical implementation of KNN is as follows.

(1) Select the hyperparameter K as the number of neighbors.
(2) Calculate the Euclidean distances of the neighbors of the data point under consideration.
(3) Find the K-nearest neighbors according to the Euclidean distances.
(4) Among these K neighbors, count the number of the neighboring data points in each category.
(5) That category that has the greatest number of neighbors is the classification of the data point.

There are two ways of using KNN for anomaly detection. In the first way, we calculate the distance of this point to the nearest center (average value of a category of data points classified with KNN) or the average of the distances to all the centers. Then, this value will be compared with a threshold to tell whether the point is an anomaly. In the second way, we calculate this distance for every sample and sort all the samples based on the distance values. The top ones with the highest values will be judged as anomalies. The distance(s) is also calculated using the Euclidean distance as in KNN.

The major advantage of KNN is that no assumption of data distribution is needed. However, the nature of KNN also makes it unsuitable for complicated data. In particular, the assessment of every point needs to involve the whole dataset. Thus, it is not suitable for big data, online applications, and high-dimensional data. Besides, this method can only find anomalous points but no group anomalies. Also, the hyperparameter K and the threshold need to be determined subjectively or/and tuned manually. In addition, this method is not suitable for tasks with lots of anomalies. We also need to note that the use of the Euclidean distance assumes data is distributed spherically and brings forth difficulties at the sphere boundaries. Finally, this method works well for global instead of local anomalies.

12.5.4 Density Based: LOF, COF, INFLO, and LoOP

The adoption of a threshold in distance-based methods endows such methods with the ability of global anomaly detection. However, it is very common that many data are not evenly distributed. This may challenge the selection of a suitable threshold and even threaten the validity of the global methods. Density-based methods can be used to process such tasks with local anomalies [141]. Methods in this category include LOF, COF, ODIN, MDEF, INFLO, LoOP, LOCI, aLOCI, etc. Basic information is provided for four typical density-based methods in the following.

Local Outlier Factor (LOF)

LOF computes the local density deviation of a given data point with respect to its neighbors. It considers outliers as the samples that have a substantially lower density than their neighbors. In LOF, we can calculate the LOF of every data point, which is defined as the ratio between the average local reachability density of the neighbors divided by the object's own local reachability density. The above definition contains several concepts that are essential in density-based methods such as DBScan for clustering. These concepts may prevent us from easily understanding the definition. Let us check these concepts before giving out the definitions of LOF.

The K-nearest neighbors of a point, e.g., A, can be denoted by $N_k(A)$. The K-distance of this point is the distance of Point A to the K-nearest neighbor. Thus, all the K-nearest neighbors should be within this distance from Point A. This distance is used to define another import concept: reachability distance rd:

$$rd_k(A, B) = \max(k - \text{distance}(B), d(A, B)) \tag{12.6}$$

where $rd_k(A, B)$ is the reachability distance of Point A from Point B, $k - \text{distance}(B)$ is the K-distance of Point B, and $d(A, B)$ is the distance between Points A and B. This quantity presents one way to measure the distance between two points. It is proposed to reduce the statistical fluctuations between all points A close to B. Increasing the value of K increases the smoothing effect. In this definition, all points belonging to the K-nearest neighbors of B (the "core" of B in DBSCAN) are considered to be "equally distant."

With this definition, we further define the local reachability distance (lrd) of Point A by

$$lrd_k(A) = \frac{1}{\frac{\sum_{B\in N_k(A)} rd_k(A,B))}{|N_k(A)|}} \tag{12.7}$$

As can be seen from the above equation, the local reachability distance (lrd) of Point A is the inverse of the average reachability distance (average of rd_k of all the $N_k(A)$ points represented by B in the above equation) of object A from its neighbors.

Then, the local outlier factor, LOF, is computed with

$$\text{LOF}_k(A) = \frac{\sum_{B\in N_k(A)} lrd_k(B)}{|N_k(A)| \cdot lrd_k(A)} \tag{12.8}$$

An LOF value of approximately 1 indicates that the object is comparable to its neighbors (and thus not an outlier). A value below 1 indicates a denser region, which would be an inlier, while values significantly larger than 1 indicate outliers. That is,

- $\text{LOF}_k(A) \sim 1$ means Point A has a similar density as neighbors.
- $\text{LOF}_k(A) \leq 1$ means Point A has a higher density than neighbors (inlier).
- $\text{LOF}_k(A) \geq 1$ means Point A has a lower density than neighbors (outlier).

We will not discuss the details of implementing the LOF from scratch. Instead, the following code (Scikit-learn) presents a simple example of using LOF.

```
from sklearn.neighbors import LocalOutlierFactor
np.random.seed(9527)
# Generate train data
X_inliers = 0.3 * np.random.randn(100, 2)
X_inliers = np.r_[X_inliers + 2, X_inliers - 2]
# Generate some outliers
X_outliers = np.random.uniform(low=-4, high=4, size=(20, 2))
X = np.r_[X_inliers, X_outliers]
```

```
n_outliers = len(X_outliers)
ground_truth = np.ones(len(X), dtype=int)
ground_truth[-n_outliers:] = -1

# Fit the model
clf = LocalOutlierFactor(n_neighbors=20, contamination=0.1)
y_pred = clf.fit_predict(X)
n_errors = (y_pred != ground_truth).sum()
X_scores = clf.negative_outlier_factor_

# Plot circles with radius proportional to the outlier scores
radius = (X_scores.max() - X_scores) / (X_scores.max() - X_scores
    .min())
plt.scatter( X[:, 0], X[:, 1], s=1000 * radius, edgecolors="r",
    facecolors="none", label="Outlier scores",)
```

As shown in Fig. 12.3, the normalized LOF was plotted as the "radius" in the data points. Accordingly, the points with the bigger circles (higher LOF) are more likely to be anomalies.

COF, INFLO, and LoOP adopt different modifications to improve some aspects of LOF.

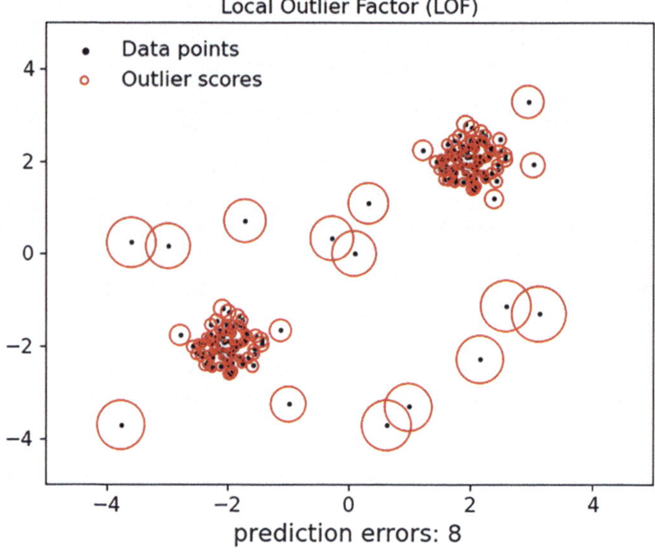

Fig. 12.3 Local outlier factor (LOF)

Connectivity-Based Outlier Factor (COF)

LOF uses the Euclidean distance and thus assumes a spherical distribution for the data, which can cause problems in some conditions. By contrast, the local densities in the connectivity-based outlier algorithm are calculated using the chaining distance instead of the Euclidean distance. The main idea of the connectivity-based outlier algorithm is to assign a degree of outlier, i.e., connectivity-based outlier factor or COF, to each data point. A high COF value of a data point represents a high probability of being an outlier.

Influenced Outlierness (INFLO)

LOF is inclined to count the points on the boundaries as anomalies. INFLO is proposed to overcome this shortcoming. INFLO computes the influenced outlierness score for the points. This score serves as the comparison between density in the neighborhood of points for outlier scoring and the density in the reverse neighborhood.

Local Outlier Probability (LoOP)

LOF adopts a square root in the equations for calculating densities in the LOF. In addition, it assumes the distributions in the neighborhood follow the Gaussian distribution.

Overall, density-based methods are superior in that they can find local anomalies in unevenly distributed data points and give a score for quantifying how anomalous a point is, e.g., higher means more anomalous. However, such methods are not suitable for high-dimensional data and can only deal with anomalous points but not group anomalies. Moreover, the assessment of every point needs to involve the whole dataset. The resultant high computing cost may make them unsuitable for big data or online applications.

12.5.5 Clustering Based

Clustering methods adopt different clustering hypotheses for anomaly detection. The computational complexity of such anomaly detection methods relies on the complexity of the corresponding clustering methods. The following three popular hypotheses can offer a rough idea of how to perform clustering-based anomaly detection.

- Hypothesis 1: A point that does not belong to any cluster is hypothesized to be an anomaly. Clustering methods that adopt this hypothesis include DBScan, SNN clustering, FindOut algorithms, and WaveCluster algorithms.

- Hypothesis 2: A point that has a large distance from its nearest cluster center is an anomaly. Clustering methods with this hypothesis include K-means, self-organizing maps (SOM), and GMM.
- Hypothesis 3: Points in sparse clusters or small clusters are anomalies. Clustering methods incorporating this hypothesis are CBLOF, LDCOF, CMGOS, etc.

It is noted that methods built on the first two hypotheses cannot identify a group of anomalies or anomalous clusters, while those that employ the third hypothesis can.

Clustering-based anomaly detection methods have unique advantages and disadvantages. As for advantages, predictions made with some clustering-based anomaly detection methods can be time efficient because such methods may only involve comparisons against a few other clusters; some methods like K-means can even be used for online applications. However, the performance of such methods heavily relies on the computational efficiency of the corresponding clustering methods. Besides, the anomaly detection task may suffer from a performance bottleneck when clustering large datasets, which can be computationally expensive.

12.5.6 Tree Based

Tree-based anomaly detection is achieved by dividing subspaces. This category of methods is featured by the selection of features, the selection of division points, and the way of tagging the subspaces. Due to its nature, especially the exclusion of distances, such methods are not constrained by the spherical neighboring areas and can identify anomalies with any geometries. Typical tree-based anomaly detection methods include iForest, SCiForest, and RRCF.

iForest

iForest or Isolation Forest can be understood as an unsupervised version of random forest [142]. The core concept is to divide data using decision trees until there is only one data point. In this process, the data points that can be classified sooner are more likely to be anomalies.

Such methods are, in particular, suitable for tasks with only a limited number of anomalies. The training process started with the creation of multiple decision trees with data sampled via bootstrap. For each tree, two conditions are employed at each node to generate two branches or leaves, depending on whether the condition is met. Such a condition, for example, can be whether the value of a randomly selected attribute is greater than a threshold. This process is repeated until the tree cannot be further divided. Then, in testing, a sample is tested with the trained trees, and the average depth of this sample is obtained. The samples with shallower depths are more likely to be anomalies.

iForest is only suitable for identifying global anomalies instead of local anomalies although efforts have been made to improve this aspect, such as "Improving iForest with Relative Mass."

SCiForest

SCiForest is an evolution or variation of iForest. As explained above, traditional iForest selects an attribute randomly. By contrast, SCiForest randomly chooses a hyperplane to split the data at any node and thus considers anomalies carried by several attributes at the same time. SCiForest adopts a split criterion that takes into account the standard deviation of the data at each node. Due to the changes, SCiForest is adapted to more complex data at the cost of higher complexity in algorithm and implementation. SCiForest, in general, exhibits enhanced performance compared with iForest. In addition, SCiForest targets clustered anomalies, while iForest targets scattered anomalies.

RRCF

RRCF can dynamically add and delete nodes from decision trees. This feature helps this algorithm treat anomalies in data flows.

Pros and Cons

Tree-based anomaly detection methods are advantageous in that every tree can be generated independently. This empowers such methods with the ability of distributed computing. Also, as introduced above, tree-based methods can be used to divide the dataset with any distributions instead of spherical distributions. The relatively good computational efficiency also makes online computing possible. However, such methods may only work well in tasks with a small number of anomalies. In addition, these methods may not be suitable for high-dimensional data, which may contain lots of irrelevant attributes.

The following code shows how to use a function from Scikit-learn to easily conduct anomaly detection with iForest.

```
import numpy as np
import matplotlib.pyplot as plt
from sklearn.ensemble import IsolationForest

rng = np.random.RandomState(9527)
# Generate normal data
X = 0.3 * rng.randn(100, 2)
X_train = np.r_[X + 2, X - 2]
# Generate abnormal data
X_outliers = rng.uniform(low=-4, high=4, size=(20, 2))
```

```
# Generate training data (normal and abnormal)
Data = np.vstack((X_train,X_outliers))

# Train the model
clf = IsolationForest(max_samples='auto', random_state=rng)
clf.fit(Data)
y_pred_train = clf.predict(X_train)
y_pred_outliers = clf.predict(X_outliers)

b1 = plt.scatter((y_pred_train==1)*X_train[:, 0], (y_pred_train
    ==1)*X_train[:, 1], c="green", s=20, edgecolor="k")
b2 = plt.scatter((y_pred_train==-1)*X_train[:, 0], (y_pred_train
    ==-1)*X_train[:, 1], c="red", s=20, edgecolor="k")
c1 = plt.scatter((y_pred_outliers==1)*X_outliers[:, 0], (
    y_pred_outliers==1)*X_outliers[:, 1], marker='s', c="green",
    s=60, edgecolor="k")
c2 = plt.scatter((y_pred_outliers==-1)*X_outliers[:, 0], (
    y_pred_outliers==-1)*X_outliers[:, 1], marker='x',s=60,c="r")
plt.legend( [b1, b2, c1, c2], ["Normal predicted as Normal", "
    Normal predicted as Anomaly", "Anomaly predicted as Normal","
    Anomaly predicted as Anomaly"],
loc="upper left",fontsize=10,frameon=False)
plt.tick_params(axis='both', labelsize=10)
```

The above code generates the results in Fig. 12.4.

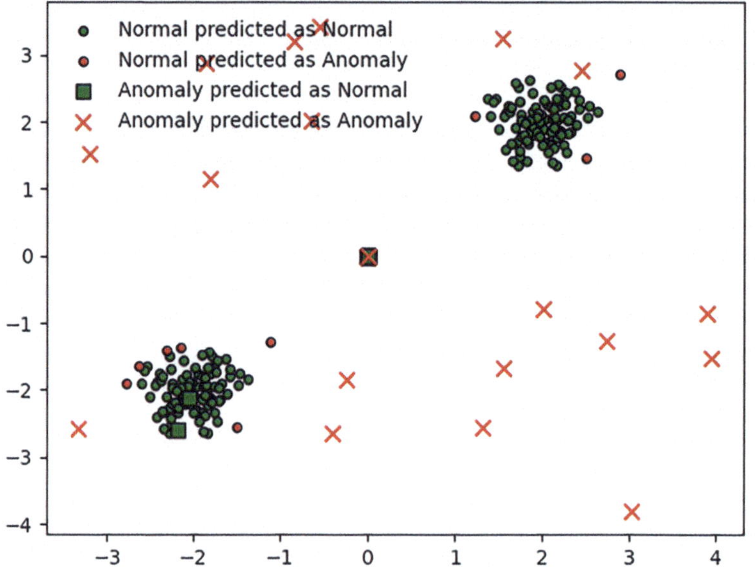

Fig. 12.4 Results of an iForest implementation

12.6 Semisupervised Learning Methods

12.6.1 Overview

The methods in this category overlap with unsupervised learning methods for anomaly detection because unsupervised methods assume that normal data samples are much more than anomalous samples. Therefore, this type of method is more suitable for datasets predominated by normal data. Brief information will be provided in the following for several popular methods in this category. Then, much more details will be provided for one of the most useful anomaly detection methods: autoencoder (or AutoEncoder).

Some common methods in this category are one-class SVM, SVDD, GMM, Naive Bayes, and AutoEncoder [143]. One-class SVM uses kernel functions to map data to a higher-dimensional space. Then, in such space, hyperplanes will be sought to achieve the maximum distance between the data and the origin. SVDD also uses kernel functions to map data to a higher-dimensional space. Then, the smallest sphere that can contain all the normal data points is sought. GMM uses normal data samples to build a model via the maximum likelihood estimation. Then, this trained model evaluates samples for anomaly detection for their probability of belonging to normal data. Naive Bayes is performed in a way similar to GMM. The model trained with normal data can give out the probability of new samples belonging to the normal data category. Autoencoder trains the data through a sequential encoder and decoder process to filter out less significant attributes. Samples whose decoding results are much different from the original sample will be determined as an anomaly.

12.6.2 Autoencoder

Introduction

Autoencoder uses a special type of artificial neural network to attract essential information from data [144]. Thus, it can be used to learn a representation (encoding) for a set of data, typically for dimensionality reduction, by training the network to ignore insignificant data ("noise"). For this purpose, we use the encoder to attract the essential information and use a decoder to convert the data with only the essential information back. In this process, as illustrated in Fig. 12.5, if the autoencoder is trained with normal data, then this trained model can retain the characteristics essential to all the normal data, and the samples that go through the process will be close to the original data. Due to the same reason, anomalies that go through such a process will generate results that are much different from the normal data.

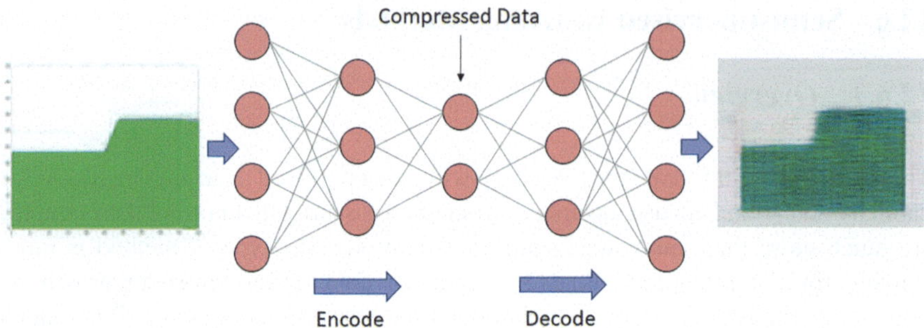

Fig. 12.5 Illustration of autoencoder

Distinct from PCA, the autoencoder gains a nonlinearity capability attributed to the use of ANNs, which helps better extract the major attributes of the normal data. This will help identify anomalies in problems with nonlinearity.

The remaining of this subsection presents code for a simple demonstration of using autoencoder for anomaly detection. In the code, an autoencoder was constructed with a simple deep neural network consisting of two parts for encoding and decoding. The MNIST dataset consisting of 50,000 training samples and 10,000 testing samples were adopted. The autoencoder was trained with the training data. 100 images out of the testing dataset were modified with random noise to serve as anomalies. The testing data was used for predictions, for which the error was obtained in terms of the sum of square root errors for each image. It can be seen that errors in the normal data and anomalies in the predicted results are much different. This indicates that the autoencoder can be used to identify the anomalies easily in this example.

Preparation: Packages and Data

First, we imported all the needed packages. A random seed was set to ensure that we would get the same results every time despite the use of random numbers.

```
import numpy as np
from keras.datasets import mnist
import ssl
ssl._create_default_https_context = ssl.
    _create_unverified_context
from keras.models import Model
from keras.layers import Dense, Input
import matplotlib.pyplot as plt
np.random.seed(9527)  # for reproducibility
```

The MNIST dataset was imported. Then, we recorded the shape of the original image for later use, which is needed because the images have to be flattened for the deep NN. 100 anomalous samples were created by randomly selecting 100 samples out of the 10,000 testing samples and then adding noise defined by a normal distribution with a mean of 0 and a standard variation of 40. Next, the data was scaled into the range between 0 and 1, which was needed for deep learning to ensure good training results and facilitate result evaluation.

```
# Load data
(x_train, _), (x_test, y_test) = mnist.load_data()
image_shape = x_train[0].shape # original image shape

# Create 100 anomalous samples (outliners) in testing data
outlier_index = np.random.randint(0,x_test.shape[0], size=100)
x_test[outlier_index] = x_test[outlier_index] + np.random.normal
    (0, 40, image_shape)

# Data preprocessing: scaling and reshaping
x_train = x_train.astype('float32') / 255.          #
    minmax_normalized
x_test = x_test.astype('float32') / 255.            #
    minmax_normalized
x_train = x_train.reshape((x_train.shape[0], -1))
x_test = x_test.reshape((x_test.shape[0], -1))
```

Model Establishment

The autoencoder model was created using deep NN. As can be seen in the following code, the NN was built layer by layer. The NN has a symmetric structure for the two parts: encoding and decoding.

```
# Built Autoencoder
input_img = Input(shape=(x_train.shape[-1],))
# Encoding
encoded = Dense(24, activation='relu')(input_img)
encoded = Dense(12, activation='relu')(encoded)
encoded = Dense(8, activation='relu')(encoded)
encoder_output = Dense(4)(encoded)
# Decoding
decoded = Dense(8, activation='relu')(encoder_output)
decoded = Dense(12, activation='relu')(decoded)
decoded = Dense(24, activation='relu')(decoded)
decoded = Dense(x_train.shape[-1])(decoded)
# Build autoencoder NN model
autoencoder = Model(inputs=input_img, outputs=decoded)
# Compile autoencoder
autoencoder.compile(loss='mse', optimizer='adam')
```

Training and Prediction

Then, we first trained the model with the training data. Next, predictions were performed with testing data, which contains the anomalies.

```
# Training
autoencoder.fit(x_train, x_train, epochs=20, batch_size=30,
    shuffle=True)

# Prediction
autoencoded_imgs = autoencoder.predict(x_test)
```

Result Evaluation and Visualization

The original data ("x_test") and the decoded data ("autoencoded_imgs") were compared to obtain the error. Then, we split the data into normal samples and anomalies for plotting.

```
Error = np.sum((x_test-autoencoded_imgs)**2,axis=1) # Sum of
    squre root errors
Normal_data = np.delete(Error,outlier_index) # Normal data
Outliners = Error[outlier_index] # Anomalies
plt.scatter(np.delete(np.arange(Error.size),outlier_index),
    Normal_data,c='g',label='Normal data')
plt.scatter(outlier_index,Error[outlier_index],c='r',label='
    Anomalies')
plt.legend(loc="middle right",fontsize=10,frameon=False)
plt.xlabel('Sample number')
plt.ylabel('Error')
```

As shown in Fig. 12.6, the errors or anomalies are much different from those of the normal samples. Thus, when new data that is much different from the normal data (used for training this autoencoder model) is processed, the autoencoder will tend to give significantly different results. These differences can be used to differentiate anomalies from normal data.

12.7 Anomaly Detection Issues

12.7.1 Data Quality

Data quality is critical to the anomaly detection practice. In particular, the selection of anomaly detection algorithms relies on the nature and quality of the data. This is because the data, to a great extent, defines what anomaly detection problem to address. The quality of the dataset is a major driver in developing accurate and usable anomaly detection models.

Common data quality issues in anomaly detection applications are listed below:

- Missing data or incomplete datasets
- Inconsistent data including formats, types, scales, etc.
- Duplicate data
- Erroneous data, including those caused by humans

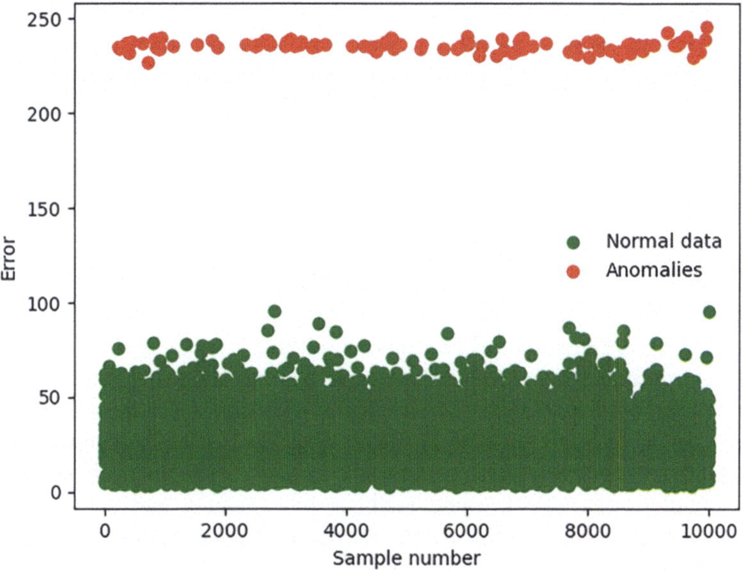

Fig. 12.6 Results for using autoencoder for anomaly detection

Common useful practices for addressing the above data quality issues in anomaly detection include but are not limited to the following ones:

- Missing or null values can be discarded or filled with the most possible values. Such values can be obtained using interpolation, mean values, or others.
- All the data can be checked for format, consistency, and completeness. Computer code for manual checks and adjustments can be employed to standardize all the data.
- Remove duplicate data based on unique sample identities or tags like time, sequence order, and ID.
- Check the scales and units of all data as well as their compatibility with the selected anomaly detection algorithms. Perform rescaling or normalization if needed.
- Observe and improve the data generated by human experts. If not possible, then try to reduce the dependency on such data.

12.7.2 Imbalanced Distributions

The imbalanced distribution of the samples is a common issue in anomaly detection. This issue is especially obvious when we use classification algorithms with labeled data for anomaly detection, which can easily suffer from the class imbalance issue.

For example, when the training set contains a very small number of anomaly samples, the classification model may yield very good accuracy. However, such a model may still be useless because the accuracy is primarily contributed by the predominant normal samples while the model's consideration of anomalies may be far from adequate.

Such issues can be addressed from different angles including data, model, and model evaluation metrics:

- Enlarge the dataset if possible.
- Over-sampling addresses class imbalance by randomly copying/duplicating minority samples (may correspond to anomalies). This technique does not abandon any information and can cause overfitting.
- Under-sampling addresses the imbalance issue by randomly removing sampling from the predominant classes. It can lose some information during the process but can improve computational efficiency and reduce memory requirements.
- Algorithms can be changed or adapted to better consider imbalanced data. In particular, ensemble learning methods like bagging and boosting can be considered. Both of them involve the resampling of data to improve the model, which better considers anomaly samples in the process.
- Evaluation metrics such as F-score and precision sometimes are better than the accuracy for considering imbalanced data.

12.7.3 High-Dimensional Data

The detection of anomalies in large datasets with high dimensionality issues has been a pressing challenge and an active research area in anomaly detection. High dimensionality not only complicates the volume, variety, velocity, and veracity of the data but also increases the levels of difficulties for the anomaly detection job. First, the increasing dimensionality can lead to data sparsity, which is difficult to analyze. Second, the insignificant or unnecessary attributes brought by the increasing dimensions can obscure or even conceal the nature of the anomalous data. Also, since outliers are defined as data instances in sparse regions, an inadequate discriminative region can be observed in almost equally sparse locations of high-dimensional space.

The increase in dimensions makes data points scattered and isolated and makes it elusive to find the global optima of the dataset. The more dimensions that are added to a dataset, the more complex it becomes, as each added dimension brings about a substantial number of false positives. The curse of dimensionality refers to a number of occurrences that emerge when analyzing and organizing data in high-dimensional spaces. The very nature of these occurrences is that, whenever there is an increase in dimensionality, the volume of the space increases proportionally, such that all other data points become sparse. This sparsity is challenging for any technique that requires a statistical value. Further, it produces numerous complications associated with other noise levels, which may be irrelevant features or unnecessary attributes. These could complicate or even conceal the data instances. This is the main reason why many algorithms struggle with high-dimensional data. Also, as the number of dimensions increases, statistical approaches such as distance measures become less useful, because the points become almost equidistant from each other due to the curse of dimensionality.

The most common strategy to address the problem of high dimensionality is to remove less significant or unnecessary dimensions (i.e., variables or attributes). This can be done by identifying the most important dimensions (a subset of features), i.e., dimension selection, or combining dimensions into a smaller set of new variables, i.e., dimensionality reduction. Many dimensionality reduction methods, such as PCA, MDS, Laplacian Eigenmaps, and diffusion maps can be used for this purpose. Statistical models, including proximity-based, cluster-based, distance-based, density-based, and classification-based models, have also been used to tackle the dimensionality problem, though such methods bring greater computational complexity as the number of dimensions increases.

12.7.4 Model Sensitivity

Model sensitivity is another issue that may need to be dealt with in anomaly detection practice. You can imagine the scenarios and consequences when false alerting constantly pops out, which may be desired in some cases but can seriously hurt users' confidence in the model in some other scenarios. You can imagine this as setting the threshold or bar for triggering an alarm. The safest way could be to make sure the anomaly detection perfectly describes the data, especially the anomalous data. However, we may need to move the bar above or below the threshold depending on our preference for different components in the confusion matrix such as false positives.

The key is the model sensitivity. For example, it may be common for a normal variance to deviate away from that baseline when the limits are too tight around the baseline. As for the false alerting issue, we can consider the following strategies:

- Enlarge the dataset to obtain better models.
- Assess the preference for different types of predictions and adjust the limits based on the preference.
- If there are too many false alerts, widen the limits by increasing the confidence interval. For example, we can adjust the alert to look for a few times (like twice) of the original limit.

12.8 Practice: Implement Typical Anomaly Detection Methods

>>> More and up-to-date course materials including practices @ AI-engineer.org <<<

1. Please develop code for three sigma, Z-score, and boxplot based on the code given in the book. Compare the results for the data generated in the following way:

```
data = np.random.randn(50000) * 20 + 20
```

2. Please modify the code for the autoencoder to process CIFAR-10 dataset. Also, find a different way to create anomalies and see how that will change the differences between the predictions for normal data and anomalies.

Chapter 13
Association Rule Learning

13.1 Overview

This chapter introduces association rule learning, a machine learning topic that could appear much different from others. Thus, to flatten the learning curve, we will first go through the definition, data format, scope, and history of association rule learning using language that can be easily understood. Next, essential concepts of association rules will be laid down in a more technical way to provide the knowledge needed for understanding representative algorithms. Based on that, three popular association rule learning algorithms, i.e., Apriori, FP Growth, and Eclat, will be explained. The procedures of performing association rule learning using these three algorithms will be detailed as we work through a typical problem using each of them.

13.2 Basics of Association Rule Learning

13.2.1 Definition

Association rule learning, which is also called association rule analysis and association rule mining in many publications [145], is a rule-based type of machine learning for discovering interesting relations between variables. This definition is not that straightforward to understand, especially considering the vague meanings of terms like "relations between variables," "rule," and "interesting." Also, association rule learning and its algorithms (or methods) contain various new parameters/concepts, such as the ones embedded in the definition. This fact makes association rule learning difficult for learners without expertise in data mining to understand the topic and implement the learning algorithms.

Z. "L." Liu, *Artificial Intelligence for Engineers*,
https://doi.org/10.1007/978-3-031-75953-6_13

In addition to these parameters and concepts, association rule learning treats data that is usually formulated in a format slightly different from data processed in other machine learning topics due to historical and practical reasons. Therefore, association rule learning may appear much different from other supervised and unsupervised machine learning topics in many aspects, which can further confuse learners.

In the remainder of this section, instead of introducing more concepts, definitions, and applications of association rule learning, we will start with the relationships between association rule learning and the machine learning methods we learned before from the perspective of data format. A brief on the history and application of association rule learning will be presented to help explain this machine learning topic for general readers. A more strict definition and explanation of this category of machine learning methods will be presented in the next section, and detailed introductions to typical methods will be provided after that.

13.2.2 Relationships with Other Machine Learning Topics

Let us first take a look at data. This is because association rule learning has a stronger data analysis favor than other machine learning topics. As we know, machine learning's application for data analysis differentiates itself from general data mining in light of automated analytical model establishment and less emphasis on analyzing the relationships within the data. In particular, in data analysis, the search for hidden relationships among data, which can also be interpreted as identifying patterns, revealing interdependency, or finding similarities, is a primary focus. This may make the appearance and nature of the data and hidden relationships to be searched for heavily dependent on the methods, especially the tasks for which such methods were developed. This is particularly obvious as we move from regular machine learning topics to association rules.

Let us take a look at the format of the data that we mostly deal with in previously introduced supervised and unsupervised learning. Such data is usually presented as an array consisting of I data points (instance), and each data point has J attributes (dimensions):

$$\bar{X} = \left[\vec{x}_1^T \ \vec{x}_2^T \ \cdots \vec{x}_I^T\right]_{1\times J}^T = \begin{bmatrix} x_{11} \ x_{12} \cdots x_{1J} \\ \vdots \quad \ddots \quad \vdots \\ x_{I1} \ x_{I2} \cdots x_{IJ} \end{bmatrix}_{I\times J} \tag{13.1}$$

where a data point $\vec{x}_i$ is a column array with J attributes as $\left[x_{i1}, x_{i2}, \cdots, x_{iJ}\right]^T$. From the perspective of statistics, each column in $\bar{X}$ is an attribute/feature and can be viewed as a random variable. In such unlabeled data represented by $\bar{X}$, there are two directions: breadth direction (i.e., horizontal in $\bar{X}$) that enumerates different attributes or features and depth direction (i.e., vertical in $\bar{X}$) that allows us to list different instances or data points.

Supervised machine learning methods, for classification or regression tasks, are mostly developed to find the relationships along the depth direction, i.e., between instances. For example, classification methods aim to find whether certain instances are more alike based on their attributes and labels so that they can be counted as one class. Regression methods try to quantify the instances based on their attributes with number(s), so that relationships between data points can be better correlated with the numeric labels and be evaluated mathematically.

In unsupervised learning, clustering is also developed to find similarities between instances, i.e., depth direction, based on similarities inferred from instances' attributes. By contrast, dimensionality reduction in unsupervised learning targets the breadth direction and tries to reduce the number of features, or the so-called dimensionality, based on the characteristics of the features.

Association rule learning is a type of unsupervised learning method. It is more like dimensionality reduction in that it focuses on the breadth direction. However, it differs from dimensionality reduction in a few obvious ways. First, dimensionality is intended to find the major dimensions or components as combinations/projections of element dimensions for dimension reduction purposes, while association rule analysis is developed for learning relations or associations between dimensions. More importantly, association rule analysis adopts a data format that is much different from that of other machine learning methods.

Figure 13.1 shows the tabular data format, which is used in most machine learning topics, and the transactional data format, which is adopted in association rule analysis.

Tabular Data

TID	A (Alligator)	B (Block)	C (Linear Crack)	D (Depression)	E (Edge)
1	0.3	0	0.2	0.1	0
2	0	0.2	0.1	0	0.2
3	0.2	0.3	0.1	0	0.05
4	0	0.25	0	0	0.15

Transactional Data

TID	A (Alligator)	B (Block)	C (Linear Crack)	D (Depression)	E (Edge)
1	1	0	1	1	0
2	0	1	1	0	1
3	1	1	1	0	1
4	0	1	0	0	1

Fig. 13.1 Engineering examples (pavement cracks) for regular tabular data (percentage of coverage) and tabular data of association rule learning (occurrence)

We can also easily observe another significant difference: the attribute values are all binary. In detail, these attributes of the data in association rule analysis all refer to the occurrence of an event: **an "item" represented by the attribute** appears (or occurs) or not. Using statistics language, this can also be described using a variable, which can be 0 or 1 with certain probabilities. Of course, one variable (attribute) and the other variable(s) may have relationships, e.g., the likelihood it will happen depends on the other one. Association rule analysis is developed to find such relationships, especially, whether the two events associated with some variables will happen together or not. Events that are most likely to happen together are of the utmost interest.

13.2.3 Understanding via History

The fact that association rule analysis takes data in this format as input stems from historical and practical reasons. While the concepts behind association rules can be traced back to much earlier applications, association rule analysis was mostly thought to be defined in the 1990s. A milestone was set when computer scientists Agrawal et al. developed an algorithm-based way to find relationships between items using point-of-sale (POS) systems, which generated transactional data as illustrated above [146]. This algorithm, which is of a machine learning nature, could discover links between different items purchased. This algorithm is mathematically defined as association rules and developed for predicting the likelihood of different products being purchased together.

For retailers, association rule analysis offers a way to better understand customer purchase behaviors. Such association rule analysis is often referred to as market basket analysis. We can understand it by looking at an example of a supermarket. In a supermarket, all products that are purchased together are put together. For example, if a customer buys bread, he most likely can also buy butter, eggs, or milk, so these products are stored on a shelf or mostly nearby. This retail origin explains why the data treated in association rule analysis may also take the transactional data, in which a separate record is made for each transaction or item instead of a complete set of associated items. It also explains why the feature/attribute is called an item and has a binary nature.

In addition to market basket analysis, association rule analysis has also been widely applied in web usage mining, continuous production, medical diagnosis, protein sequences, and census data. In engineering, association rule learning has great potential in intelligent transportation systems, construction engineering (building information modeling), and performance analysis in various learning and testing scenarios.

13.3 Essential Concepts of Association Rules

In this section, essential concepts for understanding, formulating, and implementing association rule analysis will be presented. First, we will recall the data format to introduce several key concepts that are widely used in common association rule analysis algorithms. Next, the "rule" in the definition of association rule analysis, which is a key and kind of hard to understand, will be explained. Next, we will be checking the ways of measuring the interestingness or significance of rules, which determine which associations are more predominant and will be discovered.

13.3.1 Items, Itemsets, and Rules

First, let us take a second look at the data. As pointed out in the previous section, **the feature/attribute/dimension is now mostly referred to as an "item"** in association rule analysis due to its retail origin. In a transactional dataset, each row is written as TID: itemset, in which TID is a unique identifier that we assign to the transaction record for the row and a number recording the sequential position of the transaction in a day. It is noteworthy that in some association rule analysis methods a slightly different data format, such as {item: TIDs}, may be used, which will be described when introducing such methods and thus not treated as a widely accepted case here.

Besides, association rule mining, at a basic level, involves the use of machine learning models to analyze patterns in data, especially co-occurrences, in a dataset. Therefore, the coexistence of two or more items in a dataset is of special interest. This prompts us to use the concept of k-itemset: an itemset that contains k items. For example, {Bread, Milk} is a 2-itemset. In many cases, the major work of association rule analysis is to find the most interesting k-itemset. In general, k-itemset with large k values are more interesting, because they represent higher-level associations.

In order to better discuss association, let us first take a look at the concept of "rules" in association rule analysis. The "rule" in association rule analysis refers to conditional relationships like "If...Then.." (Fig. 13.2). Here, the if-element is called the antecedent, and the then-statement is called the consequent. To better define to rule, we can describe the problem in the following way.

Let $F = \{f_1, f_2, \cdots, f_J\}$ be a set of J binary attributes/features called items.

Let $D = \{t_1, t_2, \cdots, t_I\}$ be a set of transactions called the dataset.

Each transaction in D has a unique transaction ID and contains a subset of the items in F.

Fig. 13.2 Schematic of if-then rule

Then, a rule in association rule analysis or, more specifically, an if-then rule is defined mathematically as

$$A \Rightarrow B, \text{ where } A, B \subseteq F. \tag{13.2}$$

where A and B are two items ($k = 1$) or itemsets ($k \geqslant 1$), whose association is to be analyzed. As can be seen, a rule consists of two different items (or itemsets). In the definition of the rule, A is called antecedent or left-hand side (LHS), and B is called consequent or right-hand side (RHS). The statement $A \Rightarrow B$ is often read as "if A then B," where the antecedent A is the "if" and the consequent B is the "then." This simply implies that, in theory, whenever A occurs in a dataset, B occurs as well. In a simple scenario, the goal of association rule mining can be the search for rules that will predict the occurrence of an item (Item B) based on the occurrence of other items (Item f_j) in the transaction.

13.3.2 Support, Confidence, and Lift

Associations described with such rules are usually not deterministic, that is, they are associated with probabilities. In association rule analysis, we look for the most interesting rules or rules that can best serve our practical purposes. Therefore, we need additional metrics or parameters to measure the interestingness of the rules so that we can find the rules with the best or acceptable interestingness. Among such criteria, **support** and **confidence** are the two most widely used metrics [147, 148]. **Lift** is relatively less popular but still adopted in many association rule analysis algorithms [148]. **Conviction** has also been used in some cases [149]. The meaning and definition of these four parameters will be given in the following, and a more in-depth explanation of the adoption of them will be presented afterward.

Support is a quantitative indicator of how frequently an itemset appears in the dataset. Therefore, it is the frequency of occurrence of an itemset. Then, Support(A, B) is the fraction of transactions that encompass both A and B:

$$\text{Support}(A, B) = P(A \cap B) = \frac{\text{number of transactions containing } A \text{ and } B}{\text{total number of transactions}} \tag{13.3}$$

The above is the support of the itemset $\{A,B\}$. In the example illustrated in Fig. 13.1, support for {Aligator, Block} $= 1/4 = 25\%$, it means that 25% of the transactions contain the itemset {Aligator, Block}. Note that A and B in the above equation are different from the crack abbreviations in Fig. 13.1 (i.e., A and B).

Confidence indicates the frequency of the if-then statements is found to be true. It measures how frequent the items in B appear in transactions that consist of A:

$$\text{Confidence}(A \Rightarrow B) = P(B|A) = \frac{P(A \cap B)}{P(A)}$$
$$= \frac{\text{number of transactions containing } A \text{ and } B}{\text{number of transactions containing } A} \quad (13.4)$$

As can be seen, the confidence of the association rule $A \Rightarrow B$ can also be interpreted as an estimate of the conditional probability $P(B|A)$. In the abovementioned example, confidence for Block $\Rightarrow$ Alligator $= 2 / 3 = 66\%$ means that if a block crack appears, then the probability of alligator crack presence is 66%.

Support and confidence measure how interesting a rule is. Support is an important measure because a rule that has very low support may occur simply by chance. A low support rule is also likely to be uninteresting from a business perspective because it may not be profitable to promote items that customers seldom buy together. For these reasons, support is often used to eliminate uninteresting rules. Support is also used for the efficient discovery of association rules. Confidence, on the other hand, measures the reliability of the inference made by a rule. For a given rule $A \Rightarrow B$, the higher the confidence, the more likely that B is present in transactions containing A. Confidence also provides an estimate of the conditional probability of B given A.

Association rules suggest a strong co-occurrence relationship between items and do not necessarily imply causality. For example, a rule may show a strong correlation in a dataset because it appears very often, but it may occur far less when applied. This would be a case of high support and low confidence. Conversely, a rule might not particularly stand out in a dataset, but continued analysis could show that it occurs very frequently. This would be a case of high confidence and low support. Using these measures helps analysts separate causation from correlation and allows them to properly value a given rule.

A third metric, lift, is defined as the ratio of confidence to support. This metric can be used to compare confidence to expected confidence or assess how frequent an if-then statement is expected to be found true. It is the probability that B appears when A shows up considering the popularity of B:

$$\text{Lift}(A \Rightarrow B) = \frac{P(B|A)}{P(B)} = \frac{P(A \cap B)}{P(A)P(B)} \quad (13.5)$$

The lift of the association rule of $A \Rightarrow B$ can also be understood as how the occurrence of A can contribute to the occurrence of B. If the lift is greater than 1, then the probability of the B occurring under the conditions of A is greater than the probability that B occurs. Therefore, the presence of A has a positive effect on B. The degree of such "lift" effect is given by the value of the parameter lift. If the rule has a lift of 1, then A and B are independent, and no rule can be derived from them. If the lift is lower than 1, then the presence of A will have a negative effect on B.

To use these metrics to discover rules of interest from data, we will also need to define thresholds for these parameters to select rules. This is because if the rules are built by analyzing all the possible itemsets from the data, then there would be so many rules that many of them would not have any meaning. Also, the number of “rules” will grow exponentially as the size of the dataset increases. Therefore, it is necessary to define thresholds to find those rules that are well represented by the data.

13.3.3 Association Rule Analysis Using the Concepts

Among these metrics, support and confidence are more predominantly used in association rule learning. Often, we have to satisfy a user-specified minimum support and a user-specified minimum confidence at the same time. While creating association rules, we can set minimum thresholds for support and confidence to filter out some non-interesting rules and keep only frequent itemsets and strong rules. A frequent itemset is an itemset whose support is greater than or equal to a minimum support threshold. A strong rule is a rule whose support and confidence satisfy their threshold values. Therefore, a general goal of association rule analysis is to find all association rules having **support** $\geqslant$ **support threshold** and **confidence** $\geqslant$ **confidence threshold**.

Usually, the association rule generation is split into two different steps that need to be applied:

(1) A minimum support threshold to find all the frequent itemsets in the database.
(2) A minimum confidence threshold for the frequent itemsets to create rules.

The above concepts of association rule analysis also bring some issues to its implementations. First, the number of rules to be assessed can be large. Second, it may be difficult to guarantee that the discovered rules are relevant or interesting. These two factors can lead to poor performance of association rule learning algorithms. Sometimes, some algorithms may contain too many variables and parameters, which also causes issues with understanding.

A trick we can use to reduce the number of rules to keep the problem tractable is the Apriori principle. That is, if a rule does not meet the minimum support/confidence requirement, then any superset of that rule will not meet the requirement and thus do not need to be considered. In terms of support, if any k-itemset does not meet the minimum support level, then any $k + 1$ or higher itemsets containing this itemset will not have enough support. In the consideration of confidence, for example, if the rule {A,B,C} $\Rightarrow$ {D} fails to meet the minimum confidence, then any relevant rule in which the left-hand side is a superset of {A,B,C} will not meet the requirement.

In addition, it could be tricky to find appropriate parameters and threshold settings for implementing algorithms.

13.4 Apriori

Apriori algorithm [150] is one of the most powerful algorithms used for data extraction and it is usually the first-introduced association rule analysis algorithm in a majority of the association rule learning literature. "Apriori" is used as the name because it uses prior knowledge of frequent itemset properties. It mainly aims at finding frequent itemsets and related association rules.

13.4.1 Procedure

It is a "bottom-up" approach, in which the search for frequent itemsets starts from k-itemsets and k increases as frequent subsets are extended one item at a time. This is a step called candidate generation. The generated groups of candidates are tested against preset criteria to find k-itemsets that meet the support criteria (for strong rules) and have the highest k. Apriori uses breadth-first searches, in which "breadth" refers to the breadth of the Hash tree structure that Apriori uses to count candidate itemsets and is slightly related to the previously introduced direction of the data but not the same thing.

For implementation, we will need to set a support threshold for filtering out k-itemsets that are not of interest. Then, we will repeat a process consisting of counting (scanning), filtering (trimming), and extension (forming) for k-itemsets, from 1-itemsets to the maximum k-itemsets. In this process, the first step is to scan all candidate k-itemset. The second step is to filter out k-itemset that does not meet the minimum support criterion to trim the candidate k-itemsets. The third step is to extend the current k-itemsets by adding one item that still exists in the trimmed candidate group of k-itemsets, and all possible $k+1$-itemsets will form the itemsets for canning in the next iteration of the same process. Thus, the first iteration will start from all the 1-itemsets and then collect all 1-itemsets whose support is higher than the support threshold.

13.4.2 Implementation with an Example

Figure 13.3 presents an example to illustrate the implementation of the Apriori algorithm. As shown, a support threshold of 50% is set before we start. Let us still work on the dataset given in Fig. 13.1, which has four transactions (four TIDs) and five items. In the first iteration, we count all the 1-itemsets as the first step. In the second step of the first iteration, the support values of all the 1-itemset are calculated. We can find that Item 4, i.e., D, is filtered out because it is the only candidate itemset that does not satisfy the minimum support criteria: support $\geqslant$

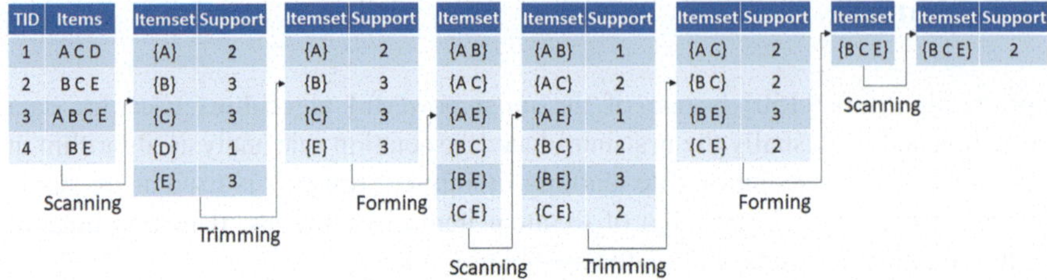

Fig. 13.3 Illustration of the implementation of Apriori with an example

50%. A trimmed candidate group of itemsets is formed as the result of the second step. In the third step, the four itemsets in the trimmed group of candidate itemsets are extended to 2-itemsets with items from the trimmed group of itemsets, leading to all the possible extended itemsets as the candidate itemsets for the next iteration. The same process is repeated for 2-itemsets until we obtain a 3-itemset {B,C,E}, which cannot be further extended to find 4-itemset. The process ends here, and we find the rules that we want in this way.

13.4.3 *Pros and Cons*

Apriori is a very classic association rule analysis algorithm. Apriori is advantageous in that its idea and implementation are easy to understand and it can also be implemented on large datasets. However, one major disadvantage needs to be recognized: it could be computationally expensive when a large number of candidate rules (candidate k-itemsets) need to be searched and their support values need to be evaluated.

Many other algorithms for association rule analysis were developed based on it, such as FP-Tree, which will be introduced in the next section. These algorithms inherit many concepts from Apriori and improve Apriori in some aspects. Direct utilization of Apriori in real-world data mining applications is not common anymore. However, it is still an algorithm worth learning to gain a good understanding of association rule analysis.

13.5 FP Growth

Frequent Pattern Growth (FP Growth) is developed based on Apriori [151]. Apriori can be computationally demanding because it needs to repeatedly scan the dataset to generate lists of k-itemsets. FP Growth adopts a treelike data structure, which can help store conditional patterns to save effort at multiple scanning processes.

13.5.1 Procedure

The procedure of performing the FP Growth algorithm consists of the following steps:

(1) Scanning and Filtering (as Apriori): Scan the transaction data and obtain the frequencies of all 1-itemsets (essentially items). Then, remove all the 1-itemset with a support value lower than the threshold.
(2) Header Table: Create a header table by sorting the remaining 1-itemsets based on their support values.
(3) Trimming and Sorting: Remove items without adequate support from the original transaction data (not header table), and sort the items in individual transactions based on the items' support values (from Step 2).
(4) FP Tree: Generate an FP tree by adding different transactions from the sorted data one by one. We start from the root node (marked as "null" or Ø) and then add nodes corresponding to different items according to their order in the transaction.
(5) Conditional Pattern Base: Search for the conditional pattern bases for different items in the header table from top to bottom. Mine the conditional pattern base recursively for frequent itemsets.
(6) Rule: Finally, we can discover rules based on the statistics of the frequent itemsets.

The above procedure is not easy to understand. Considering this difficulty, these steps will be explained in the following subsections using the simple example given in the previous section. This example shows the different observations of pavement cracks. Items $A \sim E$ represent five typical types of cracks. Association rule analysis can be employed the explore the correlation between the occurrences of different types of cracks from the observations.

13.5.2 Item Header Table

Let us work on the first three steps centered around the creation of the header table.

First, we will need to scan the original data as shown in the left part of Fig. 13.4 to obtain the frequencies of occurrence of all the items. This will generate a list of items, which can be also viewed as 1-itemsets and their corresponding frequencies. Next, we will sort these items based on frequencies: from more frequent items to less frequent ones. This is the scanning and sorting of the items in the original data.

Then, items with support values that are less than the threshold value will be removed. In the example, we use a support threshold of 50%. Thus, any item that appears less than 2 will be eliminated. In this example, Item D only appears once, which corresponds to a support of $1/4 = 25\% < 0.5\%$, and thus needs to be removed. The remaining items in the preferred order will be put together as the

Fig. 13.4 Creation of header table: support $\geqslant 2$

Data

TID	Items
1	A C D
2	B C E
3	A B C E
4	B E

Header Table

Itemset	Support
{B}	3
{C}	3
{E}	3
{A}	2

Sorted Data

TID	Items
1	C A
2	B C E
3	B C E A
4	B E

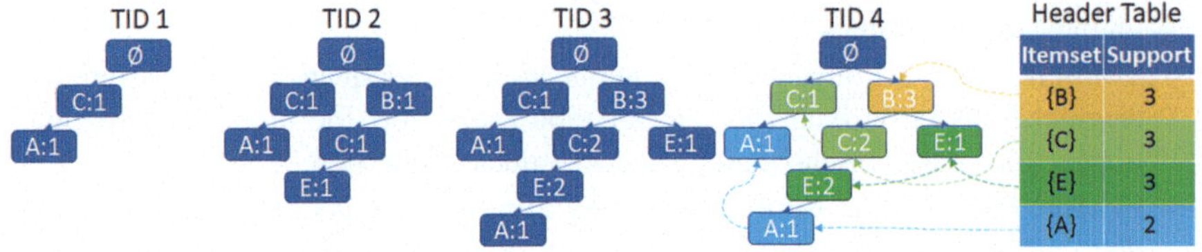

Fig. 13.5 Generation of the FP tree

header table in the middle subplot of Fig. 13.4. Now, the remaining 1-itemsets, i.e., {B}, {C}, {E}, and {A}, and their frequencies form the header table.

Next, we will sort the original data based on the order of the items in the header table. For this purpose, the items in individual transactions will be filtered and sorted. Tanking TID 1, for example, Item D will be removed because it is excluded from the header table, and then the remaining Item A and Item C will be sorted according to the order in the header table: C A.

The header table will be used in the later process for creating the FP tree. We will see that the 1-itemsets sorted in the order of support also help build up the FP tree. In particular, those more frequent items will appear close to the root node, and consequently, they would be shared by nodes corresponding to less frequent items, which appear as descendant nodes in the tree.

13.5.3 FP Tree

The FP tree is generated step by step, and each step is associated with one transaction. We can start with the first transaction in the sorted (and trimmed) transaction data: TID 1: {C A}. The number in each node indicates the frequency of that node. This will create a branch as shown in Fig. 13.5. Because the tree has no existing nodes except the root node, so we just create new nodes for each item.

Then, we work on TID 2: {B C E}. The first node, B, is different from the previous branch, so we will need to create a new branch (or sub-tree). The situation TID 3: {B C E A} is different: the first three nodes are the same as those in the branch created with TID 2. Thus, we will start the existing branch and branch out only new nodes, e.g., A in TID 3 appears. In this process, because the nodes {B C E} appear twice, i.e., TID 2 and TID 3, so we will increase the frequencies in the corresponding nodes as TID 3 is added. TID 4: {B E} can follow the first node B in the right branch, whose frequency will increase by 1 to allow for the addition of Item B from TID 4. Then, Item E will require us the create a new leaf node.

The creation of more complicated FP trees can follow the same procedure. For illustration purposes, we can also mark the different items in different colors and link the nodes corresponding to the same item with lines. This can help us better visualize the connection between the header table and the FP tree and will help us understand the mining of the FP tree for frequent items in the next subsection.

13.5.4 Mining FP Tree for Frequent Itemsets

As mentioned, the FP tree serves as a special data structure that can help us mine frequent itemsets without repeatedly scanning the dataset. This mining process starts from the bottom of the header table, which corresponds to the leaf nodes of the tree, especially those with a long path. Thus, we will mine the tree from Nodes A, E, C, and B sequentially. Usually, we do not need to try all the items, because the last items may correspond to the nodes close to the root node and thus can bring short k-itemsets.

For each item that we consider, we will need to identify the conditional pattern base. This conditional pattern base, which appears as a condensed segment of the header table, contains information that can help us easily obtain frequent itemsets made of items in it.

In the above example, we start with node "A." The conditional pattern base of Item A can be viewed as the sub-tree that uses "A" as the leaf node. In the above example, there are two "A" leaf nodes. These two "A" leaf nodes share no common nodes except the root node, so they will be considered separately. In the left branch ending with leaf node "A," there is only one item, so this sub-tree is two short and cannot provide information about frequent itemsets (1-itemset only). Thus, we do not need to consider this sub-tree.

The right branch ending with leaf node "A" presents a sub-tree with three nodes (excluding the leaf node "A"). Also, the non-leaf node in this leaf sub-tree has a frequency of 2. As a result, we obtain a conditional pattern base written as {B:2, C:2, E:2} and is illustrated in Fig. 13.6. This conditional pattern base tells that we have 2-itemsets: {B:2, C:2}, {C:2, E:2}, {B:2, E:2}, {B:1, A:1}, {C:1, A:1}, and {E:1, A:1}; 3-itemsets: {B:2, C:2, E:2}, {B:1, C:1, A:1}, and {B:1, E:1, A:1}; 4-itemsets {B:1, C:1, E:1, A:1}. The number "1" means 1 of such sets can be identified from the conditional pattern sets. You can see the convenience of using the conditional pattern base as a useful structural feature of the FP tree for easily storing and presenting information about frequent itemsets.

Let us continue the above mining process. The next item we should check is Item E. As shown in Fig. 13.7, there are two E nodes, in which the left "E" node will

A

Ø

C:1 B:3

A:1 C:2 E:1

E:2

A:1

Conditional Pattern Base

Itemset	Support
{B}	2
{C}	2
{E}	2

Fig. 13.6 Use of FP tree for association rule learning for Item A

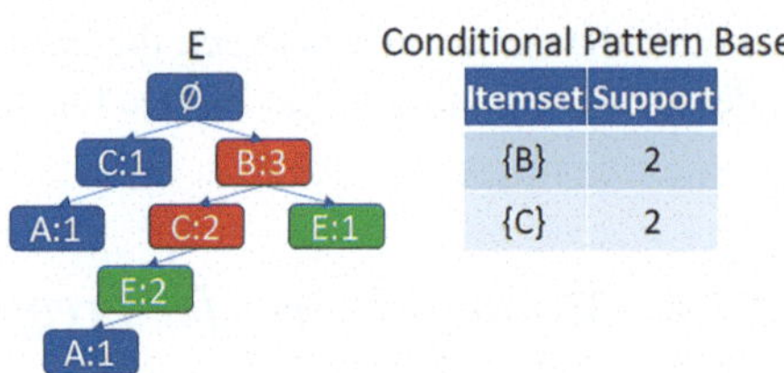

Fig. 13.7 Use of FP tree for association rule learning for Item E

also be treated as a leaf node in this case as the right A node has been considered. Because the left E node represents a longer path, so we will keep it. Then, the conditional pattern base for Item E is {B:2, C:2}, in which "2" comes from the left E node: E:2. Using this conditional pattern base, we can obtain 2-itemsets {B:2, E:2}, {C:2, E:2} and 3-itemsets {B:2, C:2, E:2}.

In the above mining process or Item E, if the C:2 node was C:1, then we need to treat it in a different way. Under this condition, the non-leaf node C has a support lower than the threshold, so we will need to remove his node. Then, we have two "E" nodes left under node "B." So, the conditional pattern base for node "E" will be {B:3}, where 3 comes from the two "E" nodes: E:2 and E:1, as $1 + 2 = 3$.

We can continue the mining process. However, in this example, the mining for Item C and Item B will not give us any frequent itemsets longer than 1, so we will stop the mining process.

13.6 Eclat

The Eclat algorithm [152] is possibly the most popular and powerful scheme for association rule mining. The full name of Eclat is equivalence class clustering and bottom-up lattice traversa. Different from Apriori and FP Growth, Eclat utilizes depth-first searches, which makes it run faster than Apriori. Also, set intersections are used to calculate the support of the candidate items while avoiding the generation of subdivisions that are not present in the prefix tree.

Fig. 13.8 Two data formats: horizontal in Apriori and FP Growth and vertical in Eclat

Regular (horizontal)

TID	A	B	C	D	E
1	1	0	1	1	0
2	0	1	1	0	1
3	1	1	1	0	1
4	0	1	0	0	1

Eclat (vertical)

Item	TID Set
A	{1,3}
B	{2,3,4}
C	{1,2,3}
D	{1}
E	{2,3,4}

13.6.1 Procedure

It is not difficult to find out that the above introduction contains a key in the distinction between Eclat and Apriori and its variations: depth first vs. breath first. This distinction is actually associated with the different data formats adopted by them. As shown in Fig. 13.8, Apriori and P-growth use TID or horizontal (alignment of items/features) data format, i.e., {TID: itemset} for mining frequent itemsets. On the contrary, Eclat adopts the vertical data format, i.e., {item: TID_set}. This employment of the vertical data format allows Eclat to function in a much different way.

The typical procedure for performing association rule analysis with Eclat is as follows:

(1) Scan the data and convert the data format from horizontal to vertical.
(2) The length of a TID_set reflects the support (more accurately, it is the frequency) of the item. Then, remove items that do not have enough support.
(3) Starting from $k = 1$, construct $(k + 1)$-itemsets from k-itemsets by obtaining the intersections of the TID sets of the corresponding itemsets.

The last step is not easy to understand without an example. It will be further explained in the following subsection.

13.6.2 Implementation

Let us check how to apply Eclat for association rule mining. We will be mining vertically formatted data for frequent itemsets using the same example.

First, we converted the original data from the horizontal format to the vertical format. This generated data in the leftmost subplot in Fig. 13.9. This data also includes Item D, which has a support lower than 50% (or a frequency of 2). So, we removed Item D. The product is the 1-itemset that we need. Again, the length of

1-itemset

Item	TID Set
A	{1,3}
B	{2,3,4}
C	{1,2,3}
D	{1}
E	{2,3,4}

2-itemset

Item	TID Set
{A,B}	{3}
{A,C}	{1,3}
{A,D}	{1}
{A,E}	{3}
{B,C}	{2,3}
{B,E}	{2,3,4}
{C,D}	{1}
{C,E}	{2,3}

3-itemset

Item	TID Set
{A,B,C}	{3}
{A,B,E}	{3}
{A,C,E}	{3}
{B,C,E}	{2,3}

4-itemset

Item	TID Set
{A,B,C,E}	{3}

Fig. 13.9 Identification of k-itemsets with Eclat

the TID set is the frequency. Thus, the TID set of Item A has a length of 2, so this 1-itemset {A} has a frequency of 2.

Next, we create 2-itemsets based on the 1-itemsets that we just obtained. As shown in the second subplot in Fig. 13.9, we can obtain eight possible 2-itemsets. 2-itemset like {B,D} does not exist, so it was not listed. 3-itemset {B, E} has the highest frequency: 3.

It is noteworthy that {D}, which should be removed from the 1-itemsets due to its unaccepted support, was still used when generating 2-itemsets. But, all 2-itemsets involving {D} still have inadequate support. This is actually the Apriori principle: if an item/itemset is infrequent, supersets containing that item/itemset will also be infrequent. That is, we can remove infrequent items and itemsets (k) before constructing supersets or itemsets with a greater length ($k + 1$). This can save us much time in the mining process.

13.7 Practice: Perform Association Rule Learning with Eclat

>>> More and up-to-date course materials including practices @ AI-engineer.org <<<

Please obtain the frequent itemsets in the following dataset with the given code for Eclat. Based on it, search for a meaningful yet simple dataset online and perform association rule learning with the data. Compare the results obtained with Eclat and the frequent itemsets that you obtain manually.

```
import numpy as np

# Function for generating a list with unique values (remove
    redundant items)
def unique(list1):
```

```
    # Initialize a null list
    unique_list = []
    # Loop over all elements
    for item in list1:
        # Check if it exists in unique_list or not
        if (not (all(x in unique_list for x in [item]))) or (item
     not in unique_list): # Item is a list or a non-list object
            unique_list.append(item)
    return(unique_list)

# Initialize global parameters
X = [[1,['A','C','D']],[2,['B','C','E']],[3,['A','B','C','E'
    ]],[4,['B','E']]]
X_ECLAT = []
X_ECLAT_items = []
X_ECLAT_itemss_unique = []
Itemsets = []
support = 2

# Prepare ECLAT input based on regular association rule analysis
    input data (like Apriori)
X_ECLAT_items.extend(X[j][1] for j in range(len(X))) # Collect
    all the items into a 2D list
X_ECLAT_items = list(np.concatenate(X_ECLAT_items).flat) #
    Flatten the 2D list into 1D
X_ECLAT_items.sort()
X_ECLAT_itemss_unique = unique(X_ECLAT_items)
X_ECLAT.extend([[X_ECLAT_itemss_unique[j]],[]] for j in range(len
    (X_ECLAT_itemss_unique)))

for i in range(len(X_ECLAT_itemss_unique)):
    for j in range(len(X)):
        if X_ECLAT_itemss_unique[i] in X[j][1]:
            X_ECLAT[i][1].append(X[j][0]) # Add the TID to the
    value of X_ECLAT of the item (as key)

# Remove itemset in k-itemsets with support less than the
    threshold
def support_filter(itemsets, support):
    # itemsets is the k-itemsets list

    # itemsets may contain k+1 or higher itemsets, obtain the
    average for later identification of these higher itemsets
    k_average = 0
    for j in range(len(itemsets)):
        k_average = j/(j+1)*k_average + 1/(j+1)*len(itemsets[j
    ][0])
        # print(k_average)

    n = len(itemsets)
    i = 0
    while i < n: # Cannot use for-loop because the size of the
    itemsets will change as we remove itemset without enough
    support
```

```
        if len(itemsets[i][1]) < 2 and n > 1 and len(itemsets[i
    ][0])<=k_average:
        # "n > 1" is to keep the highest k-itemsets regardless of
     its support, e.g., even support = 1
        # len(itemsets[i][0])<=k_average enable to keep itemsets
    with k+1 or a greater length
            itemsets.pop(i)
            n = n - 1 # Reduce the total number of itemsets by 1
        else:
            i = i + 1 # Move to (i+1) itemset only when (i)th
    itemset is not removed

# Find the common elements in two lists
def common_member(a, b):
    a_set = set(a)
    b_set = set(b)

    if (a_set & b_set):
        return list(a_set & b_set)

# Generate all the frequent k-itemsets
k = 0
while True:
    # Step 1: Forming
    # First iteration (k=1) does no extra work (1-itemsets are
    there); the later iterations need the following forming step
    if k == 0:
        Itemsets.append(X_ECLAT.copy()) # 3D list, [[[1-itemset 1
     ,1-itemset 1 total number],[1-itemset 2,1-itemset 2 total
    number],...],[[2-itemset 1 ,2-itemset 1 total number],[]]]
        # support_filter(Itemsets[k], support)
    else: # k > 0:
        Itemsets.append([]) # Add a new list in list for new k-
    itemsets as k increases by 1
        for i in range(len(Itemsets[k-1])):
            # start_index = X_ECLAT_itemss_unique.index(max(
    Itemsets[k-1][i][0])) # Find the maximum index of the current
     itemset; needed to avoid repetition
            for j in range(i+1,len(Itemsets[k-1])):
                Combined_items_unique = unique( Itemsets[k-1][i
    ][0] + Itemsets[k-1][j][0] ) # New (k+1) k-itemsets
                Combined_items_unique.sort()
                Common_TIDs = common_member(Itemsets[k-1][i][1],
    Itemsets[k-1][j][1])
                # print([ Combined_items_unique,Common_TIDs ])
                if Common_TIDs: # if not empty
                    Itemsets[k].append( [ Combined_items_unique,
    Common_TIDs ] )
    print(k)
    Itemsets[k] = unique(Itemsets[k])
    support_filter(Itemsets[k], support)

    if not Itemsets[k]:
```

```
        Itemsets.pop(-1) # Remove [] if k-itemsets is empty for
    the last (highest) k
        break
    k += 1
```

Chapter 14
Value-Based Reinforcement Learning

14.1 Overview

This chapter presents the basics of reinforcement learning (RL) and, based on that, introduces value-based RL as one of the two major categories of RL algorithms. For this goal, the basic RL concepts, including Markov decision process and essential RL terms, like environment, state, action, value, reward, and policy, will be explained first. Next, we will have an RL demonstration using simple virtual environments. Then, the Bellman equation, which casts the basis for the "learning" or improvement of an RL learning agent, will be discussed. This includes the Bellman equation's mathematical formulations, deduction, and use. After going through the basics, we will start the introduction to value-based RL algorithms with an overview of popular RL algorithms. Following that, typical value-based RL algorithms including two time-difference algorithms, i.e., Q-learning and Sarsa, and one episode-update algorithm, i.e., Monte Carlo, will be introduced. Details will be provided to cover both theories and guidelines for the implementation of these algorithms.

14.2 Basics of Reinforcement Learning

14.2.1 Basic Concepts

In a loose sense, any learning problem can be conceptualized as the interaction between a learning agent and the world whereby the agent performs learning and application of the learned knowledge. In supervised and unsupervised learning, all the data, no matter whether labeled or unlabeled, is ready and carries knowledge about the world. By contrast, data in reinforcement learning is not available before

Z. "L." Liu, *Artificial Intelligence for Engineers*,
https://doi.org/10.1007/978-3-031-75953-6_14

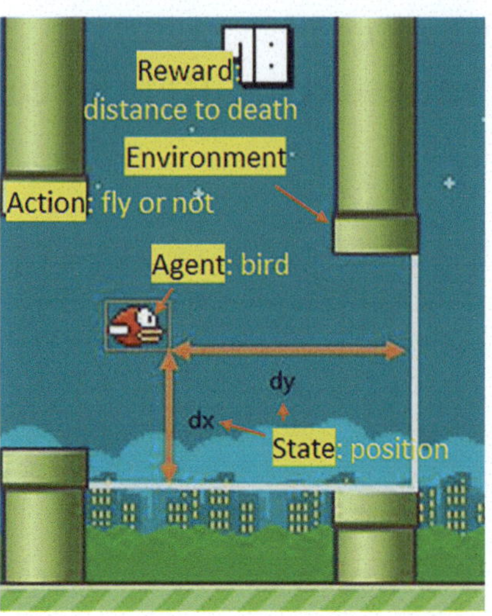

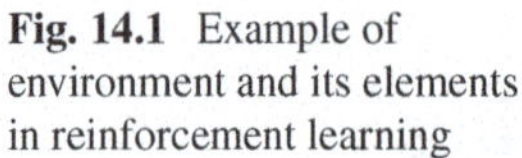
Fig. 14.1 Example of environment and its elements in reinforcement learning

a learning process starts. Instead, the agent obtains the data as the learning agent interacts with the environment. The data can be understood as "labeled" because we need to figure out what to do (described as "action" in RL; can be viewed as labels) under what conditions (described as "state" in reinforcement learning; can be viewed as data without labels). The model will be continuously improved as more data is obtained in the training process. The model's performance also needs to be evaluated with metrics in reinforcement learning. Such metrics are so-called reward functions that tell the goodness of specific actions for certain states.

Based on the above introduction, we can generally conceptualize reinforcement learning as illustrated in Fig. 14.1. The world outside of a learning agent is modeled as an *environment*. In this flappy bird game, the agent is the bird or the player controlling the agent, and the environment is the space consisting of the pipe obstacles. In order to describe the learning process, we need to know the current status of the learning agent, i.e., *state*, and the *action* the learning agent will take to interact with the environment. In this example, the state is the location of the bird, while the action is to fly or not. The action may bring the agent to a different (new) state and get *reward* from the environment that leads to beneficial results, e.g., survive longer in the game. The goal of any reinforcement learning process is to maximize the rewards so that the learning agent can gain a better performance. The Markov decision process (MDP) to be introduced will present a more strict model for conceptualizing such sequential decision processes [153, 154].

14.2.2 Markov Decision Process

The studies of MDP stemmed from the realm of optimal control. In 1957, a US researcher, Richard Bellman [155] first proposed the discrete time MDP via an optimal control model. Between 1960 and 1962, Howard [156] and Blackwell [157] proposed a dynamic programming algorithm for solving MDP problems. The MDP and its solution methods were later adopted for applications in various areas, such as autonomous control, suggestion systems, and reinforcement learning. In addition to the above essential components like state, action, and reward, MDP is characterized by the Markov property: the state at the current time is only related to the state and action at the previous time and independent of the state and action at other time points. This can be described using the following probability equation:

$$P[s_{t+1}|s_t] = P[s_{t+1}|s_1, a_1, s_2, a_2 \ldots, s_t] \tag{14.1}$$

This equation means that the consideration of earlier states does not make a difference in MDP.

Sequential decision problems based on MDP can be grouped into two categories: model based and model-free. Dynamic programming was proposed for model-based problems. That is, the influence of the learning agent's action on the environment—the probability $P^a_{ss'}$ and reward $r^a_{ss'}$ of entering a new state, s', after taking an action, a, in the current state, s—is known. By contrast, reinforcement learning was proposed for model-free problems, but it can also be used to solve model-based problems. More information about the classification of RL algorithms, including model-based and model-free categories in a broad sense, will be presented before we start working on the first major category of RL algorithms, i.e., value-based RL algorithms, in this chapter.

The Bellman equation is viewed as the core of dynamic programming. Based on it, a solution to MDP problems can be derived. Because both dynamic programming and reinforcement learning are intended for sequential decision problems, they share many concepts in common including the Bellman equation. Due to this reason, many people also perceive the Bellman equation as the core of reinforcement learning. However, reinforcement learning usually includes additional materials in addition to the sole use of the Bellman equation.

14.2.3 Policy Function, State Function, State-Action Function, and Reward Function

A policy determines how an agent selects an action in a state of the environment. This policy can be deterministic, that is, one action will be selected in a specific state. Or, the policy can also be stochastic, that is, in a specific state, the agent can select an action from different candidate actions (discrete or continuous) with different probabilities. In RL, it can be more specifically referred to as the learning policy, while in applications like control tasks, it may be called the control policy. Every policy can be formulated using a policy function:

$$\pi(s) : S \mapsto A \tag{14.2}$$

With the above equation, when a state, s, is entered as the input, the policy function will give out an action, a, as the output.

Value functions, in a broad sense, contain both V and Q. But in a narrow sense, we call V a value function and Q an action-value function. Both V and Q are widely used in RL algorithms. These two functions are closely related, and they can be converted to each other. Thus, in some RL algorithms, such as Q-learning and Sarsa, you may only see one of them, such as Q. These two functions can be confusing. Some aspects of these two types of functions are clarified here.

First, both of the two functions are usually associated with a specific learning policy. For example, we write them as $V^{\pi}(s)$ and $Q^{\pi}(s, a)$, where the superscript π marks the policy. However, in algorithms such as Q-learning and Sarsa, the policy is continuously improved as more data is processed by the agent to improve the policy. Thus, π may also be omitted to avoid confusion.

Second, from the symbol, we can see that $V(s)$ is the value of a state (without mentioning any specific action), while $Q(s, a)$ is the value associated with an action in a state, in which both an action and a state are needed. These two types of functions can be defined as follows. $V^{\pi}(s)$ is the expected value of rewards following policy π forever when the agent starts the learning process from the state s: $V^{\pi}(s) = \mathbb{E}(R)$. Mathematically, this is formulated as

$$V^{\pi}(s, \pi) = \mathbb{E}_{\pi}(R|s, \pi) = \mathbb{E}_{\pi}\left[\sum_{k=0}^{\infty} \gamma^{k} r_{t+k+1} | s_0 = s\right] \tag{14.3}$$

where R is the long-term reward. R at the current step t can be obtained by summing up the rewards at different steps as $R_t = \sum_{k=0}^{\infty} \gamma^{k} r_{t+k+1}$. This equation can also be written as

$$V^{\pi}(s, \pi) = \mathbb{E}_{\pi}(R_t|s, \pi) = \mathbb{E}_{\pi}\left[\sum_{k=0}^{\infty} \gamma^{k} r_{t+k+1} | s_t = s\right] \tag{14.4}$$

By contrast, $Q_{\pi}(s, a)$ is the expected value of rewards yielded by first taking action a in state s and then following the policy π forever. Therefore, Q can be related to rewards as

$$Q^{\pi}(s, \pi) = \mathbb{E}(R|s, a, \pi) \tag{14.5}$$

Based on the definition, the two types of functions can be converted to each other using the following equations:

$$
\begin{aligned}
V^{\pi}(s) =&\mathbb{E}_a[Q^{\pi}(s,a)] \\
&=\sum_a \pi(a|s)Q^{\pi}(s,a)
\end{aligned}
\tag{14.6a}
$$

$$
Q^{\pi}(s,a) = \sum_{s'} P^a_{ss'}[r^a_{ss'} + \gamma V^{\pi}(s')]
\tag{14.6b}
$$

where $\pi(a|s)$ is the probability of selecting action a if we follow policy π (policy can be determinate but here a more general definition with indeterminate policy is adopted), $P^a_{ss'}$ is the probability of moving from state s to s' if we take action a, and $r^a_{ss'}$ is the reward gained as we move from s to s'. In Q-learning and Sarsa, random factors were added via ϵ-greedy, which corresponds to $P^a_{ss'}$. In the literature, we can also see $r^a_s = \sum_{s'} P^a_{ss'} r^a_{ss'}$.

It is also common to use * to represent the optimal policy, which leads to the highest long-term reward. Accordingly, $V^*(s)$ is the maximum possible value of $V^{\pi}(s)$:

$$
V^*(s) = \max_{\pi} V^{\pi}(s)
\tag{14.7}
$$

The theory of MDP states that if π^* is an optimal policy, we will act optimally. Therefore, we will take the optimal action by choosing the action from $Q^*(s,\cdot)$, and this generates the optimal $V^*(s)$:

$$
V^*(s) = \max_a Q^*(s,a).
\tag{14.8}
$$

It is worthwhile to point out that, in policy-based algorithms, we usually evaluate the policy, i.e., $\pi(a|s)$, which can be viewed as a probability distribution of actions. By contrast, in value-based algorithms, we usually use and improve the optimal functions V^* or/and Q^*, which can be viewed as V or/and Q values that can best help us identify the optimal action (e.g., action a with the highest $Q(s,a)$ value in s).

14.2.4 Implementation of RL Environment

In this section, brief information is provided to show how to work with an RL environment. Two different ways will be illustrated. The first way involves the use of environments created by RL packages such as OpenAI Gym [158]. OpenAI Gym provides functions for users to easily employ a variety of RL environments. These environments can be used to test RL algorithms that are developed by the users, so that the user can focus on the RL algorithms instead of wasting time to find,

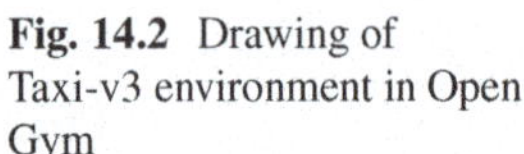

Fig. 14.2 Drawing of Taxi-v3 environment in Open Gym

implement, or calibrate RL environments. The second refers to the development of RL environments from scratch.

Implementation with OpenAI Gym

We use an environment called "Taxi-v3" from OpenAI Gym as an example. As illustrated in Fig. 14.2, the taxi needs to start at a random location, pick up, and drop off a passenger at two locations randomly selected from the four special locations marked with four colors. There are six actions: 0, move south; 1, move north; 2, move east; 3, move west; 4, pickup passenger; and 5, drop off passenger. The map has 25 cells (5 rows by 5 columns). Therefore, there are 500 states: 5 rows * 5 columns * 5 passenger positions (at four special locations + in taxi) * 4 destinations = 500 states.

The following code shows the use and check of the environment. The first step to use this environment is to import the package and initialize the environment using "gym.make('Taxi-v3')." We can also check or/and use the size of the action and state spaces to get a better understanding of the environment. Such information will also likely be needed in later use of the environment for RL.

```
import numpy as np
import gymnasium as gym # Old versions use "import gym"

# Initialization
env = gym.make("Taxi-v3", render_mode="human") # Call the "Taxi-
    v3" environment from gym with the name "env"
action_size = env.action_space.n # Obtain the size of the action
    space: how many actions. There are six here
state_size = env.observation_space.n # Obtain the size of the
    state space: 5 rows * 5 columns * 5 passenger positions * 4
    destinations =500 states
```

A key step in the use of such RL environment is to make one step in an episode. In this step, a new state will be obtained. Also, a reward value will be obtained after taking the action. Another variable is the signal to indicate whether the episode ends. In OpenAI Gym, this signal is represented using a variable called "done." This can be seen in the following code:

```
 # Update state
        state_new, reward, done, _, _ = env.step(action)
```

When an episode is finished, i.e., some conditions for ending the episode are met, we usually need to stop the RL iterations. This can be done as follows:

```
# Stop an episode when the environment tells an episode ends
    if done == 1:
        print('This is the #%i training episode' % n)
        break
```

After each episode or before we start testing, we will need to reset the environment. In testing, we also need to use loops by iteratively calling the "env.step()" function to add more steps within each episode. In addition, an "env.render()" function is provided in OpenAI Gym to illustrate the results. This is shown in the following code:

```
# Testing
state = env.reset()[0]
while True:
    # Choose an action based on the Q table only
    action = np.argmax(Table[state]) # Choose an action based on
    the Q table
    state, reward, done, _, _ = env.step(action)
    env.render() # Green is passenger; Red is destination
    if done == 1:
        break
```

Implementation from Scratch

In this part, the construction of a very simple environment that can be used in a way similar to OpenAI Gym environments is illustrated in the following code. Because the code for building the environment serves as Python functions that can be imported by other Python code, we used object-oriented programming to develop the code. The code includes a few key parts similar to OpenAI Gym: initialize variables and functions, develop functions to reset the environment, develop a function to make one step in the episode (including assigning rewards and stopping the episode), and develop a render function to illustrate the state when using the episode.

```
import numpy as np

class treasure1D:
    def __init__(self,N):
```

```
        self.n_cell = N # The greater this number, the higher
    n_episode is needed
        self.Reward = np.zeros(self.n_cell);
        self.Reward[-1]=1.0 # n_cell cells, only the rightmost
    has reward (end of game)
        self.States = range(self.n_cell)
        self.state = np.random.choice(self.States)
        self.reward = 0
        self.done = False
        self.observation_size = 1 # Number of elements in
    observation (there is only one in this env: state/position)

    def reset(self):
        self.state = np.random.choice(self.States)
        self.reward = 0
        self.done = False
        return self.state

    def step(self,action):
        # Update state
        reward = 0
        if action == 'left' or action == 0:
            if self.state == self.States[0]: # Leftmost
                state_new = self.state # Does not move
                self.done = True
                self.reward = -10
            else:
                state_new = self.state - 1 # Obtain new state s^
    prime
        else:
            if self.state == self.States[-1]: # rightmost
                state_new = self.state # Does not move
                self.done = True
                self.reward = 10
            else:
                state_new = self.state + 1 # Obtain new state s^
    prime
        self.state = state_new
        return self.state, self.reward, self.done, []

    def render(self):
        print ("The state is ", self.state, "The reward is ",self
    .reward)
```

This environment simulates a very simple 1D treasure hunt game. In each episode, an agent will be placed at a random cell of a 1D map consisting of multiple cells. The agent needs to find the treasure in the rightmost cell. Rewards/penalties are only assigned to the cells on the two ends of the map. The agent cannot move out of the map from the left. Thus, it will stay in the leftmost cell when it tries to move left from there, and a reward of -10 (penalty of 10) is applied when that happens. An episode ends when the agent reaches the rightmost cell, i.e., finding the treasure, and a reward of 10 is made.

14.3 Bellman Equation

The Bellman equation lays out a relationship between the state value function or the action-state value function in the current state and that in the following state. Therefore, this equation appears in a recursive format and can be used iteratively to obtain V or/and Q in an episode. In this subsection, we will first give out the Bellman equation in different forms. Then, details will be offered to show how the Bellman equation in typical forms can be derived to understand the essence of learning in RL.

14.3.1 Formulations of Bellman Equation

The Bellman equation can be formulated with the state value function and the action-state value function. The Bellman equation formulated with the state value function describes the values function in the current state s and the value function in the following state s':

$$V^{\pi}(s) = \sum_{a} \pi(a|s) Q^{\pi}(s, a) = \sum_{a} \pi(a|s) \sum_{s'} P_{ss'}^{a} \left[r_{ss'}^{a} + \gamma V^{\pi}(s') \right]. \tag{14.9}$$

Likewise, the Bellman equation formulated with the action-state value function describes the action-state value function in the current state s and that in the following state s':

$$\begin{aligned} Q^{\pi}(s, a) &= \sum_{s'} P_{ss'}^{a} \left[r_{ss'}^{a} + \gamma V^{\pi}(s') \right] \\ &= \sum_{s'} P_{ss'}^{a} \left[r_{ss'}^{a} + \gamma \sum_{a'} \pi(a'|s') Q^{\pi}(s', a') \right]. \end{aligned} \tag{14.10}$$

Based on the above equations and the relationship in Sect. 14.2.3, we can obtain the following Bellman equation for the optimal state value function and action-state value function:

$$V^{*}(s) = \max_{a} Q^{*}(s, a) = \max_{a} \sum_{s'} P_{ss'}^{a} \left[r_{ss'}^{a} + \gamma V^{\pi}(s') \right]. \tag{14.11a}$$

$$Q^{*}(s, a) = \sum_{s'} P_{ss'}^{a} \left[r_{ss'}^{a} + \gamma V^{*}(s') \right] = \sum_{s'} P_{ss'}^{a} \left[r_{ss'}^{a} + \gamma \max_{a'} Q^{*}(s', a') \right]. \tag{14.11b}$$

14.3.2 Deduction of Bellman Equation

In the following, we will show the deduction of the V Bellman equation and Q Bellman equation one by one.

V Bellman Equation:

$$
\begin{aligned}
V^{\pi}(s) =& \mathbb{E}_{\pi}\left[r_{t+1} + \gamma r_{t+2} + \gamma^{2} r_{r+3} + \cdots | s_t = s\right] \\
=& \mathbb{E}_{\pi}\left[\sum_{k=0}^{\infty} \gamma^{k} r_{t+k+1} | s_t = s\right] \\
=& \mathbb{E}_{\pi}\left[r_{t+1} + \gamma \sum_{k=0}^{\infty} \gamma^{k} r_{t+k+2} | s_t = s\right]
\end{aligned}
\tag{14.12}
$$

A key step in the deduction is to recall the definition of the expectation of immediate reward:

$$
\mathbb{E}_{\pi}\left[r_{t+1} | s_t = s\right] = \sum_{a} \pi(a|s) \sum_{s'} P_{ss'}^{a} r_{ss'}^{a}. \tag{14.13}
$$

Similarly, the expectation of the sum of the rewards in the following steps can be written as

$$
\begin{aligned}
\mathbb{E}_{\pi}\left[\sum_{k=0}^{\infty} \gamma^{k} r_{t+k+2} | s_t = s\right] =& \sum_{a} \pi(a|s) \sum_{s'} P_{ss'}^{a} \mathbb{E}_{\pi}\left[\sum_{k=0}^{\infty} \gamma^{k} r_{t+k+2} | s_{t+1} = s'\right] \\
=& \sum_{a} \pi(a|s) \sum_{s'} P_{ss'}^{a} \mathbb{E}_{\pi}\left[V^{\pi}(s')\right]
\end{aligned}
\tag{14.14}
$$

Substituting the above two equations into the equation for defining $V^{\pi}(s)$, we obtain

$$
V^{\pi}(s) = \sum_{a} \pi(a|s) Q^{\pi}(s, a) = \sum_{a} \pi(a|s) \sum_{s'} P_{ss'}^{a}\left[r_{ss'}^{a} + \gamma V^{\pi}(s')\right]. \tag{14.15}
$$

Q Bellman Equation:
Similar to the deduction of the V Bellman equation, we can start from the definition of the Q and then split it into two parts:

$$\begin{aligned} Q^\pi(s,a) =& \mathbb{E}_\pi\left[r_{t+1} + \gamma r_{t+2} + \gamma^2 r_{r+3} + \cdots | s_t = s, a_t = a\right] \\ =& \mathbb{E}_\pi\left[\sum_{k=0}^{\infty} \gamma^k r_{t+k+1} | s_t = s, a_t = a\right] \\ =& \mathbb{E}_\pi\left[r_{t+1} + \gamma \sum_{k=0}^{\infty} \gamma^k r_{t+k+2} | s_t = s, a_t = a\right] \end{aligned} \tag{14.16}$$

Next, we need to deal with the two terms in the above equation based on the definition of the expectation of rewards in a way similar to what we did for the V Bellman equation:

$$\mathbb{E}_\pi\left[r_{t+1} | s_t = s, a_t = a\right] = \sum_{s'} P_{ss'}^a r_{ss'}^a. \tag{14.17}$$

$$\begin{aligned} \mathbb{E}_\pi\left[\sum_{k=0}^{\infty} \gamma^k r_{t+k+2} | s_t = s, a_t = a\right] =& \sum_{s'} P_{ss'}^a \mathbb{E}_\pi\left[\sum_{k=0}^{\infty} \gamma^k r_{t+k+2} | s_{t+1} = s'\right] \\ =& \sum_{s'} P_{ss'}^a \left[\sum_{a'} \pi(a'|s') \mathbb{E}_\pi \sum_{k=0}^{\infty} \gamma^k r_{t+k+2} | s_{t+1}\right. \\ & \left. = s', a_{t+1} = a'\right] \\ =& \sum_{s'} P_{ss'}^a \left[\sum_{a'} \pi(a'|s') Q^\pi(s', a')\right] \\ =& \sum_{s'} P_{ss'}^a \left[V^\pi(s', a')\right] \end{aligned} \tag{14.18}$$

It is noted that the above two equations are different from those for the V function because the action a is determined in the definition of Q. As a result, the sum weighted by $\pi(a|s)$ is not needed anymore.

Combining the above equations, we obtain

$$\begin{aligned} Q^\pi(s,a) =& \sum_{s'} P_{ss'}^a \left[r_{ss'}^a + \gamma V^\pi(s')\right] \\ =& \sum_{s'} P_{ss'}^a \left[r_{ss'}^a + \gamma \sum_{a'} \pi(a'|s') Q^\pi(s', a')\right]. \end{aligned} \tag{14.19}$$

14.3.3 Use of Bellman Equation in Reinforcement Learning

The use of the Bellman equation in reinforcement learning is slightly different from that in dynamic programming. In dynamic programming, the Bellman equation is implemented in iterations to seek solutions to MDP problems. In reinforcement learning, two significant differences are the use of the optimal value functions and the learning rate.

The first difference is marked by the use of the greedy method in the search for a better policy. Instead of always utilizing the best action in the current policy (exploitation purpose), nonoptimal actions that may lead to better policies will also be taken at certain probabilities. The search for the optimal policy can be performed via the improvements in the action-state value function or/and state value function or the whole policy represented by a distribution $\pi(s, a)$. The former is usually adopted in value-based algorithms, while the latter is used in policy-based algorithms.

In value-based RL algorithms, we usually use ϵ-greedy method to reach a balance between exploration and exploitation. That is, in most conditions, i.e., at a probability of $1-\epsilon$, we use the greedy policy: adopting an action, a, with the optimal action-value function $Q^*(s, a)$, and select a random action in other conditions, i.e., at the probability of ϵ. Therefore, in the former case $(1 - \epsilon)$, we can use the relationship between the optimal state value function and the optimal action-state value function:

$$V^*(s) = \max_a Q^*(s, a). \tag{14.20}$$

Usually, in such value-based RL algorithms, such as Q learning [159], the transfer between states is fixed. That is, we do not need to consider $P^a_{ss'}$ anymore. With the above two considerations, the Bellman equation for the optimal Q function, i.e., Eq. 14.19, becomes

$$Q^*(s, a) = r^a_{ss'} + \gamma V^*(s', a') = r^a_{ss'} + \gamma \max_{a'} Q^*(s', a'). \tag{14.21}$$

"Greedy" is primarily reflected via the max function.

The second difference resides in the use of a learning rate. A learning rate is used to integrate both the old value from the history (e.g., existing Q table) and the new value obtained with the Bellman equation when making updates:

$$Q^*(s, a) = (1 - \alpha)Q^*(s, a) + \alpha \left[r^a_{ss'} + \gamma \max_{a'} Q^*(s', a') \right]. \tag{14.22}$$

where $Q^*(s, a)$ on the left-hand side of the equation is the new Q function in state s when taking action a; $Q^*(s, a)$ on the right-hand side of the equation is the $Q^*(s, a)$ from the history, e.g., the previous episode, and the second term on the right-hand side of the equation is the new $Q^*(s, a)$ obtained purely with the Bellman equation. The learning rate α gives out the weights to sum up these two values so that the update will not experience abrupt changes. That is, if $\alpha = 1$, then the update will be completed with the new value (from the Bellman equation) only. As a result, the update process may not be stable because the new Q is irrelevant to the historical ones. On the contrary, if $\alpha = 0$, the new value (in the current episode) will always be equal to the old value (from the previous episode), and consequently, the learning will stop.

14.4 Value-Based RL

14.4.1 Overview of RL Algorithms

The classification of machine learning algorithms can be complicated and is still evolving quickly due to the fast development of RL. Figure 14.3 presents a classification that includes the most classic and popular algorithms in the state of the practice. Currently, the RL based on MDP is predominant. Both model based and model-free RL within this category gained success. However, more research and breakthroughs have been made in the time-difference model-free RL, which can be further categorized into value-based and policy-based algorithms. In more recent studies, both value-based and policy-based algorithms have been integrating deep neural networks to enable impactful deep reinforcement learning [160, 161].

The remaining of this chapter will use value-based RL as an entry for introducing reinforcement learning. We will show how to construct the classic Q learning (or Q-learning) [159] and Sarsa [162], from theories, procedure, to pseudo-code. In the next chapter, we will focus on policy-based RL algorithms. A classic gradient-

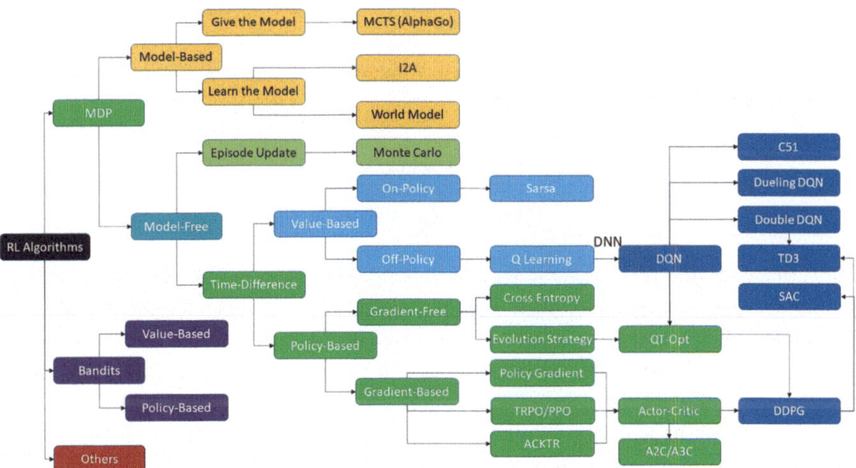

Fig. 14.3 Overview of popular RL algorithms

based algorithm, policy gradient, will be presented in detail. Based on that, more information will be provided for other gradient-based policy-based RL algorithms.

14.4.2 Q Learning and Sarsa

The essence of Q learning is to use the average value of $Q(s, a)$ to estimate $Q(s, a)$: a strategy called time-difference. A much different but easier to understand strategy is to collect all (or a large amount of) the data, e.g., one episode, and then calculate the average. This is the strategy adopted by episode-update methods like Monte Carlo. Therefore, distinct from episode update methods, time-difference methods like Q-learning update the average value, e.g., $Q(s, a)$, immediately after a new data point (or sample, e.g., s, a, r) is collected instead of after an episode is finished. The math underlying this time difference can be derived with the following equation.

$$
\begin{aligned}
u_k =& \frac{1}{K}\sum_{k=1}^{K} x_k \\
=& \frac{1}{K}\left(x_K + \sum_{k=1}^{K-1} x_k\right) \\
=& \frac{1}{K}[x_K + (K-1)u_K] \\
=& u_{K-1} + \frac{1}{K}(x_K - u_{K-1})
\end{aligned}
\tag{14.23}
$$

In the above equation, the average value at the current step u_K is obtained by adding $\frac{1}{K}(x_K - u_{K-1})$ to the average of the previous step u_{K-1}. $\frac{1}{K}$ is the learning rate for a constant number of steps. Then, we can use α, a generalized learning rate, to reformulate the above equation as

$$
\begin{aligned}
u_k =& u_{K-1} + \alpha(x_K - u_{K-1}) \\
=& (1-\alpha)u_{K-1} + \alpha x_K
\end{aligned}
\tag{14.24}
$$

Thus, this equation provides one way of understanding the world, i.e., characterized by the average or expected value, by continuously taking more measurements and using such measurements to update the average or approach the expected value.

In Q learning, we want to select the action in the current state that can lead to the highest cumulative rewards. Accordingly, we define $Q^{\pi}(s, a)$ as the expected value of first taking action a in state s and then following the policy π forever. In Q learning, we are improving our understanding of the environment and, consequently, the policy by updating the $Q(s, a)$ values. Using the above equation for updating the expected value, $Q(s, a)$ can be updated as follows:

$$Q(s, a) = (1 - \alpha)Q(s, a) + \alpha[r + \gamma \max_{a'} Q(s', a')] \quad (14.25)$$

where r is the reward gained as we move from state s to the next state s' after taking action a and $\max_{a'} Q(s', a')$ is the maximum of the Q values of all the actions in the state s'. $\gamma \max_{a'} Q(s', a')$ represents the experience from previous learning. For example, $\max_{a'} Q(s', a')$ can inform the agent in the current state if the agent previously (in an early learning process or episode) gained a high reward in state s' after taking action a'. As can be seen, this equation is the same as the Q update equation 14.22 using the Bellman equation, disregarding the difference in symbols.

Q learning is an off-policy learning, whereby the agent can learn from the data generated beforehand instead of learning as the agent generates data in an environment. An RL algorithm that is very close to Q learning but with a distinct on-policy nature is Sarsa. In Sarsa, the learning agent must participate in the learning process as it improves the policy. This is achieved by replacing $\max_{a'} Q(s', a')$ with $Q(s', a')$. Thus, instead of selecting the action leading to the maximum reward in the next state (s' part of the Q table is from history only (off the policy) at this moment), the agent in Sarsa first uses the policy derived from Q (still on the same policy, with ϵ-greedy) to get $Q(s', a')$ (next state) and then use this $Q(s', a')$ to update the $Q(s, a)$ in the current state:

$$Q(s, a) \leftarrow (1 - \alpha)Q(s, a) + \alpha[r + \gamma Q(s', a')] \quad (14.26)$$

The detailed procedures for implementing Q learning and Sarsa are exemplified using the following pseudo-code.

Q learning:

Initialize $Q(s, a)$ arbitrarily
Repeat for each episode:
  Initialize s (e.g., choose a s randomly)
  Repeat for each step of the episode:
    Select a from s using a policy derived from Q (e.g., ϵ-greedy[1])
    Take action a, and obtain r and s' based on the environment
    Update Q: $Q(s, a) \leftarrow (1 - \alpha)Q(s, a) + \alpha[r + \gamma \max_{a'} Q(s', a')]$
    Update s: $s \leftarrow s'$
  Until s is terminal

Sarsa:

Initialize $Q(s, a)$ arbitrarily
Repeat for each episode:
  Initialize s (e.g., choose a s randomly)
  Select a from s using a policy derived from Q (e.g., ϵ-greedy)
  Repeat for each step of the episode:

[1] Select a that has the highest Q at the s with probability of ϵ. and select a random action with a probability of $1 - \epsilon$.

Take action a, and obtain r and s' based on the environment
Select a' from s' using a policy derived from Q (e.g., ϵ-greedy
Update Q: $Q(s, a) \leftarrow (1 - \alpha)Q(s, a) + \alpha[r + \gamma Q(s', a')]$
Update s: $s \leftarrow s'$
Until s is terminal

Example code for implementing Q learning to address the Taxi-v3 problem is given in the following:

```
import numpy as np
import pandas as pd
import gym

# Initialization
env = gym.make("Taxi-v3") # Call the "Taxi-v3" environment from
    gym with the name "env"
action_size = env.action_space.n # Obtain the size of the action
    space: how many actions. There are six here
state_size = env.observation_space.n # Obtain the size of the
    state space: 5 rows * 5 columns * 5 passenger positions * 4
    destinations =500 states
Table = np.zeros((state_size, action_size)) # Initialize the Q
    table

# Hyperparameter
epsilon = 0.9 # Parameter for epsilon-greedy
alpha = 0.1 # Learning rate
gamma = 0.8 # Decay of rewards

# Training
n_episode = 10000 # Number of episodes for training
for n in range(n_episode):
    state = env.reset() # Initialize the environment for each
    episode
    while True:
        # Choose an action based on the Q table with epsilon
    greedy
        action_Q = np.argmax(Table[state]) # Choose an action
    based on the Q table
        action_greedy = np.random.choice(Table[state].size) #
    Choose an action randomly for use in the epsilon-greedy
    method
        action = np.random.choice([action_Q,action_greedy],size
    =1,p=[epsilon,1-epsilon])[0]
        # Update state
        state_new, reward, done, _ = env.step(action)
        # Update the table
        Table[state,action] = (1-alpha)*Table[state,action] +
    alpha*(reward+gamma*max(Table[state_new]))
        # Prepare for the next step
        state = state_new
        # Stop an episode when the environment tells an episode
    ends
```

```
        if done == 1:
            print('This is the %i the episode' % n)
            break

# Testing
state = env.reset()
while True:
    # Choose an action based on the Q table only
    action = np.argmax(Table[state]) # Choose action based on Q
    table
    state, reward, done, _ = env.step(action)
    env.render() # Green is passenger; Red is destination
    if done == 1:
        break
```

14.4.3 Monte Carlo Method

In the current practices of reinforcement learning, the use of episodic update methods represented by the Monte Carlo method is less popular than step-update (time-difference) methods. Compared to time-difference methods such as Q learning and Sarsa, the Monte Carlo method is also value based: the learning is carried out by improving the value functions. However, instead of updating the value functions step by step in an episode, the Monte Carlo method first finishes all the steps in the episode and then updates the value. Accordingly, for each episode, a series of states will be generated randomly: $s_1, a_1, r_2, s_2, a_2, r_3, \ldots, s_t, a_t, r_{t+1}, \ldots, s_T, a_T$.

Then, let us recall the definition of the value function in the MDP. However, a minor difference is made so that the equation can be used to calculate the value of every state in a series of states:

$$V^{\pi}(s) = \mathbb{E}^{\pi}(G_t | s_t = s) = \mathbb{E}^{\pi}\left[\sum_{i=t}^{T} \gamma^{i-t} r_{i+1}\right] \tag{14.27}$$

Therefore, for every step in an episode, we will have one state s (and choose one action a), which corresponds to a G_t value in the above function. Using the average value update equation, we can update the value function of every state s that appears in a series (episode):

$$V(s) = (1 - \alpha)Q(s) + \frac{1}{N} G_t \tag{14.28}$$

where N is the number of times that this state appears in the whole learning process (e.g., all the episodes that have been tested). It is possible that a state can appear multiple times in an episode (series). To deal with such a situation, we can choose to consider only the reward gained (G_t) in the first appearance of this state (called first visit) or use the rewards from every appearance of the state (called every visit). The second method usually requires more computational effort but may get better results when the number of series is small.

In fact, we usually update the action-value function (Q table). This update can be performed in a similar way:

$$Q(s,a) = (1-\alpha)Q(s,a) + \frac{1}{N}G_t \tag{14.29}$$

To use the equation, we need to record the N number for every state for updating the state value and for every action in every state for updating the action-state value. This will consume a lot of memory. A convenient alternative is to replace $\frac{1}{N}$ with a learning rate α.

14.5 Practice: Solve RL Problem Using Q Learning

>>> More and up-to-date course materials including practices @ AI-engineer.org <<<

Put together code to implement Q learning to solve a simple problem.

1. Please understand the "Taxi-v3" environment from OpenAI Gym (now gymnasium, search online). Please try the environment based on the code given and explained in the book. Test the key functions for using the environment.
2. Please train an RL model using Sarsa for the Taxi-v3 environment. You can start from the following example code for Q learning:

```
# Hyperparameter
epsilon = 0.9 # Parameter for epsilon-greedy
alpha = 0.1 # Learning rate
gamma = 0.8 # Decay of rewards

# Training
n_episode = 10000 # Number of episodes for training
for n in range(n_episode):
    state = env.reset()[0] # Initialize the environment for
    each episode
    while True:
        # Choose an action based on the Q table with epsilon
    greedy
        action_Q = np.argmax(Table[state]) # Choose an action
    based on the Q table
        action_greedy = np.random.choice(Table[state].size) #
    Choose an action randomly for use in the epsilon-greedy
    method
        action = np.random.choice([action_Q,action_greedy],
    size=1,p=[epsilon,1-epsilon])[0]
        # Update state
```

```
        state_new, reward, done, _, _ = env.step(action) # Old
     version Gym just need 1 "_"
        # Update the table
        Table[state,action] = (1-alpha)*Table[state,action] +
    alpha*(reward+gamma*max(Table[state_new]))
        # Prepare for the next step
        state = state_new
        # Stop an episode when the environment tells an
    episode ends
        if done == 1:
            print('This is the #%i training episode' % n)
            break
env.close()
```

Hints: You will need to initialize a Q table, for example, with 0's. The shape of the Q table is related to the state size and the action size. Try to understand the relevant line(s) in the above Q learning code and the code for Taxi-v3 in the book to correctly perform the initialization.

3. Please use the trained model from Part 2 for testing (one episode is enough). You can refer to the sample code for testing, which is given in the book. You will need to import and initialize the environment before testing.
   Hints: You will need to use "env = gym.make("Taxi-v3", render_mode='human')" before run "env.render()."

Chapter 15
Policy-Based Reinforcement Learning

15.1 Overview

This chapter introduces another major category of reinforcement learning algorithms: policy-based RL. First, we will try to smoothly transition from value-based RL by discussing issues with value-based RL and how such issues can be addressed with policy-based RL. Next, we will get familiar with the basic concepts in policy-based RL and reveal the major steps of policy-based RL studies: objective function construction, policy definition, training via policy improvements, and algorithm improvements. With the understanding of these steps, we will first introduce an objective function to derive the policy gradient. Based on it, the deduction of the policy gradient theorem will be presented. Then, we will show the Monte Carlo implementation with the derived policy gradient theorem. Next, we will show the issues with the simple Monte Carlo implementation, which can be improved in two different directions: policy and value. As for policy, we will introduce more forms of objective functions and show a more widely accepted policy gradient theorem. For value, we will discuss more policy evaluation methods for reducing the variance. Various classic policy gradient algorithms including REINFORCE, REINFORCE with baseline, Actor-Critic (TD(0) and TD(λ)), and so on will be introduced.

15.2 Policy-Based RL vs. Value-Based RL

Value-based algorithms such as Q learning and deep Q learning [163] have some issues:

(1) A minor change in the value function can lead to a change in the selection of actions. Such noncontinuous changes are an important reason why it is hard for value-based methods to get converged results.

Z. "L." Liu, *Artificial Intelligence for Engineers*,
https://doi.org/10.1007/978-3-031-75953-6_15

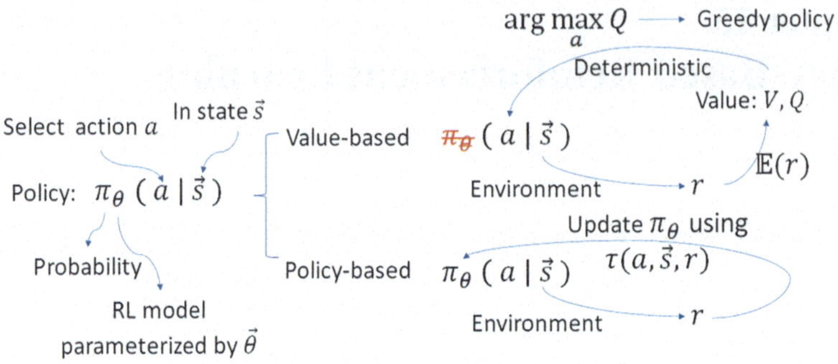

Fig. 15.1 Value-based RL vs. policy-based RL

(2) The search for the optimal Q value is difficult in a continuous space. This issue is very obvious when there are high-dimensional or continuous action spaces. For example, in autonomous driving applications, the number of states corresponding to different road and vehicle conditions is huge.
(3) Value-based algorithms cannot learn stochastic policies. Due to this reason, it is easy for a value-based learning agent to get stuck in a specific state. One example is a two-player game like rock-scissors-paper; if one player acts deterministically, the other player can develop countermeasures to win.

Policy-based algorithms can be used to address these issues by adjusting the policy instead of the value functions (V and Q) as in value-based RL algorithms. In value-based RL, both of these value functions can determine the selection of actions in different states. As shown in Fig. 15.1, the state values determine the selection of actions via a greedy policy: select the action with the highest action-state value (i.e., the highest expected cumulative rewards). To enable learning, stochastic factors were added by modifying the greedy policy with ϵ in the training process (in testing, only the greedy policy is used), which represents the probability of exploring better actions that are not suggested by the greedy policy but can achieve higher long-term rewards.

By contrast, policy-based algorithms directly determine the selection of actions in different states via the policy (Fig. 15.2). Such a policy can be viewed as a probability distribution of different actions at different state variables. This distribution tells the probability of selecting any action in the given state. Thus, when a state with different state variables is determined, we will know the probabilities of selecting different actions, though only one action will be selected based on the probabilities. To select an action, the Softmax function is usually used for discrete actions, while the Gaussian distribution is commonly used for continuous actions. The greedy policy (not ϵ greedy) can be viewed as a special policy that corresponds to a Dirac delta distribution at the point where the maximum action-state function value is located: only the action corresponding to the maximum action-state value

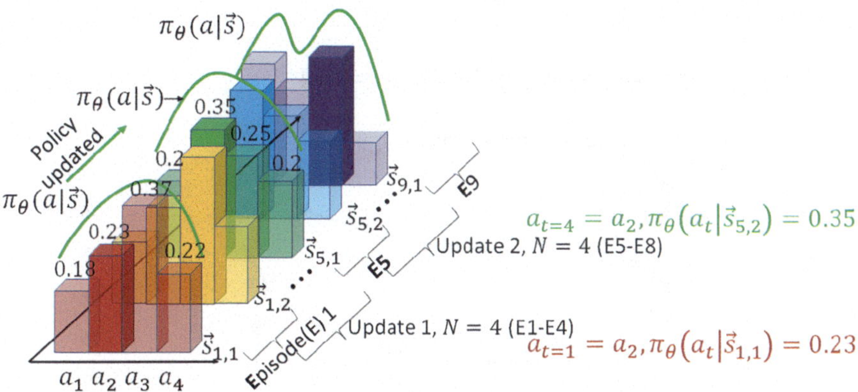

Fig. 15.2 Illustration of policy update in policy-based RL

will be selected at a probability of 100%. In the training process of a policy-based learning algorithm, this distribution representing the policy will be adjusted to improve the policy.

15.3 Basic Concepts

The probability distribution used to represent a policy can be written as $\pi_\theta(a|s)$, which describes the probability of selecting an action a in the state s. This function is parameterized by θ, which could be one number or an array of numbers.

Figure 15.2 illustrates the updates of $\pi_\theta(a|s)$, in which the policy is updated every step with epsodic data. As shown, in Step 1 of Episode 1 represented by a state $\vec{s}_{1,1}$, we selected action a_2, which has a probability of 0.23. The generated awards were used to update $\pi_\theta(a|s)$. This led to $\vec{s}_{1,2}$, in which *the updated*$\pi_\theta(a|s)$ can be simply viewed as the new probabilities of different actions. This process continued as more episodes of data was used.

When an agent acts in an environment using a policy $\pi_\theta(a|s)$, a series of states and actions can be generated. This series is called a trajectory, which is illustrated in Fig. 15.3. Such a trajectory can be formulated as follows:

$\tau(s_1, a_1, s_2, a_2, \cdots, s_t, a_t)$

The stochastic nature of policy-based algorithms makes their theories appear different from those of value-based algorithms [164]. One difference is the widespread use of probability functions and expectations in the theories of policy-based algorithms, which is not necessary in value-based algorithms. One policy corresponds to one probability distribution. As a result, different trajectories can be generated even if an agent starts from the same state following one policy. Therefore, in order to evaluate a policy, expectations of rewards and state functions are commonly used in the theories of policy-based theories. It is noted that many introductions to RL are

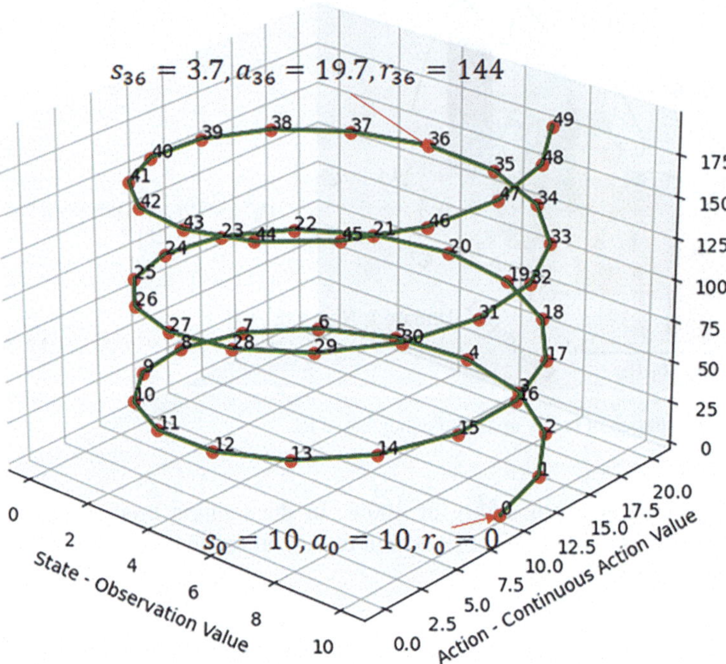

Fig. 15.3 Trajectory of an episode in RL learning

written up using frameworks considering the probabilities and expectations, which present a comprehensive description suitable for both value-based and policy-based algorithms, whereas some other introductions are laid down without probability for simplicity, which is more intended for value-based algorithms. This could be confusing to many people who are new to RL.

Reinforcement learning is similar to supervised learning in a few aspects. That is, we are looking for a mapping from states to actions. But here, it is no longer a simple one-to-one mapping. The rewards and their functions, such as the state value function and action-state value function, serve as the "objective/loss function" in training and "evaluation metrics" in testing for RL. The training process of RL is a process of maximizing the objective function. Thus, such an objective function is usually constructed with the reward and policy. In that way, the probability of actions can be adjusted to obtain more rewards (or higher objective function value). Because the policy is parameterized by θ, the objective function is also parameterized by θ. The training process aims to obtain the optimal parameter θ^*, which can maximize the objective function:

$$\theta^* = \arg\max_{\theta} J(\theta) \tag{15.1}$$

In the following, we will see that the development, implementation, and improvement of policy gradient methods can be divided into four steps or components:

(1) Propose an objective function to assess the effectiveness of the policy. We will see that many different objective functions can be used.
(2) Define a policy mathematically. For training, we will need to find out how to improve the policy by maximizing the objective function: the policy gradient theorem.
(3) Construct and implement an algorithm for training. That is, we will develop a procedure to improve the policy in the process of maximizing the objective function.
(4) Further improve the algorithm. Many issues caused difficulties in getting good training results with policy gradient methods. Different improvements were proposed, leading to different methods/algorithms of policy gradients.

The rest of this chapter is organized to introduce these four components sequentially.

15.4 Objective Function and Policy Gradient Theorem

15.4.1 Objective Function

The objective function can be constructed in different ways. In the following, we will first show a very simple way to construct the objective function and derive the policy gradient theorem with it. Three more objective functions will be presented after that, based on which various policy gradient algorithms will be described.

In a simple way, we can use the expectation of the sum of rewards as the objective function. This expectation can be obtained by letting the learning agent generate a large number of trajectories. When this number is large enough, we can use the average of the total rewards to approximate this expectation:

$$
\begin{aligned}
J(\theta) &= \mathbb{E}_{\tau \sim \pi_{\theta(\tau)}}[r_0 + r_1 + \cdots + r_t] = \mathbb{E}_{\tau \sim \pi_{\theta(\tau)}}\left[\sum_t (r(\tau))\right] \\
&\approx \frac{1}{N}\sum_i^N \left[\sum_t r(s_{i,t}, a_{i,t})\right]
\end{aligned}
\tag{15.2}
$$

where i is the number of τ, which is the trajectory obtained in episode i, and t is the number of the step in this episode.

The above equation can be written in a continuous form:

$$
J(\theta) = \mathbb{E}_{\tau \sim \pi_{\theta(\tau)}}\left[\sum r(\tau)\right] = \int_\tau r(\tau)\pi_\theta(\tau)d\tau
\tag{15.3}
$$

With the above definition, the training process in policy-based reinforcement learning tasks can be formulated as

$$\arg\max_{\theta} J(\theta) = \arg\max_{\theta} \mathbb{E}_{\tau\sim\pi_{\theta(\tau)}} \left[\sum r(\tau)\right] = \arg\max_{\theta} \int_{\tau} r(\tau)\pi_{\theta}(\tau)d\tau \tag{15.4}$$

The RL implementation with the above equation is similar to the search for the minimum loss or maximum objective function in supervised learning. We can still use optimization methods such as gradient descent. Notwithstanding, there is one major difference: we look for a maximum objective function determined by the rewards in reinforcement learning instead of a minimum objective function determined by the difference between the predicted and true labels in supervised learning. In gradient *ascent*, the parameters that enable the maximum of the objective function are achieved via the following equation:

$$\theta = \theta + \alpha\nabla_{\theta} J(\theta) \tag{15.5}$$

15.4.2 Policy Gradient Theorem

Now, the key is to derive an equation that can be easily used to calculate $\nabla_{\theta} J(\theta)$. This equation can be derived as follows:

$$\begin{aligned}
\nabla_{\theta} J(\theta) &= \int \nabla_{\theta} \left[\pi_{\theta}(\tau) \cdot r(\tau)\right] d\tau \\
&= \int \pi_{\theta}(\tau) \cdot \frac{\nabla_{\theta}\pi_{\theta}(\tau)}{\pi_{\theta}(\tau)} \cdot r(\tau)d\tau \\
&= \int \pi_{\theta}(\tau) \cdot \nabla_{\theta} \log \pi_{\theta}(\tau) \cdot r(\tau)d\tau \\
&= \mathbb{E}_{\tau\sim\pi_{\theta}(\tau)} \left[\nabla_{\theta} \log \pi_{\theta}(\tau) \cdot r(\tau)\right]
\end{aligned} \tag{15.6}$$

where log is the natural log, i.e., ln.

In the above equation, the operator is placed on $\pi_{\theta}(\tau) \cdot r(\tau)$, but it only takes effect on $\pi_{\theta}(\tau)$, because $r(\tau)$ is not a function of θ. The key work in the above deduction is to take a $\pi_{\theta}(\tau)$ out from the gradient operation, which enables the construction of an expectation.

Next, we need to recall the essential characteristics of MDP, i.e., Markov property [165], so that τ in the above equation can be related to individual action-state sets for implementations. In this property, the current state is only related to the state and action at the previous time and independent of the states and actions at other times. Then, the probability of a trajectory τ can be formulated as

$$\pi_\theta(\tau)=P_\theta(s_0,a_0,s_1,a_1,\cdots,s_T,a_T,s_{T+1})=P(s_0)\prod_{t=0}^{T}[\pi_\theta(a_t|s_t)\cdot P(s_{t+1}|s_t,a_t)] \tag{15.7}$$

Besides, the sum of the rewards obtained in the trajectory τ can be formulated as

$$r(\tau)=\sum_{t=0}^{T}r(s_t,a_t) \tag{15.8}$$

Please be aware that, at this moment, we just sum up the rewards from one trajectory to get $r(\tau)$, and the discount factor has not been considered yet. The discount factor will be discussed in a later section for advanced policy gradient algorithms.

Substituting the above two equations into the $\nabla_\theta J(\theta)$ equation, we obtain

$$\begin{aligned}\nabla_\theta J(\theta)=&\mathbb{E}_{\tau\sim\pi_\theta(\tau)}\left[\nabla_\theta\left(\log P(s_0)+\sum_{t=0}^{T}\log\pi_\theta(a_t|s_t)+\log P(s_{t+1}|s_t,a_t)\right)\right.\\&\left.\cdot\left(\sum_{t=0}^{T}r(s_t,a_t)\right)\right]\\=&\mathbb{E}_{\tau\sim\pi_\theta(\tau)}\left[\nabla_\theta\left(\sum_{t=0}^{T}\log\pi_\theta(a_t|s_t)\right)\cdot\left(\sum_{t=0}^{T}r(s_t,a_t)\right)\right]\\=&\mathbb{E}_{\tau\sim\pi_\theta(\tau)}\left[\left(\sum_{t=0}^{T}\nabla_\theta\log\pi_\theta(a_t|s_t)\right)\cdot\left(\sum_{t=0}^{T}r(s_t,a_t)\right)\right]\end{aligned} \tag{15.9}$$

The above equation is called *the policy gradient theorem*. This equation for the calculation of the gradient of the objective function contains an expectation, which cannot be calculated directly. Therefore, in implementations, we can only approximate it with an average of enough samplings of τ:

$$\nabla_\theta J(\theta)\approx\underbrace{\frac{1}{N}\sum_{i=1}^{N}\left[\underbrace{\left(\sum_{t=0}^{T}\nabla_\theta\log\pi_\theta(a_{i,t}|s_{i,t})\right)}_{\text{Policy: probability of }\tau}\cdot\underbrace{\left(\sum_{t=0}^{T}r(s_{i,t},a_{i,t})\right)}_{\text{Value: rewards of }\tau}\right]}_{\text{Use mean to approximate expectation}} \tag{15.10}$$

In addition to the use of the mean to approximate the expectation, there are *two key parts: policy and reward (value)*. The former determines what action will be selected, and the latter tells what the value of the action (or reward generated by the

Table 15.1 Different value evaluation functions in common policy gradient algorithms

Equation for value evaluation	Algorithm
$G_t = \sum_{i=t}^{T} \gamma^{i-t} r_{i+1}$	REINFORCE version 2
$G_t - V_w(s_t)$[a]	REINFORCE with baseline
$Q_w(s_t, a_t)$	Q Actor-Critic
$A_w(s_t, a_t) = Q_w(s_t, a_t) - V_w(s_t)$	Advantage Actor-Critic
$\delta = r_t + V_w(s_{t+1}) - V_w(s_t)$	Temporal-Difference Actor Critic
$\delta(\lambda)$[b]	Eligibility Actor Critic

[a] V_w and Q_w are an approximation to V^π and Q^π, respectively, which are parameterized by w, such as linear functions as $Q(s, a) = \phi(s, a) \cdot w$ and a deep neural network, and will be improved in the training process

[b] $\delta(\lambda)$ is the δ obtained using a TD(λ) scheme, which will be introduced later

action) is. For the policy, we will need to find a way to mathematically formulate π_θ and, based on it, obtain a simple way to obtain the gradient of (the log of) the policy. For the reward (value), in a simple way, we can directly implement the above equation by adding up the rewards for all the steps of a τ. The part for evaluating the value (reward) in the above equation is $\sum_{t=0}^{T} r(s_{i,t}, a_{i,t})$. This part can be replaced by other equations for valuation evaluations as listed in Table 15.1. In a later subsection, we will discuss more ways of evaluating the value, which leads to more advanced policy gradient algorithms.

15.4.3 Simple Episodic Monte Carlo Implementation of Policy Gradient: REINFORCE V1

In the remaining part of this subsection, we will give out a simple Monte Carlo algorithm to implement the above theory. This algorithm is called REINFORCE in many places [166] but is named REINFORCE version 1 here to differentiate it from some other variations to be introduced in later subsections. The following is the pseudo-code for REINFORCE version 1.

REINFORCE version 1

Initialize the policy parameter θ at random.
Repeat for each update (N episodes):
  Generate N samples of τ with the policy π_θ
  Calculate $\nabla_\theta J(\theta) \approx \frac{1}{N} \sum_{i=1}^{N} \left[\left(\sum_{t=1}^{T} \nabla_\theta \log \pi_\theta(a_{i,t}|s_{i,t}) \right) \cdot \left(\sum_{t=1}^{T} r(s_{i,t}, a_{i,t}) \right) \right]$
  Update policy parameters: $\theta \leftarrow \theta + \alpha \nabla_\theta J(\theta)$
  Until the criterion for stopping the training (updating θ) is met.

This pseudo-code is presented to illustrate a procedure for simply implementing a policy gradient method with the above basic form of the policy gradient. If you try it, you will find that it is relatively difficult to make the policy to converge. This is because the updates on the policy calculated with $\nabla_\theta J(\theta)$ have high variance, which, in some cases, may be too high to ensure convergence to the optimal policy within the allotted time.

15.4.4 Strategies for Improving Policy Gradient Implementation

Different strategies have been explored to improve the performance of policy gradient methods, some of which led to popular algorithms. These strategies are grouped into the following types, which correspond to the different components in the above basic form of the policy gradient theorem (Eq. 15.10):

(1) The first type of strategy is to improve the calculation of the expectation $\mathbb{E}$. A simple and straightforward way is to increase the number of episodes per gradient update. More advanced ways propose special procedures to implement the expectation operation, such as TRPO and PPO [167], which can prevent the policy from changing too rapidly.
(2) The second type of strategy was developed from the perspective of the policy. We can propose more flexible and suitable functions for the policy. One typical example is the use of a deep neural network to replace traditional logistic, Softmax, and Gaussian policy functions [168].
(3) The third type is devoted to the value part. This can be the reformulation of the cumulative rewards by adding the discount factor or the removal of the rewards before the current step, which will be introduced in the next subsection. Reward normalization by subtracting the mean and/or dividing by the standard deviation can also effectively reduce the variance. Among this type, one major subcategory of strategies is the use of new functions for the value part, which generates lots of policy gradient algorithms to be introduced (see Table 15.1). In addition to new value functions, better estimations of these new values, such as a generalized advantage estimation, can also help control the variance.

Section 15.5 will be devoted to Item 2. More information about Items 1 and 3 can be found in Sect. 15.6.

15.5 Policy Function

As mentioned above, the Softmax function is usually used for discrete actions or, more accurately, the selection of discrete actions based on different state variables. Let us first recall this equation:

$$f(x) = \frac{1}{1 + e^{-x}} \tag{15.11}$$

Next, let us use a problem with two actions (i.e., a_0 and a_1) and four state variables (or called observations; thus, each state is described by four state variables/observations) to illustrate the use of the Softmax function to construct a policy function $\vec{\pi}_\theta(a|s)$:

$$\vec{\pi}_\theta(a|\vec{s}) = \begin{bmatrix} \pi_\theta(a = a_0|\vec{s}) \\ \pi_\theta(a = a_1|\vec{s}) \end{bmatrix} = \begin{bmatrix} \frac{1}{1+e^{-\vec{\phi}(\vec{s},a_0)^T \cdot \vec{\theta}}} \\ 1 - \frac{1}{1+e^{-\vec{\phi}(\vec{s},a_0)^T \cdot \vec{\theta}}} \end{bmatrix} \tag{15.12}$$

where $\vec{\phi}(\vec{s}, a_0)^T \cdot \vec{\theta}$ is used considering that we usually need to use multiple parameters to define a model. To facilitate the following deduction, we use $\vec{\theta}$ to represent a column array that contains all the RL model parameters.

15.5.1 *Linear Policy Function for Discrete Actions: Formulation 1*

The four state variables $\vec{s}_{4\times1} = [s_1, s_2, s_3, s_4]^T$ and the associated $\vec{\theta}$ and $\vec{\phi}(s, a)$ can be viewed as a vector or a vertical array and written formally as

$$\vec{s}_{4\times1} = \begin{bmatrix} s_0 \\ s_1 \\ s_2 \\ s_3 \end{bmatrix}, \vec{\phi}(s, a_0)_{8\times1} = \begin{bmatrix} s_0 \\ s_1 \\ s_2 \\ s_3 \\ 0 \\ 0 \\ 0 \\ 0 \end{bmatrix}, \vec{\phi}(s, a_1)_{8\times1} = \begin{bmatrix} 0 \\ 0 \\ 0 \\ 0 \\ s_0 \\ s_1 \\ s_2 \\ s_3 \end{bmatrix}, \vec{\theta}(s)_{8\times1} = \begin{bmatrix} \theta_0 \\ \theta_1 \\ \theta_2 \\ \theta_3 \\ \theta_4 \\ \theta_5 \\ \theta_6 \\ \theta_7 \end{bmatrix}$$

In Step t, the probability of selecting action a_t in state $\vec{s}$, i.e., $\pi_\theta(a_t|\vec{s})$, is written as $\pi_\theta(a_t, \vec{s})$ in some places, which is not correct. This is because, strictly speaking, we should have $\pi_\theta(a_t, \vec{s}) = \pi_\theta(a_t|\vec{s}) \cdot P(\vec{s})$. It is noted that, in Eq. 15.12, we have $\sum \pi_\theta(a_t|\vec{s}) = \pi_\theta(a_t = a_0|\vec{s}) + \pi_\theta(a_t = a_1|\vec{s}) = 1$.

We can employ the Softmax function for the policy, which can used for two or more actions:

$$\pi_\theta(a_t|s) = \frac{e^{\vec{\phi}(s,a_t)^T \cdot \vec{\theta}}}{\sum_j e^{\vec{\phi}(s,a_j)^T \cdot \vec{\theta}}} \tag{15.13}$$

where a_t is the action selected in step t and $a_t \in \{a_0, a_1\}$ in this example has two possible actions.

The policy gradient then can be calculated using the following equation, for which the derivation is not given here:

$$\nabla_{\vec{\theta}} \ln \vec{\pi}_\theta(a_t|\vec{s})_{8\times 1} = \vec{\phi}(\vec{s}, a_t) - \mathbb{E}[\vec{\phi}(\vec{s}, \cdot)] = \vec{\phi}(\vec{s}, a_t) - \sum_i [\vec{\phi}(\vec{s}, a_i)^T \cdot \pi_\theta(a_i|s)]$$
$$= \vec{\phi}(\vec{s}, a_t) - \sum_i \left[\vec{\phi}(\vec{s}, a_i) \cdot \frac{e^{\vec{\phi}(\vec{s},a_i)^T \cdot \vec{\theta}}}{\sum_j e^{\vec{\phi}(\vec{s},a_j)^T \cdot \vec{\theta}}} \right] \tag{15.14}$$

The above equation is for the update in a step. As illustrated, in Step t, a_t is the selected action in the state described by $\vec{s}$. $\nabla_{\vec{\theta}} \ln \vec{\pi}_\theta(a_t|\vec{s})$ is also known as the score function. The update on $\vec{\theta}$, which parameterizes the policy π_θ, is performed with this score function as follows:

$$\vec{\theta}_{8\times 1} = \vec{\theta} + \alpha \cdot [\nabla \ln \vec{\pi}_\theta(a_t|\vec{s}) \cdot r(\tau)] \tag{15.15}$$

where the formula for $r(\tau)$ can be found in the above context, and more (improved) formulae will be introduced in the next subsection.

15.5.2 Linear Policy Function for Discrete Actions: Formulation 2

As mentioned, the purpose of a linear policy function is to construct a function to predict the policy (coefficient $\vec{\theta}$) with the linear combination of state (variables) and actions. Thus, there are different ways to achieve this goal.

The following is another way, in which both the vectors and the equations of policy and policy gradient need to be reformulated correspondingly.

$$\vec{s}_{4\times 1} = \begin{bmatrix} s_0 \\ s_1 \\ s_2 \\ s_3 \end{bmatrix}, \vec{\phi}(s)_{4\times 1} = \vec{s}_{4\times 1} = \begin{bmatrix} s_0 \\ s_1 \\ s_2 \\ s_3 \end{bmatrix}, \vec{\theta}(\vec{s}, \vec{a})_{4\times 2} = [\vec{\theta}_{a_1}, \vec{\theta}_{a_2}] = \begin{bmatrix} \theta_0, \theta_4 \\ \theta_1, \theta_5 \\ \theta_2, \theta_6 \\ \theta_3, \theta_7 \end{bmatrix}$$

The policy function is constructed using the Softmax function in the same way:

$$\pi_\theta(a_t|s) = \frac{e^{\vec{\phi}(s)^T \cdot \vec{\theta}_{a_t}}}{\sum_j e^{\vec{\phi}(s)^T \cdot \vec{\theta}_{a_j}}} \tag{15.16}$$

The gradient of the policy function, or the score function, can be derived to obtain the following equation:

$$\nabla_{\vec{\theta}_{a_k}} \ln \vec{\pi}_\theta(a_t|\vec{s})_{4\times 1} = \vec{\phi}(\vec{s}) \cdot \delta_{kt} - \mathbb{E}[\vec{\phi}(\vec{s}, \cdot)] = \vec{\phi}(\vec{s}) \cdot \delta_{ij} - \vec{\phi}(\vec{s}) \cdot \pi_\theta(a_t|\vec{s})$$
$$= \vec{\phi}(\vec{s})_{4\times 1} \delta_{kt} - \vec{\phi}(\vec{s})_{4\times 1} \cdot \frac{e^{\vec{\phi}(\vec{s})^T \cdot \vec{\theta}_{a_t}}}{\sum_j e^{\vec{\phi}(\vec{s})^T \cdot \vec{\theta}_{a_j}}} \tag{15.17}$$

where δ_{kt} is 1 when $k = t$ and 0 when $k \neq t$. The above equation can be used to calculate the score functions for different model parameters $\vec{\theta}_{a_k}$ ($a_k \in \{a_0, a_1\}$), which is different from Formulation 1. These different score functions can be adopted to update the corresponding parameters:

$$\vec{\theta}_{a_k} = \vec{\theta}_{a_k} + \alpha \cdot \left[\nabla_{\vec{\theta}_{a_k}} \ln \vec{\pi}_\theta(a_t|\vec{s})_{4\times 1} \cdot r(\tau)_{1\times 1} \right] \tag{15.18}$$

15.5.3 *Policy Function for Continuous Actions*

The Gaussian policy is suitable for continuous actions. In such cases, an action will be selected from a range according to a policy represented by a Gaussian distribution $\mathcal{N}(\vec{\phi}(s)^T \cdot \vec{\theta}, \sigma^2)$, in which $\mu(s) = \vec{\phi}(s)^T \cdot \vec{\theta}$. In the CartPole environment [169], $\vec{s}$, $\vec{\phi}$, and $\vec{\theta}$ can be constructed as follows:

$$\vec{s}_{4\times 1} = \begin{bmatrix} s_0 \\ s_1 \\ s_2 \\ s_3 \end{bmatrix}, \vec{\phi}(\vec{s})_{4\times 1} = \vec{s}_{4\times 1} = \begin{bmatrix} s_0 \\ s_1 \\ s_2 \\ s_3 \end{bmatrix}, \vec{\theta}_{4\times 1} = \begin{bmatrix} \theta_0 \\ \theta_1 \\ \theta_2 \\ \theta_3 \end{bmatrix}$$

The mathematical equation for the policy is

$$\pi_\theta(a_t|\vec{s}) = \frac{1}{\sqrt{2\pi}\sigma} e^{-\frac{[a_t - \vec{\phi}(\vec{s})^T \cdot \vec{\theta}]^2}{2\sigma^2}} \tag{15.19}$$

The mathematical formulation of the policy gradient is

$$\nabla_{\vec{\theta}} \ln \pi_\theta(a_t|\vec{s})_{4\times 1} = \frac{[a_t - \mu(\vec{s})]\vec{\phi}(\vec{s})_{4\times 1}}{\sigma^2} = \frac{[a_t - \vec{\phi}(\vec{s})^T \cdot \vec{\theta}]\vec{\phi}(\vec{s})}{\sigma^2} \tag{15.20}$$

The above score can be used to update the model parameters as

$$\vec{\theta}_{4\times 1} = \vec{\theta} + \alpha \cdot \left[\nabla_{\vec{\theta}} \ln \vec{\pi}_\theta(a_t|\vec{s}) \cdot r(\tau) \right] \tag{15.21}$$

15.6 Common Policy Gradient Algorithms

The previous subsection provides a simple way of constructing the objective function and deriving the policy gradient theorem. This subsection shows more objective functions, which could lead to a slightly different equation for the policy gradient theorem and a more widely accepted algorithm for implementing a Monte Carlo policy gradient method. Based on that, more advanced policy-based algorithms will be introduced.

15.6.1 More Objective Function Formulations

- When whole episodes can be generated, we can use the start value as the objective function:

$$J_1(\theta) = V^{\pi_\theta}(s_1) = \mathbb{E}_{\pi_\theta}[v_1] \tag{15.22}$$

- In a continuous environment, we can use the *average value* as the objective function:

$$J_{avV}(\theta) = \sum_s d^{\pi_\theta}(s) V^{\pi_\theta}(s) \tag{15.23}$$

- We can also use the average reward per time-step as the objective function:

$$J_{avR}(\theta) = \sum_s d^{\pi_\theta}(s) \sum_a \left[\pi_\theta(s,a) R_s^a\right] \tag{15.24}$$

In the above objective functions, $d^{\pi_\theta}(s)$ is the probability of generating s in policy π_θ, and $R_s^a = \sum_{s'} P_{ss'}^a R_{ss'}^a$. For J_1, $J_{avV}/(1+\gamma)$, and J_{avR}, we can strictly prove that the same equation for policy gradient, i.e., Eq. 15.25, can be derived as long as $\pi_\theta(s,a)$ is differentiable:

$$\nabla_\theta J(\theta) = \mathbb{E}_{\pi_\theta}\left[\nabla_\theta \log \pi_\theta(s,a) Q^{\pi_\theta}(s,a)\right] \tag{15.25}$$

Instead of deriving this equation strictly, which is relatively complicated, here we use another simple yet less strict way to show how to move from the basic form of the policy gradient theorem introduced in the previous subsection to the above equation. Let us first recall the basic form of the policy gradient theorem:

$$\begin{aligned}
\nabla_\theta J(\theta) =& \mathbb{E}_{\tau \sim \pi_\theta(\tau)}\left[\nabla_\theta \log \pi_\theta(\tau) \cdot r(\tau)\right] \\
\approx& \frac{1}{N} \sum_{i=0}^{N} \left[\underbrace{\left(\sum_{t=0}^{T} \nabla_\theta \log \pi_\theta(a_{i,t}|s_{i,t})\right)}_{\text{Policy}} \cdot \underbrace{\left(\sum_{t=1}^{T} r(s_{i,t}, a_{i,t})\right)}_{\text{Value: Policy Evaluation}} \right]
\end{aligned} \tag{15.26}$$

If we compare Eqs. 15.26 and 15.25, we can observe differences in both the policy part (episodic score function vs. step score function) and value part (pure sum of episode vs Q function a state-action set). Let us first take a look at the value part to see how to move from the basic form for episodic updates to the new form for stepwise updates. If we look at the basic form, we can see that the rewards of the whole trajectory $r(\tau)$ serve as a multiplier for every grad-log-prob term in the sum over the episode. That is, the reward obtained over the whole episode is used to change the probabilities of each action taken during the episode. This means that those rewards obtained before an action will still affect that action via the gradient update. This is not desirable because we want to attribute an action only to those (and future) rewards received after taking that action. With this change, the basic form changes to

$$\nabla_\theta J(\theta) = \mathbb{E}_{\tau\sim\pi_\theta(\tau)}\left[\left(\sum_{t=0}^{T}\nabla_\theta \log \pi_\theta(a_{i,t}|s_{i,t})\right)\cdot\left(\sum_{t=t'}^{T} r(s_{i,t'}, a_{i,t'})\right)\right] \tag{15.27}$$

Another general observation is that adding a discount factor to the consideration of the value will improve the issue of high variance that we discussed before. Therefore, we can add discount factors to the following steps to construct R_t:

$$\begin{aligned}\nabla_\theta J(\theta) =&\mathbb{E}_{\tau\sim\pi_\theta(\tau)}\left[\left(\sum_{t=0}^{T}\nabla_\theta \log \pi_\theta(a_{i,t}|s_{i,t})\right)\cdot\left(\sum_{t=t'}^{T} \gamma^{t'-t} r(s_{i,t'}, a_{i,t'})\right)\right]\\ =&\mathbb{E}_{\tau\sim\pi_\theta(\tau)}\left[\left(\sum_{t=0}^{T}\nabla_\theta \log \pi_\theta(a_{i,t}|s_{i,t})\right)\cdot R_t\right]\end{aligned} \tag{15.28}$$

Again, R_t is expressed with G_t in some literature. Next, we apply the expectation operation on the two terms on the right-hand side of the equation separately:

$$\begin{aligned}\nabla_\theta J(\theta) =&\mathbb{E}_{\tau}\left[\left(\sum_{t=0}^{T}\nabla_\theta \log \pi_\theta(a_{i,t}|s_{i,t})\right)\right]\cdot \mathbb{E}_\tau[R_t]\\ =&\mathbb{E}_{s_0,a_0,\ldots,s_T,a_T}\left[\left(\sum_{t=0}^{T}\nabla_\theta \log \pi_\theta(a_{i,t}|s_{i,t})\right)\right]\cdot \mathbb{E}_{s_{t+1},r_{t+1},\ldots,s_T,r_T}[R_t]\\ =&\mathbb{E}_{s_0,a_0,\ldots,s_T,a_T}\left[\left(\sum_{t=0}^{T}\nabla_\theta \log \pi_\theta(a_{i,t}|s_{i,t})\right)\right]\cdot Q(s_t,a_t)\\ =&\mathbb{E}_{\tau}\left[\left(\sum_{t=0}^{T}\nabla_\theta \log \pi_\theta(a_{i,t}|s_{i,t})\right)\cdot Q(s_t,a_t)\right]\\ =&\mathbb{E}_{\tau}\left[\sum_{t=0}^{T}\nabla_\theta \log \pi_\theta(a_{i,t}|s_{i,t})\cdot Q(s_t,a_t)\right]\end{aligned} \tag{15.29}$$

In the above deduction, we employed the definition of the Q function: $Q(s_t, a_t) = \mathbb{E}_{s_{t+1},r_{t+1},\ldots,s_T,r_T}[R_t]$ or more strictly as

$$Q^{\pi_\theta}(s,a) = \mathbb{E}_{\pi_\theta}[r_t|s_t, a_t] \tag{15.30}$$

where R_t is the cumulative reward when taking an action a_t in a state s_t. Here, we use subscript t to emphasize that they are a specific action/state (as a value of a variable) in a trajectory/episode, whereas the small-letter a and s are for any given action/state (as a variable). R_t was rewritten as G_t in many reinforcement learning literature. Then, the above policy gradient equation can be reformulated as

$$\nabla_\theta J(\theta) = \mathbb{E}_{\pi_\theta}\left[\sum_{t=0}^{T} \nabla_\theta \log \pi_\theta(s_t, a_t) R_t\right] \tag{15.31}$$

In the above equation, we can see that both $Q(s_t, a_t)$ and R_t can be used for the value part, though Q could be more accurate in some contexts. However, R_t is better for application since Q is an expectation.

15.6.2 Simple Stepwise Monte Carlo Implementation of Policy Gradient: REINFORCE V2

In the last step, let us take a look at $\nabla_\theta J(\theta)$, which is used to calculate the increment of the policy. As shown in Eq. 15.31, a large increment can be obtained by obtaining the sum of $\nabla_\theta \log \pi_\theta(a_{i,t}|s_{i,t}) \cdot Q(s_t, a_t)$ for one episode. Therefore, we can imagine that a small increment could be equal or proportional to $\nabla_\theta \log \pi_\theta(a_{i,t}|s_{i,t}) \cdot Q(s_t, a_t)$ directly. Therefore, it is possible to remove the summation operator in the above policy gradient equation to obtain an equation for a smaller increment of $\nabla_\theta J(\theta)$:

$$\nabla_\theta J(\theta) = \mathbb{E}_{\pi_\theta}\left[\nabla_\theta \log \pi_\theta(s_t, a_t) R_t\right] = \mathbb{E}_{\pi_\theta}\left[\nabla_\theta \log \pi_\theta(s_t, a_t) Q(s_t, a_t)\right] \tag{15.32}$$

This equation can be implemented using the following pseudo-code, which is also called REINFORCE version 2.

REINFORCE version 2

Initialize the policy parameter θ at random.
Repeat for each update (1 episodes):
  Generate 1 sample of τ with the policy π_θ
  for $t = 1, 2, \cdots, T$:
    Calculate $R_t = \sum_{k=0}^{T-t} \gamma^k r_{t+k+1}$
    Calculate $\nabla_\theta J(\theta) = \left[\nabla_\theta \log \pi_\theta(s_t, a_t) R_t(s_t, a_t)\right]$
    Update policy parameters: $\theta \leftarrow \theta + \alpha \nabla_\theta J(\theta)$ (or $\theta \leftarrow \theta + \alpha \gamma^t \nabla_\theta J(\theta)$)
  Until the criterion for stopping the training (updating θ) is met.

Compared with version 1, this version of REINFORCE seemingly updates θ after every step in an episode instead of every N episode. However, the policy is essentially updated every episode because (1) an episode/sample needs to be generated before starting the policy update for that episode and (2) the updated policy will take effect when generating the next episode. Thus, this is still a Monte Carlo method.

15.6.3 Actor-Critic

REINFORCE exhibits some major issues:

- An agent will take many actions in one episode. It is difficult to tell whether an action is better. That is, this method has a high variance.
- The time required for reaching converged results is long.
- It can only be used in an episodic environment.

Many efforts have been made to address the above issues, especially for reducing the high variance. One good way of reducing the variance is the use of discount factor as shown in Version 2 of REINFORCE ($\theta \leftarrow \theta + \alpha\gamma^t \nabla_\theta J(\theta)$). Another good example of such efforts is the Actor-Critic algorithm [170].

As mentioned, the high variance is attributed to the fact that it is hard to evaluate the action. To address this issue, the Actor-Critic algorithm adds a step to the policy gradient algorithm for evaluating actions. The original policy gradient part in REINFORCE—updating the policy by adjusting θ so that actions leading to more rewards will have a higher probability of being selected—acts more like an actor for the policy update. The newly added part is intended to evaluate the value of different actions like what we do in value-based methods, which can be viewed as a critic for policy evaluation. In this way, the Actor-Critic integrates and expedites the learning process by enhancing the evaluation ability of the algorithm. Such algorithm integration benefits from both policy-based and value-based methods. We can also find that such RL algorithms also include the two essential steps in dynamic programming: policy update and policy evaluation.

The critic works by using a policy evaluation function to approximate the real action-state function:

$$Q_w(s, a) \approx Q^{\pi_\theta}(s, a) \tag{15.33}$$

As a result, the Actor-Critic algorithm has two sets of parameters.

Actor: Update the policy parameters, θ.

Critic: Update the parameters of the action-state value function, w.

Because $Q_w(s, a)$ is just an approximation of the real action-state function, the Actor-Critic can be viewed as an approximation instead of the exact policy gradient method:

$$\nabla_\theta J(\theta) \approx \mathbb{E}_{\pi_\theta}\left[\nabla_\theta \log \pi_\theta(s, a) Q_w(s, a)\right] \tag{15.34}$$

For the critic, a simple treatment is to use a linear function to approximate the action-state value function:

$$Q_w(s, a) = \vec{\phi}(s, a)^T \cdot \vec{w} \tag{15.35}$$

Then, we can update the parameter w using

$$\vec{w} = \vec{w} + \Delta\vec{w} = \vec{w} + \vec{\phi}(s, a) \cdot \Delta Q_w(s, a) \tag{15.36}$$

In the above equation, it is assumed that $\vec{\phi}(s, a)^T = \vec{\phi}(s, a)^{-1}$. Next, to obtain $\Delta Q(s, a)$, we can recall the optimal Q value update equation in value-based learning algorithms:

$$Q(s, a) \leftarrow (1 - \alpha)Q(s, a) + \alpha[r + \gamma Q(s', a')] = Q(s, a) + \Delta Q \tag{15.37}$$

Based on the equation, ΔQ can be obtained as the difference between the current and next Q values within one update:

$$\Delta Q(s, a) = \alpha \cdot \delta = \alpha \cdot [r + \gamma Q(s', a') - Q(s, a)] \tag{15.38}$$

where δ is the error of time difference (TD error).

To sum up, we have

$$\vec{w} = \vec{w} + \Delta\vec{w} = \vec{w} + \alpha \cdot \vec{\phi}(s, a) \cdot \delta = \vec{w} + \alpha \cdot \vec{\phi}(s, a) \cdot [r + \gamma Q(s', a') - Q(s, a)] \tag{15.39}$$

Then, the Actor-Critic can be implemented using the following pseudo-code.

Actor-Critic

Initialize the policy parameter $\vec{\theta}$ and $\vec{w}$ at random.
Repeat for each update (1 episode):
  Generate 1 sample of τ with the policy π_θ
  for $t = 1, 2, \cdots, T$:
    Calculate $R_t = \sum_{k=0}^{T-t} \gamma^k r_{t+k+1}$
    Calculate $\delta = \cdot[r + \gamma Q(s', a') - Q(s, a)]$
    Update policy parameters: $\vec{\theta} \leftarrow \vec{\theta} + \alpha\nabla_{\vec{\theta}} J(\theta)$
    Update value function parameters $\vec{w}$: $\vec{w} = \vec{w} + \alpha \cdot \vec{\phi}(s, a) \cdot \delta = \vec{w} + \alpha \cdot \vec{\phi}(s, a) \cdot [r + \gamma Q(s', a') - Q(s, a)])$
  Until the criterion for stopping the training (updating $\vec{\theta}$) is met.

In the Actor-Critic method, we use $Q_w(s, a)$ to approximate $Q^{\pi_\theta}(s, a)$. Thus, this method adopts an approximate policy gradient. This approximation can introduce a bias, which can defeat the convergence to a suitable policy. It has been proven that a well-designed $Q_w(s, a)$ that satisfies the following two conditions can remove the bias. This is called the compatible function approximation theorem:

- $\nabla_{\vec{w}} Q_w(s, a) = \nabla_{\vec{\theta}} \log \pi_\theta(s, a)$
- $\epsilon = \mathbb{E}_{\pi_\theta}\left[(Q^{\pi_\theta}(s, a) - Q_w(s, a))^2\right]$

where ϵ is a very small constant.

15.6.4 Actor-Critic with Baseline

Besides the introduction of a critic, we can also reduce the variance by subtracting a baseline [171]. In detail, we can introduce a baseline function $B(s)$, which will be taken off from the policy gradient. It is intentionally designed to be a function of only states and irrelevant to action. As a result, this function will not change the policy gradient:

$$\begin{aligned}\mathbb{E}_{\pi_\theta}\left[\nabla_\theta \log \pi_\theta(s,a)B(s)\right] &= \sum_s d^{\pi_\theta}(s)\sum_a \left[\nabla_\theta \log \pi_\theta(s,a)B(s)\right] \\ &= \sum_s d^{\pi_\theta}(s)B(s)\sum_a \left[\nabla_\theta \log \pi_\theta(s,a)\right] = 0\end{aligned} \tag{15.40}$$

A good choice for $B(s)$ is the state value function $V^{\pi_\theta}(s)$. Based on this perception, we introduce an advantage function $A^{\pi_\theta}(s,a)$, which is defined as

$$A^{\pi_\theta}(s,a) = Q^{\pi_\theta}(s,a) - V^{\pi_\theta}(s) \tag{15.41}$$

Then, we will use this advantage function to replace Q^{π_θ} when calculating the policy gradient:

$$\nabla_{\vec{\theta}} J(\theta) = \mathbb{E}_{\pi_\theta}\left[\nabla_{\vec{\theta}} \log \pi_\theta(s,a)A^{\pi_\theta}(s,a)\right] \tag{15.42}$$

The next question is how to obtain $A^{\pi_\theta}(s)$. Now, the equation for $A^{\pi_\theta}(s)$ has two functions Q and V. Previous experience indicates that, no matter whether we use linear functions just like what we introduced for the basic Q Actor-Critic method or use deep neural networks to approximate such equations, we do not need to deal with two functions. To address this issue, we can utilize the Bellman equation:

$$Q(s_t,a_t) = \mathbb{E}\left[r_{t+1} + \gamma V(s_{t+1})\right] \tag{15.43}$$

So, we can rewrite the advantage value function as

$$A^{\pi_\theta}(s,a) = Q^{\pi_\theta}(s,a) - V^{\pi_\theta}(s) = r_{t+1} + \gamma V(s_{t+1}) - V(s_t) \tag{15.44}$$

As a result, we usually use the TD error to approximate $A^{\pi_\theta}(s,a)$, because it is an unbiased estimate of $A^{\pi_\theta}(s,a)$:

$$\delta^{\pi_\theta} = r + \gamma V^{\pi_\theta}(s') - V^{\pi_\theta}(s) \tag{15.45}$$

where $t' = t+1$, and here we carry the superscript π_θ to be accurate, which can be removed for simplicity and approximated using w (e.g., approximate policy in terms of linear functions or neural networks parameterized with w) in implementations. The expectation of this TD error is

$$\begin{aligned}\mathbb{E}_{\pi_\theta}[\delta^{\pi_\theta}|s,a] =& \mathbb{E}_{\pi_\theta}[r+\gamma V^{\pi_\theta}(s')|s,a] - V^{\pi_\theta}(s) \\ =& Q^{\pi_\theta}(s,a) - V^{\pi_\theta}(s) = A^{\pi_\theta}(s,a)\end{aligned} \tag{15.46}$$

Hence, we have

$$\nabla_{\vec{\theta}} J(\theta) = \mathbb{E}_{\pi_\theta}\left[\nabla_{\vec{\theta}} \log \pi_\theta(s,a) \cdot \delta^{\pi_\theta}\right]. \tag{15.47}$$

15.6.5 More Policy Gradient Algorithms

The above TD scheme is TD(0), in which we only go forward a step to calculate the TD difference. To improve the performance, we can also go with TD(λ). The Advantage Actor-Critic method is abbreviated as A2C algorithms in many places. When allowing more actors to gain experience asynchronously, we can get A3C [172].

In fact, we can see that the baseline method can also be used with an MC implementation of the policy gradient theorem (which is called vanilla PG [173]) and also with Actor-Critic, which is a TD implementation of the policy gradient theorem. The difference between these two types is analogous to that between the MC implementation of value-based methods and Sarsa. In the former, we obtained all the stepwise rewards r and can calculate the V function ($G_t - V = B$), while in the latter, we will need to use the current strategy to explore the next step and then calculate the TD error (A) to update the policy.

Finally, deep neural networks can be used for both the policy function (e.g., actor) and the evaluation function (e.g., critic). This can enable various deep reinforcement learning methods.

15.7 Practice: Understand and Modify Policy Gradient Code for Addressing RL Problem

>>> More and up-to-date course materials including practices @ AI-engineer.org <<<

Please work on the following items one by one.

1. Please understand a simple "Treasure1D" environment, which was explained in the chapter for value-based RL.

```
import numpy as np

class treasure1D:
    def __init__(self,N,treasure_loc):
        self.n_cell = N # Number of cells; the greater this
    number, the higher n_episode is needed
```

```
        self.Reward = np.zeros(self.n_cell);
        self.Reward[treasure_loc]=1.0 # n_cell cells, only the
    rightmost has reward (end of game)
        self.States = range(self.n_cell)
        self.state = np.random.choice(self.States)
        self.reward = 0
        self.done = False
        self.action_space = np.array(['left','right'])
        self.state_size = N # Gym environments provide this
    variable
        self.observation_size = 1 # Number of elements in
    observation (there is only one in this env: state/position)

    def reset(self):
        self.state = np.random.choice(self.States)
        self.reward = 0
        self.done = False
        return self.state

    def step(self,action):
        # Update state
        reward = 0
        if action == 'left' or action == 0:
            if self.state == self.States[0]: # Leftmost
                state_new = self.state # Does not move
                self.done = True
                self.reward = -100
            else:
                state_new = self.state - 1 # Obtain new state s^
    prime
                self.reward = 1
        else:
            if self.state == self.States[-1]: # rightmost
                state_new = self.state # Does not move
                self.done = True
                self.reward = 1000
            else:
                state_new = self.state + 1 # Obtain new state s^
    prime
                self.reward = 1
        self.state = state_new
        # self.reward = self.Reward[state_new]
        return self.state, self.reward, self.done, [], []

    def render(self):
        print ("The state is ", self.state, "The reward is ",self
    .reward)
```

2. Please understand the following code for policy gradient (using Discrete Action Formulation 2 in the textbook). Please run it to get the result and explain what happens. Next, please try to use a different seed number, i.e., 123, and see if you can still get good results. If not, think about why and find a way to address the issue.

```
import numpy as np
import pandas as pd
from envs_liu2 import treasure1D

# Hyperparameter
alpha = 0.00001 # Learning rate
gamma = 0.8 # Decay of rewards; Not used.

# Initialization
rs = np.random.RandomState(9527) # Please try seed 123 (effects
    on the initialization of Theta)
n_cell = 10 # The greater this number, the higher n_episode is
    needed
treasure_loc = -1 # Treasure state/position/cell

env = treasure1D(n_cell,treasure_loc)
state = env.reset()

Actions = env.action_space
States = np.arange(n_cell)
S = np.zeros((env.observation_size,1))
# Option 1 (Seed like 123 will not converge because it rarely
    moves to the rightmost; seed like 9627 can converge)
Theta = rs.rand(env.observation_size,Actions.size)
# Option 2 ?

# Function for Sampling
def sampling(States,Actions,Theta):
    Sample = []
    # Choose an initial action randomly
    state = np.random.choice(States)
    S = np.array(state).reshape(env.observation_size,1)

    while True:
        # Choose action following the policy function (Softmax)
        action = np.random.choice(Actions,size=1,p=(np.exp(S.
    T@Theta)/np.sum(np.exp(S.T@Theta))).flatten() )[0]
        print(state,Theta)
        # Update state
        state_new,reward,done,_,_ = env.step(action)
        Sample.append([state,action,reward])

        state = state_new
        S = np.array(state).reshape(env.observation_size,1)
        # Stop an episode when the treasure is found
        if state == States[treasure_loc]:
            break
    return Sample

N_update = 100#100
N = 20 # Number of sampling to approach the expectation of Del(J)

for n in range(N_update):  # n_update iterations for updating
    theta
```

```
    Del_J = np.zeros(Theta.shape)

    for n_ in range(N): # N iterations for obtaining Del(J):
    using average of N episodes to approach expectation
        Sample = sampling(States,Actions,Theta)
        Tau = np.array(Sample)
        Tau_states = np.resize(np.float32(Tau[:,0]),(Tau.shape
    [0],1))
        Tau_actions = Tau[:,1]
        Tau_rewards = np.resize(np.float32(Tau[:,2]),(Tau.shape
    [0],1))
        Sum_del_log_pi = np.zeros(Theta.shape)
        Sum_r = 0
        for t in range(Tau.shape[0]):
            delta_ik = np.zeros((1,Actions.size))
            delta_ik[:,np.where(Actions == Tau_actions[t])[0][0]]
     = 1
            S = Tau_states[t].reshape(env.observation_size,1)
            del_log_pi = S * delta_ik - S @ np.exp(S.T @ Theta) /
     np.exp(S.T @ Theta).sum(axis=1)
            Sum_del_log_pi = Sum_del_log_pi + del_log_pi
        Sum_r = Tau_rewards.sum()
        del_Theta = Sum_del_log_pi * Sum_r
        Del_J = Del_J + 1/N * del_Theta
    Theta = Theta + alpha*Del_J
    print(n, Theta)

# Testing
state = env.reset() # state = 3 # or state = env.reset()
S = np.array(state).reshape(env.observation_size,1)
step = 1
while True:
    # Choose action based on Softmax probabilities
    action = np.random.choice(Actions,size=1,p=(np.exp(S.T@Theta)
    /np.sum(np.exp(S.T@Theta))).flatten() )[0]
    # Update state
    state_new,reward,done,_,_ = env.step(action)
    print(step,state,'--->',state_new)
    step = step + 1
    state = state_new
    S = np.array(state).reshape(env.observation_size,1)
    # Stop an episode when the treasure is found
    if state == States[treasure_loc]:
        break
```

Hints: The seed number affects the initialization of Theta, which is related to the initial probabilities of selecting different actions ("left" or "right").

3. *Please use advanced algorithms like deep Q learning to solve the problem defined in the "CartPole-v" environment from Gymnasium.

Appendix A
Appendices

A.1 Overview

This chapter presents three appendices for three categories of auxiliary materials in AI study and implementations: mathematics for machine learning, optimization, and evaluation metrics. For math, the three most AI-relevant math topics, i.e., statistics, information theory, and array operations, will be provided to bridge possible knowledge gaps for learners and serve as reference materials for experienced practitioners. Next, optimization as an essential step to the solution of machine learning problems will be introduced. In the final, basic evaluation metrics, which can be used to assess the performance of the models in both the training, e.g., loss function, and later testing stages, will be presented. Readers can study the topics in the appendices systematically or selectively based on their needs.

A.2 Mathematics for Machine Learning

Machine learning can be performed from either a theoretical or a practical perspective, both of which have been approached by many learners and practitioners. However, no matter which way to take, mastery of some mathematical knowledge is needed. In particular, statistics and linear algebra are essential in many machine learning algorithms. In addition, some knowledge from tensor analysis, such as tensor operations and tensor notations, is highly related to array operations, which are very common in machine learning, and to the mathematical descriptions of many engineering problems. Although such knowledge is not emphasized in traditional machine learning literature, it can be very helpful in understanding, communicating, and implementing machine learning, especially for engineering problems. This section is presented to bridge the abovementioned possible knowledge gaps.

Z. "L." Liu, *Artificial Intelligence for Engineers*,
https://doi.org/10.1007/978-3-031-75953-6

A.2.1 Statistics

Statistics takes different roles in machine learning, whose significance can vary among different machine learning practitioners. On one hand, many machine learning methods can be learned and implemented without knowing much statistics. That is why many AI practitioners completely excluded statistics except for methods that are directly built on statistics like Bayesian algorithms. On the other hand, effort has been made by statisticians to reinterpret and cast a more solid statistics foundation for most machine learning methods.

Therefore, the role of statistics can be minimized or maximized, depending on the background and intention of the machine learning learners/users. This fact also casts light on the overlap and difference between machine learning and applied statistics: machine learning algorithms process data to achieve a goal (i.e., mapping $f(x)$ from input x to output y), while statistics more focuses on the statistical characteristics of the data (e.g., relationships between random variables (data attributes) and distributions of data and their attributes). There is no correct or wrong about the choice, which more depends on the purpose.

This subsection first discusses some basic yet essential statistics concepts: the two types of probabilities, random variables, conditional probability, chain rule, independence, expectation/variance/deviation/covariance, and common probability distributions. Related topics such as surrogate functions (logistic, Softmax) will also be covered. Then, some important information from another highly relevant applied statistics area, i.e., information theory, will be provided to cover concepts such as information and probability, amount of information (self-information), and differences between distributions (Kullback-Leibler divergence and cross-entropy).

Random Variables

Probability theory is a mathematical framework for analyzing and describing uncertainty. It is used to derive algorithms and analyze the behavior of AI models. The core concept, i.e., probability, can be studied from either the frequentist or Bayesian perspective. In both, probability is associated with specific random variables. Each random variable has its own distribution function, which quantifies the possibility that this random variable takes a specific value. A relevant concept is information, which quantifies the amount of uncertainty in a probability distribution.

In machine learning, a typical example of a random variable is an attribute in a dataset. In such a dataset, different samples can have different values for this attribute, and such values are associated with different probabilities that are determined by the distribution function of this random variable or attribute. If the samples in this dataset only have one attribute, then there will be a univariate distribution function. However, it is more common that there are multiple attributes. This fact leads to the prevalence of multivariate distribution functions or multiple univariate distribution functions, in which the possibilities of different attributes are related and mutually dependent, respectively.

Thus, we can understand the sample generation process, i.e., obtaining lab measurements or collecting samples from a pool of existing data, as a process in which values are generated for new samples' different attributes. Such attributes thus can be viewed as random numbers by assuming that different values for each attribute have different probabilities of occurrence. That is why we can usually use random numbers generated according to certain probability distributions to describe the attributes.

In some machine learning literature, X is employed to represent the variable, while x is adopted to represent the value that it takes. $f(x; a, b)$ means a distribution function of the random variable x parameterized by a and b. It is common to use $f(x)$ or $p(x)$ to represent the probability of x in a continuous distribution function and $P(x = x_i)$ for the probability of a discrete function that describes the probability of x at x_i.

Probabilities

As mentioned above, there are two mainstreams of probability theories. The first is the frequentist probability, which is obtained by measuring the frequencies of events without any presumptions or prior knowledge. Thus, this type of probability starts from the measurements of frequencies of events, for example, the number of times that different events, like the two sides of a coin, would appear. Then, the probability of a more complicated event can be calculated based on the probabilities of the basic events. The second is the Bayesian probability, which infers quantitative levels of uncertainty based on a degree of belief. Hence, some prior knowledge is needed. A typical example is the process of a doctor diagnosing a patient. The diagnosis of a disease is usually performed based on some symptoms, each of which is associated with a certain probability of a specific disease. In this example, the probability of a disease is inferred based on the prior knowledge about the related symptoms and their probabilities of indicating this disease.

Conditional Statement

In probability theories, conditional probability is a measure of the probability that an event occurs under the condition that another event has already occurred. If the event of interest is X and the event Y is known or assumed to have occurred, "the conditional probability of X given Y," or "the probability of X under the condition Y," is usually written as $P(X|Y)$ and occasionally as $P_Y(X)$. This can also be understood as the fraction of probability Y that intersects with X:

$$P(X \mid Y) = \frac{P(X \cap Y)}{P(Y)} \tag{A.1}$$

$P(X|Y)$ may or may not be equal to the unconditional probability of X: $P(X)$. If $P(X|Y) = P(X)$, then events X and Y are considered to be independent.

Bayes' Theorem

In probability theory and statistics, Bayes' theorem, or called Bayes' law or Bayes' rule, describes the probability of an event based on prior knowledge of conditions related to the event. Bayes' theorem can be derived from the definition of conditional probability:

$$P(X \mid Y) = \frac{P(X \cap Y)}{P(Y)}, \text{ if } P(Y) \neq 0, \tag{A.2}$$

where $P(X \cap Y)$ is the probability of both X and Y being true. This so-called joint probability is also frequently written as $P(X, Y)$. Similarly, we can also have the following conditional probability equation:

$$P(Y \mid X) = \frac{P(X \cap Y)}{P(X)}, \text{ if } P(X) \neq 0, \tag{A.3}$$

Obtaining $P(X \cap Y)$ from the above equation and substituting it into Eq. A.2 for $P(X \mid Y)$ yields Bayes' theorem:

$$P(X \mid Y) = \frac{P(Y \mid X)P(X)}{P(Y)}, \text{ if } P(Y) \neq 0. \tag{A.4}$$

The following is a complete explanation of different terms in the above equation.

X and Y are events, and $P(Y) \neq 0$.

$P(X \mid Y)$ is a conditional probability: the probability of event X occurring given that Y is true. It is also called the posterior probability of X given Y.

$P(Y \mid X)$ is also a conditional probability: the probability of event Y occurring given that X is true. The probability is often be interpreted as the *likelihood* of Y given a fixed X.

$P(X)$ and $P(Y)$ are the probabilities of observing X and Y, respectively, without any given conditions. They are also frequently referred to as the prior probability or marginal probability.

Law of Total Probability

The law of total probability is another useful tool that can extend Bayes' theorem to consider more events. To understand this theorem, let us assume that, in its discrete case, $\{Y_j : j = 1, 2, 3, \ldots\}$ is a finite or countably infinite partition of a sample space, and each event Y_j is measurable. That is to say, $\{Y_j : j = 1, 2, 3, \ldots\}$ is a set of pairwise disjoint events whose union is the entire sample space. Then, for any event X of this probability space, we have

$$P(X) = \sum_j P(X \cap Y_j). \tag{A.5}$$

The above law of total probability can also be alternatively formulated with the inclusion of conditional probabilities:

$$P(X) = \sum_j P(X \mid Y_j)P(Y_j), \tag{A.6}$$

or as

$$P(X) = \sum_j P(Y_j \mid X)P(X), \tag{A.7}$$

Substituting the law of total probability into the Bayesian theorem, we can obtain the format of the Bayesian theorem that we frequently use in Bayesian classifiers:

$$P(X \mid Y) = \frac{\sum_{j=1}^{J} P(Y_j \mid X)P(X)}{P(Y)} \tag{A.8}$$

As can be seen, the law of total probabilities can serve as a link between marginal probabilities and conditional probabilities to allow for more events. In fact, the law of total probability breaks up probability calculations into distinct parts: from one event to multiple events comprising that event.

However, it also needs to be pointed out that the above example for defining the total probabilities is only a very special condition in which multiple events exist in one sample space. Besides, these events are mutually independent, or in other words, have no overlaps. When there are multiple events that can possibly overlap with each other, we will need to turn to the chain rule to be introduced next. Despite this fact, the above extended formulation of Bayes' law is still very useful in constructing many machine learning algorithms.

Use of Probability in Machine Learning

Relationship Between Events: Interdependent Occurrence of Variable Values

In machine learning, we will more frequently resort to the Bayesian probability than the frequentist probability. In particular, conditional probability is an essential piece of knowledge that links prior knowledge to what we need to infer. The conditional probability described using the Bayesian theorem is a statistics topic that we would encounter a lot in machine learning.

In machine learning, it is more common to write the conditional probability as $P(Y = x|X = x)$. The meaning of this formula is the probability of $Y = y$ (variable Y takes a value y) given $X = x$. A notion underneath this formulation is that an event occurs when a random variable takes a specific value, and there are relationships like interdependencies between events. Conditional probability describes one type of such relationship:

$$P(Y = x|X = x) = \frac{P(Y = x, X = x)}{P(X = x)} \tag{A.9}$$

where $P(Y = x, X = x)$ is the joint probability, which is the probability that both $Y = y$ and $X = x$ occur.

In more complicated conditions, we may need to employ the chain rule, especially when we have multiple events that may have overlap or random variables that contain interdependency. The chain rule can be applied to conditional probabilities to connect different conditional probabilities. The following is a general equation for the chain rule, which contains two equivalent formulas:

$$P(X_1, X_2, \cdots, X_j) = \prod_{j=1}^{J} P(X_j|X_{j+1}, \cdots, X_J) = \prod_{j=1}^{J} P(X_j|X_1, \cdots, X_{j-1}) \tag{A.10}$$

This above chain rule equation can be demonstrated with the following simple example:

$$P(a, b, c) = P(a|b, c) \cdot P(b, c) = P(a|b, c) \cdot P(b|c) \cdot P(c) \tag{A.11}$$

or equivalently as

$$P(a, b, c) = P(c|a, b) \cdot P(a, b) = P(c|a, b) \cdot P(b|a) \cdot P(a) \tag{A.12}$$

The independence of variables also needs to be discussed for the co-occurrence of multiple events, which sometimes can help us simplify the discussion. For example, if two variables X and Y are independent, then the probability of two events, i.e., $X = x$ and $Y = y$, occurring simultaneously, is simply formulated as

$$P(X = x, Y = y) = P(X = x) \cdot P(Y = y) \tag{A.13}$$

Statistics of Events: Statistics of Variable Values

Besides the above inter-event relationships, in order to understand each random variable, we usually need to assess statistical measurements like expectation, deviation, and (co-) variance.

The expectation or the expected value of a function $f(x)$ with respect to a probability distribution $p(x)$ is the average or mean value that f takes on when x is drawn from p. If X is a discrete variable, we have

$$\mathbb{E}_{X\ p}[f(x)] = \sum_{x} P(x) f(x) \tag{A.14}$$

If X is a continuous variable, we have

$$\mathbb{E}_{X\ p}[f(x)] = \int p(x) f(x) \tag{A.15}$$

Variance is a measure of how much the values of a random variable X vary as we sample different values of x from its probability distribution:

$$\mathrm{Var}(f(x)) = \mathbb{E}[(f(x) - \mathbb{E}[f(x)])^2] \tag{A.16}$$

When Var is low, the values of $f(x)$ cluster near the expectation. A closely related parameter, standard deviation, is also frequently encountered in machine learning. The standard deviation can be simply calculated as the positive square root of the variance:

$$\sigma = \sqrt{\mathbf{Var}(f(x))} \tag{A.17}$$

Covariance is needed when there are multiple (random) variables. The covariance provides a measure of how much two variables are linearly related:

$$\mathrm{Cov}(f(x), g(y)) = \mathbb{E}[(f(x) - \mathbb{E}[f(x)]) \cdot (g(y) - \mathbb{E}[g(y)])] \tag{A.18}$$

It is noted that independence is a distinct property from covariance. Independence denotes the exclusion of any mutual dependency like linear and nonlinear relationships. By contrast, covariance only provides a measure of linear relationships between two random variables. Thus, low covariance implies a weak linear relationship but not necessarily independence.

We also often use the covariance matrix when we have multiple random variables that can be organized as an array/vector. A covariance matrix, or called dispersion matrix, variance matrix, or variance-covariance matrix, is a square matrix that presents the covariance between any pair of elements in an array (or vector). When we have an array $\vec{x}$, the covariance will become a matrix as follows:

$$\mathrm{Cov}(\vec{x})_{i,j} = \mathrm{Cov}(x_i, x_j) = \mathbb{E}[(x_i - \mathbb{E}[x_i]) \cdot (x_j - \mathbb{E}[x_j])] \tag{A.19}$$

Probability Distributions

The following are probability distributions that are commonly used in machine learning.

Bernoulli

Bernoulli distribution is a univariate discrete probability distribution in which there are only two values for the random variable: 0 and 1. Therefore, we usually write it as $X \subset [0, 1]$. Accordingly, these two values or outcomes are associated with two probabilities as follows:

$$P(X = 1) = \phi \tag{A.20a}$$

$$P(X = 0) = 1 - \phi \tag{A.20b}$$

Then, the probability density function p can be formulated as

$$p(X = x) = \phi^x \cdot (1 - \phi)^{1-x} \tag{A.21}$$

where x is the frequency or the ratio of numbers when $X = 1$ appears.

The mean and variance of the random variable that has a Bernoulli distribution are

$$\mathbb{E}_X[X] = \phi \tag{A.22a}$$

$$\mathrm{Var}_X[X] = \phi \cdot (1 - \phi) \tag{A.22b}$$

Multinoulli/Categorical Distribution

The Multinoulli distribution, which is also known as the categorical distribution, is a multivariate discrete distribution that generalizes the Bernoulli distribution. The discrete variable X has J states (or called events or outcomes), and P_j is the probability of the jth state. The probability mass function is

$$p_X(x_1, x_2, \cdots, x_J) = \prod_{j=1}^{J} p_j^{x_j} \tag{A.23}$$

The expected value of X is $\mathbb{E}[X] = \vec{p} = [P_1, P_2, \cdots, P_j]^T$. Thus, the expected value of X_i, which is the jth member of X corresponding to the condition wherein the jth event occurs, is equal to the probability of the corresponding event.

The covariance matrix of X is

$$\mathrm{Var}[X] = \bar{\Sigma} = \vec{p} \otimes \bar{1}_{1\times J} - \vec{p} \otimes \vec{p} \tag{A.24}$$

Or, we can write the elements of $\bar{\Sigma}_{ij}$ as

$$\bar{\Sigma}_{ij} = \begin{cases} p_i(1 - p_j) & \text{if } i = j \\ -p_i(p_j) & \text{if } i \neq j \end{cases} \tag{A.25}$$

Gaussian/Normal Distribution

The Gaussian or normal distribution is possibly the most widely used probability distribution in machine learning. The mathematical formulation of this distribution function is as follows:

$$\mathcal{N}(x; \mu, \sigma^2) = \sqrt{\frac{1}{2\pi\sigma^2}} \cdot \exp\left(-\frac{(x - \mu)^2}{2\sigma^2}\right) \tag{A.26}$$

where μ is the mean, which marks the center of the normal distribution curve, and σ is the standard deviation, which measures the distance from the center to the inflection point of the curve.

The multivariate normal distribution generalizes a normal distribution from $\mathbb{R}$ to $\mathbb{R}^n$:

$$\mathcal{N}(\vec{x}; \vec{\mu}, \bar{\Sigma}^2) = \sqrt{\frac{1}{(2\pi)^n \det(\bar{\Sigma})}} \cdot \exp\left(-\frac{1}{2}(\vec{x} - \vec{\mu})^T \bar{\Sigma}^{-1}(\vec{x} - \vec{\mu})\right) \tag{A.27}$$

where $\det(\bar{\Sigma})$ is the determinant of $\bar{\Sigma}$. $\bar{\Sigma}$ gives the covariance matrix of the distribution, which should be positive, definite, and symmetric.

Exponential Distribution

The exponential distribution function can be used to describe the probability distribution of the time between events that occur continuously and independently at a constant average rate, which is termed a Poisson point process. The probability density function of the standard exponential distribution is formulated as

$$p(x) = \exp(-x) \tag{A.28}$$

where $x \geqslant 0$. Accordingly, the probability is 0 for all $x \leqslant 0$. It is also common to use a general formula for the probability distribution function of the exponential distributions as follows:

$$p(x, \lambda, \mu) = \lambda \cdot \exp[-\lambda(x - \mu)], \qquad x \geqslant \mu, \ \lambda \geqslant 0 \tag{A.29}$$

Laplace Distribution

The Laplace distribution is intended for the distribution of the difference between two exponential random variables. Due to this reason, it is also frequently referred to as the double exponential distribution. One typical example is the Brownian motion evaluated at an exponentially distributed random time:

$$p(x; \mu, \lambda) = \frac{1}{2\gamma} \exp\left(-\frac{x - \mu}{\gamma}\right) \tag{A.30}$$

The Laplace distribution allows us to place a sharp peak of probability mass at a point μ.

Dirac Delta Distribution

The Dirac delta distribution, or written as δ distribution at some places, is proposed to describe a unit impulse. The value of this function is zero everywhere except at one point, while the integral over the entire real domain equals 1. The density function of the Dirac delta distribution is

$$p(x) = \delta(x - \mu) \tag{A.31}$$

Empirical Distribution

The empirical distribution (function) is used to describe a collection of measured values of a given variable. The value of the function at a point x equals the percentage of the measured values that are less than or equal to x in the collection. When we have I measured values, this distribution function can be formulated as

$$p(x) = \frac{1}{I} \sum_{i=1}^{I} H(x - x^{(i)}) \tag{A.32}$$

where $H(\cdot)$ is the Heaviside step function, whose value is 1 when $x - x^{(i)} > 0$ and 0 otherwise.

Logistic/Sigmoid Function

In a broad sense, the sigmoid function describes "S"-shaped, monotonically increasing curves, while the logistic function is a special case of the sigmoid function. However, in the context of machine learning, the sigmoid function is cited in a narrow sense for interchangeable use with the logistic function, which has the following formulation:

$$\sigma(x) = \frac{1}{1 + e^{-x}} \tag{A.33}$$

One major use of the sigmoid function in machine learning is to convert a continuous function into a discrete space. For example, when the output of a model is a continuous number, the sigmoid function can squeeze the value into a range between 0 and 1, which can be treated as the probability belonging to a specific class. Usually, with the assistance of a threshold, e.g., 0.5, we can turn the continuous output into a discrete label for classification, e.g., Class 0 or Class 1.

Softmax Function

The Softmax function can be viewed as a generalization of the logistic function from two classes to multiple classes/dimensions (or multiclass classification). As a normalized exponential function, it can normalize the output of a model to a probability distribution over predicted output classes. In the simplest way, the Softmax function is written as

$$\sigma(\vec{z})_k = \frac{e^{z_k}}{\sum_{l=1}^{K} e^{z_l}} \tag{A.34}$$

where $\vec{z}$, which has a shape of $K \times 1$, is the input vector for the function, and $k, l \in [1, \cdots, K]$.

The above equation can be slightly modified for machine learning. Let us assume a machine learning model represented by a vector (array of numbers) $\vec{w}$, then the probability of sample $\vec{x}$ belonging to the kth class out of K possible classes is

$$P(y = k|\vec{x}) = \frac{e^{\vec{x}^T \cdot \vec{w}_k}}{\sum_{l=1}^{K} e^{\vec{x}^T \cdot \vec{w}_l}} \tag{A.35}$$

This equation calculates the possibility of a sample $\vec{x}$ belonging to the kth class, that is, label $y = k$, based directly on the sample's attribute values. But in typical machine learning applications, we usually have a model f that converts the sample's input $\vec{x}$ to the output $\vec{z}$: $\vec{z} = f(\vec{x})$. But this $\vec{z}$, as the "intermediate" output, needs to be converted to the final output, e.g., y as the classification label. In this case, we will need to slightly modify the above equation as

$$P(y = k|\vec{z}) = \frac{e^{\vec{z}^T \cdot \vec{w}_k}}{\sum_{l=1}^{K} e^{\vec{z}^T \cdot \vec{w}_l}}. \tag{A.36}$$

A.2.2 *Information Theory*

Information theory is a branch of applied mathematics that focuses on the quantification, storage, and communication of information. In the area of AI especially machine learning applications, we frequently borrow knowledge from information theory for information quantification, which measures or describes some characteristics of data. In the following, we will introduce several information concepts (or called measures) that we frequently employ in AI.

One basic intuition behind information theory is that the occurrence of an unlikely event is more informative than that of a likely event. For example, a fact that an event occurs on certain days provides more information than a fact that an event occurs every day. For a more quantitative description, we will need to measure the amount of information (or uncertainty). Thus, a high level of uncertainty corresponds to more information. Self-information or information content is a measure that we can use to measure the amount of information associated with the occurrence of an event:

$$I(x) = -\ln P(x) \tag{A.37}$$

where $I(x)$ is the self-information of an event x, e.g., when a random variables take a value x, and $P(x)$ is the probability of it. This equation tells that an event with low probability has a large amount of information.

The above definition of self-information also implies that this measure is more for the calculation of individual random variable value (or a single event). We also have the need for quantifying the total amount of information contained in a random variable, X, that follows a specific distribution, $p(x)$. This amount of information can be measured using the information entropy, which is also called Shannon entropy in many places:

$$H(x) = \mathbb{E}_{X\sim P}[I(x)] = -\mathbb{E}_{X\sim P}[\ln P(x)] = -\sum_{x\in X} P(x)\ln P(x). \tag{A.38}$$

It is noted that the use of a log base of e in the definitions of the above is not mandatory. Instead, other choices of the base, e.g., 2 and 10, are also possible. The use of different log bases leads to different units for measuring the information amount: the base 2 corresponds to the unit of bits or "Shannons," the base e corresponds to "nat" ("natural units"), while the base 10 corresponds to the units of "dits," "bans," or "hartleys." However, such units are declared in information theory yet rarely emphasized in AI.

Another useful measure is available for quantifying the difference between two probability distributions. Such a difference can be computed using the Kullback-Leibler (KL) divergence, which is also called the relative entropy:

$$D_{KL}(P \parallel Q) = \mathbb{E}_{X\sim P}\left[\ln \frac{P(x)}{Q(x)}\right] = \mathbb{E}_{X\sim P}\left[\ln P(x) - \ln Q(x)\right] \tag{A.39}$$

where P and Q are two different distribution functions. A KL divergence value of 0 indicates that the two distributions have identical quantities of information or no difference.

In many cases, we will need to discuss the difference between two probability distributions. A measure called cross-entropy $H(P, Q)$, which is similar to KL divergence, can be used for this purpose:

$$H(P, Q) = -\mathbb{E}_{X\sim P(x)}[\ln Q(x)] = -\sum_{x} P(x)\ln Q(x) \tag{A.40}$$

For example, we need to measure how far the predicted values are from the true labels in a classification task. This actually denotes the difference between two distributions corresponding to the predictions and true labels. Such a task thus can employ cross-entropy, which quantifies the difference or distance. As a result, cross-entropy is by nature a good candidate for loss function, which is called log loss due to the use of log in its definition.

This cross-entropy is frequently used with the Softmax function in multiclass classification tasks in deep learning. In such tasks, the loss function ℓ is usually constructed as

$$\ell = -\sum_{i} y_{ik}\ln P_{ik} = -\sum_{i} y_{ik}\ln \frac{e^{z_{ik}}}{\sum_{l=1}^{K} e^{z_{il}}} \tag{A.41}$$

where y_{ik} is the one-hot label for class k of sample i, in which 1 is used for true and 0 for false, and z_{ik} is the output for the kth class of sample i.

A.2.3 Array Operations

As introduced in Chap. 1, the AI domain has been predominated by numeric AI, which features the manipulation of data with machine learning algorithms. Such data exists either in terms of numbers or symbols, which human beings can easily understand, e.g., in ASCII format, or their corresponding machine code, e.g., in binary format, which is designed to be read by computers thus not legible to us. But in both cases, we can understand data as an organized body of numbers or/and letters. In real computation, it is common to convert everything including letters, which can be used to register attribute values and labels, to numbers, except for a few classic machine algorithms that were developed originally with symbolic systems such as decision trees.

Such an organized body of data can appear as a row of numbers, a stack of multiple rows of numbers (as a rectangular layer with multiple rows and columns), or a three-dimensional data grid consisting of multiple such stacks, which correspond to 1D, 2D, and 3D data structures, respectively. The dimension can be further increased as we generate an nth order data structure by stacking multiple $(n-1)$th order data structures. In fact, we usually use "arrays" as a general term to describe these data structures with different orders or dimensions, for example, 1D array for a row of numbers. So, the training and testing in machine learning can be viewed as a sequence of array operations, which move, store, and transform data according to certain rules determined by the algorithms and array operation rules. That is why many people believe arrays play a central role in data science. Array operations and associated rules are thus vital in AI. Considering these facts, array operations, especially 2D arrays, which we can easily visualize, will be introduced in the following.

Matrix is another term that is frequently used when describing data. From a math perspective, a matrix is used to refer to a 2D array. However, it is noted that matrix and array may have different notions when these two terms are used to refer to data structure type in some packages, such as the "matrix" used in NumPy (not recommended in new versions). Also, 1D array, as a general data structure, cannot be mixed with specific data structures in programming languages or packages. Typical examples are the list and tuple in Python, which are not merely a row of numbers or letters but also come with specific operations, functions, or methods.

Tensor is also very frequently used to describe data, especially in physics and engineering mechanics. As a math term, tensor provides a notation for data so that we can easily formulate data stored in diverse and complicated data structures using math symbols. In addition, tensor analysis [174], which includes convenient operations using specific symbolic deduction rules, can be used to conveniently handle the operations of tensors to significantly facilitate deductions. Tensors in different orders correspond to arrays with the corresponding numbers of dimensions. For example, a second-order tensor can be used to represent a 2D array. Tensors are widely used in engineering and have been adopted in modern deep learning packages such as TensorFlow and the tensor data type in PyTorch. Basic knowledge about tensor operations will be presented in this chapter. Array operations that are more specific to certain AI topics such as the convolution and pooling in deep learning are introduced in the corresponding chapters.

Matrix Operations

Let us first take a look at common matrix operations that we are more familiar with. More general array operations will be introduced based on them.

Matrix Multiplication

Matrix multiplication does not perform elementwise multiplication of the two matrices like matrix addition and subtraction. Instead, for a matrix $A \in \mathbb{R}^{I\times K}$ and $B \in \mathbb{R}^{K\times J}$, the product of their matrix multiplication $C = AB$ is a matrix $C \in \mathbb{R}^{I\times J}$ where

$$C_{ij} = \sum_{k=1}^{K} A_{ik} B_{kj}. \tag{A.42}$$

As can be seen, this operation requires that the number of columns in A must equal the number of rows in B. The following rules are valid for matrix multiplication operations:

- Matrix multiplication is associative: $(AB)C = A(BC)$. Thus, the order in which we do the multiplications does not matter, though some orders can be more computationally efficient than others.
- Matrix multiplication is distributive: $A(B + C) = AB + AC$.
- Matrix multiplication is *not* commutative: $AB \neq BA$.

Transpose

The transpose of a matrix is obtained by flipping its rows and columns. That is, if $A \in \mathbb{R}^{I\times J}$, then its transpose, A^T, is a matrix $C \in \mathbb{R}^{J\times I}$ where

$$C_{ij} = A_{ji}. \tag{A.43}$$

Identity Matrix

The identity matrix $\mathbf{I} \in \mathbb{R}^{J\times J}$, which is written as $\bar{I}$ in other places of this book, is a square matrix with 1's on the diagonal and 0's everywhere else.

$$\mathbf{I} = \begin{bmatrix} 1 & 0 & \cdots & 0 \\ 0 & 1 & \cdots & 0 \\ \vdots & \vdots & \ddots & \vdots \\ 0 & 0 & \cdots & 1 \end{bmatrix}_{J\times J}. \tag{A.44}$$

Here $\mathbf{I}$ is used to differentiate from the symbol I for the dimension. The above identity matrix has a shape of $J \times J$ and can be written as $\mathbf{I}_{J\times J}$.

Identify matrix has a unique property: for any matrix $A \in \mathbb{R}^{I\times J}$, we have

$$A\mathbf{I}_{J\times J} = \mathbf{I}_{I\times I}A = A \tag{A.45}$$

where the two $\mathbf{I}$ matrices have different sizes, i.e., the first one is $J \times J$ and the second is $I \times I$. To denote the difference, it is common to use $\mathbf{I}_J$ to explicitly denote the size of $\mathbf{I}$.

Matrix Inverse

For a square matrix $A \in \mathbb{R}^{J\times J}$, the matrix inverse $A^{-1} \in \mathbb{R}^{J\times J}$ is the unique matrix such that

$$A^{-1}A = AA^{-1} = \bar{I}. \tag{A.46}$$

It is worthwhile to mention that not all the matrices have an inverse, which depends on the linear independence between rows/columns of A.

The matrix inverse provides an immediate method to solve a system of linear equations. For example, the solution to $Ax = b$ can be found by multiplying both sides of the equation by A^{-1}:

$$A^{-1}Ax = A^{-1}b \Longrightarrow x = A^{-1}b. \tag{A.47}$$

Transpose and Inverse of Matrix Product

The transpose of a matrix product is the product of the transposes in the reverse order:

$$(AB)^T = B^T A^T \tag{A.48}$$

The inverse of a matrix product also follows a similar rule: the inverse of a matrix product is the product of the inverses in the reverse order:

$$(AB)^{-1} = B^{-1}A^{-1} \tag{A.49}$$

Please note that the matrix operations in this subsection were explained using the regular matrix notation (with or without expanded indices), which is used in most data mining and machine learning literature. In the rest of this book, we use a mixed notation between the matrix notation and the tensor notation. Thus, instead of writing AB to represent the matrix product, we adopt $\bar{A} \cdot \bar{B}$.

General Array Operations

Only common array operations are presented in this subsection. Array operations that are not used too much in AI, such as cross product, scalar triple product, and vector triple product, are excluded.

Addition and Subtraction
Array addition and subtraction, which can be applied to two arrays of the same shape, are applied elementwise over the matrices. Taking 2D arrays as an example, if $\bar{A} \in \mathbb{R}^{I\times J}$ and $\bar{B} \in \mathbb{R}^{I\times J}$, then their sum $\bar{C} = \bar{A} + \bar{B}$ is another matrix of the same size $\bar{C} \in \mathbb{R}^{I\times J}$ where

$$C_{ij} = A_{ij} + B_{ij}. \tag{A.50}$$

The addition and subtraction of matrices briefly mentioned in the previous subsection are just a special case of this.

Elementwise Product
Any two arrays of the same shape can have their elementwise product, in which the corresponding elements in the two arrays can be multiplied to get the corresponding element in the product array:

$$C_{ij} = A_{ij} \times B_{ij}. \tag{A.51}$$

In NumPy, this product is applied using the symbol "*". In this book, $\odot$ is used to denote the elementwise product to avoid possible confusion. Thus, the elementwise product written in the hybrid tensor notation in this book is $\bar{C} = \bar{A} \otimes \bar{B}$.

Inner Product
Inner or dot product is widely used between two arrays with any dimensions (or tensors with any orders). The inner product is used interchangeably with the dot product on many occasions. Their difference lies in that the inner product generalizes the dot product to abstract vector spaces over a field of scalars, either real or complex numbers.

Here, we use the simplest case: two 1D arrays $\vec{u}, \vec{v} \in \mathbb{R}^I$ to explain what happens during this operation. The dot product or scalar product of two column vectors, e.g., $\vec{u}$ and $\vec{v}$, is written as $\vec{u} \cdot \vec{v}$ in the context of tensor analysis. However, it is written as $\vec{u}^T \cdot \vec{v}$ in this book. This notation enables the development of computer code, in which a major difficulty is to determine the shape of arrays (including shape change due to transpose) and the operators between arrays, to be directly implemented based on the equation. The quantity can be expressed as

$$\vec{u}^T \cdot \vec{v} = \sum_{i=1}^{I} u_i v_i \tag{A.52}$$

If we use matrices and assume $I = 3$, then we can formulate the operations as

$$\vec{u}^T \cdot \vec{v} = \begin{bmatrix} u_1 \; u_2 \; u_3 \end{bmatrix} \begin{bmatrix} v_1 \\ v_2 \\ v_3 \end{bmatrix} = u_1 v_1 + u_2 v_2 + u_3 v_3 \tag{A.53}$$

From a geometrical perspective, the dot product projects one vector onto the direction of the other one and then scales the former vector by the length of the second one. This can be written as

$$\vec{u}^T \cdot \vec{v} = |\vec{u}| \cdot |\vec{v}| \cos(\vec{u}, \vec{v}), \tag{A.54}$$

where $(\vec{u}, \vec{v})$ denotes the angle between the directions of $\vec{u}$ and $\vec{v}$.

In NumPy, the dot product can be easily applied using the symbol "@".

Outer Products

The outer or tensor product is a type of operation between tensors of any order. In linear algebra, the term outer product is typically used to refer to the tensor product of two vectors. In the dyadic context, dyadic product, outer product, and tensor product all share the same meaning and thus are used synonymously. However, the tensor product is the most general and abstract term among them.

There are several equivalent terms and notations for this product:

1. The dyadic product of two vectors $\vec{u}$ and $\vec{v}$ is denoted by their juxtaposition.
2. The outer product of two column vectors $\vec{u}$ and $\vec{v}$ is denoted and defined as $\vec{u} \otimes \vec{v}$.
3. The tensor product of two vectors $\vec{u}$ and $\vec{v}$ is denoted by $\vec{u} \otimes \vec{v}$.

To avoid confusion, this book writes the tensor product as $\vec{u} \otimes \vec{v}$.

The result of the outer product or tensor product of $\vec{u}_{3\times1}$ and $\vec{v}_{3\times1}$ is a 3×3 matrix as

$$\vec{u} \otimes \vec{v} = \vec{u}\vec{v}^{\mathrm{T}} = \begin{bmatrix} u_1 \\ u_2 \\ u_3 \end{bmatrix} \begin{bmatrix} v_1 \; v_2 \; v_3 \end{bmatrix} = \begin{bmatrix} u_1 v_1 & u_1 v_2 & u_1 v_3 \\ u_2 v_1 & u_2 v_2 & u_2 v_3 \\ u_3 v_1 & u_3 v_2 & u_3 v_3 \end{bmatrix} \tag{A.55}$$

Tensor product is associative and distributive but not commutative.

As for the associative law, we have

$$(\vec{u} \otimes \vec{v}) \otimes \vec{w} = \vec{u} \otimes (\vec{v} \otimes \vec{w}) = \vec{u} \otimes (\vec{v} \otimes \vec{w}) \tag{A.56}$$

The associative law is compatible with scalar multiplication for any scalar α:

$$(\alpha\vec{u}) \otimes \vec{v} == \alpha(\vec{u} \otimes \vec{v}) = \vec{u} \otimes (\alpha\vec{v}) \tag{A.57}$$

In the latest version of NumPy, the tensor product can be performed using "numpy.tensordot(u,v,axes=0)," in which the option "axes=0" means "no contraction," which will be introduced with more detail next.

Contraction and More General Product Operations for Arrays
If we compare the inner and outer products, we can easily see that one major difference between the two operations is that, while the outer product is to "stack" all the dimensions/axes (of arrays) or orders (of tensors) together, the inner product will "merge" the two dimensions/axes right next to the dot product sign. That is, this merging process will reduce the number of dimensions/axes by 2 because the last dimension of the tensor before the product sign and the first dimension of the tensor after the sign is "merged." This "merging" is called "contraction" and is implemented by getting the summation of the products of corresponding elements from the two dimensions/axes. For example, the inner products of two 1D arrays will produce an array with ($1 + 1 - 2 = 0$) dimensions. In this process, the (only) dimension of the 1D array before the dot sign and that of the 1D array after the dot sign "contracted" into a scalar (0th dimension), e.g., $\vec{u}^T \cdot \vec{v} = \sum_{i=1}^{I} u_i v_i$.

The outer product can be performed by "numpy.tensordot" with no contraction (of axes). But it is worth mentioning that "numpy.tensordot" can be employed to apply outer products that include the contraction of any one or more pairs of axes. For example, "numpy.tensordot(A,B,axes=[[0,1],[1,3]])" contracts the first (0) and second (1) axes of tensor $\bar{A}$ and the second (1) and fourth (3) axes of tensor $\bar{B}$, i.e., A0 contracted with B1 and A1 with B3.

Broadcasting
Broadcasting is an operation of matching the dimensions of differently shaped arrays so that further operations can be applied on those arrays. This is usually performed by comparing the shapes of two arrays elementwise. Taking NumPy as an example, broadcasting works from the last/tailing (i.e., rightmost) dimension and moves to the left dimension one by one. It will be helpful and adequate to know about the broadcasting rules in NumPy: two dimensions are compatible when they are equal or one of them is 1. If the rule is valid for every dimension (pair), then the two arrays are "broadcastable." The following are examples of shapes that cannot be broadcasted:

A (1d array): 3
B (1d array): 4 # trailing dimensions do not match

A (2d array): 2 x 1
B (3d array): 8 x 4 x 3 # second from last dimensions mismatched

If two arrays are broadcastable, broadcasting can be performed by changing every dimension value of "1" to the higher value of the corresponding dimension in the other array. The following examples exhibit the execution of the operation:

A (2d array): 5 x 4
B (1d array): 4
Result (2d array): 5 x 4

A (4d array): 8 x 1 x 6 x 1
B (3d array): 7 x 1 x 5
Result (4d array): 8 x 7 x 6 x 5

A (3d array): 15 x 3 x 5
B (3d array): 15 x 1 x 5
Result (3d array): 15 x 3 x 5

Vector Norms

Vector norms are tools for measuring "distances." They are thus of great use when constructing bounds, evaluation metrics, regularization terms, and other criteria. The most common one is the second norm or norm-2:

$$\|\vec{x}\|_2 = \sqrt{\vec{x}^T \cdot \vec{x}} = \sqrt{\sum_{i=1}^{n} x_i^2} \tag{A.58}$$

This norm is used to compute the Euclidean distance and thus called the Euclidean (also called ℓ_2) norm of $\vec{x}$.

Besides, we may occasionally also resort to the ℓ_1 norm:

$$\|\vec{x}\|_1 = \sum_{i=1}^{n} |x_i| \tag{A.59}$$

or the ℓ_∞ norm

$$\|\vec{x}\|_\infty = \max_{i=1,\dots,n} |x_i|. \tag{A.60}$$

Complexity of Operations

The big-O complexity of the different matrix operations can help us understand the challenges and computing demands of array operations. Therefore, brief information is provided here. Assuming $\bar{A}, \bar{B} \in \mathbb{R}^{I \times I}$ and $\vec{x}, \vec{y} \in \mathbb{R}^I$, let us use the following example to illustrate the idea of gauging complexities:

- Inner product $\vec{x}^T \cdot \vec{y}$: $O(I)$
- Matrix-vector product $\bar{A} \cdot \vec{x}$: $O(I^2)$
- Matrix-matrix product $\bar{A} \cdot \bar{B}$: $O(I^3)$
- Matrix inverse $\bar{A}^{-1}$ and matrix solve $\bar{A}^{-1} y$: $O(I^3)$. Note: As introduced below, these two operations are actually done differently, though they have the same big-O complexity.

There are very different complexities for computing the exact same term in different ways. For example, suppose we want to compute the matrix products $\bar{A} \cdot \bar{B} \cdot \vec{x}$. We could compute this as $(\bar{A} \cdot \bar{B}) \cdot \vec{x}$ (computing the $\bar{A} \cdot \bar{B}$) product first and then multiplying with $\vec{x}$); this approach would have complexity $O(I^3)$, as the matrix-matrix product would dominate the computation. Alternatively, if we compute the product as $\bar{A} \cdot (\bar{B} \cdot \vec{x})$ by first computing the vector product $\bar{B} \cdot \vec{x}$, which produces a vector, and then multiplying this by A), the complex is only $O(I^2)$, as we just have two matrix-vector products. In summary, the order of operations can significantly affect the time complexity of linear algebra operations.

Array Calculus

Calculus for arrays is very common in different applications [175]. It appears as matrix calculus in statistics and machine learning and as tensor calculus in physics and engineering mechanics. In both cases, calculus is discussed for multivariate problems, i.e., variables have multiple dimensions (or called axes or attributes). However, statistics and machine learning usually involve the differential calculus between first- and second-order tensors, which can be easily represented by 1D arrays and 2D arrays (matrices), while physics and engineering mechanics can involve both differential and integral calculus for tensors including tensors higher than the second order. Moreover, the former needs to write all the numbers in formats that can be easily understood, i.e., 1D arrays and matrices, because statistics and machine learning deal with data, while the latter needs to use the tensor notation because the physical relationship expressed in simple math symbols are needed to facilitate formulation and deduction. This subsection introduces the basic knowledge of array calculus.

Let us use two scalars, x and y, two 1D arrays, $\vec{x}$ and $\vec{y}$, and two 2D arrays, $\bar{X}$ and $\bar{Y}$, to illustrate the operations and rules in array calculus.

$$\vec{x} = \begin{bmatrix} x_1 \\ x_2 \\ \vdots \\ x_I \end{bmatrix}_{I \times 1} \tag{A.61}$$

$$\bar{X} = \begin{bmatrix} x_{11} & x_{12} & \cdots & x_{1J} \\ \vdots & & \ddots & \vdots \\ x_{Ii} & x_{I2} & \cdots & x_{IJ} \end{bmatrix}_{I \times J} \tag{A.62}$$

$$\vec{y} = \begin{bmatrix} y_1 \\ y_2 \\ \vdots \\ y_J \end{bmatrix}_{J \times 1} \tag{A.63}$$

$$\bar{Y} = \begin{bmatrix} y_{11} & y_{12} & \cdots & y_{1J} \\ \vdots & & \ddots & \vdots \\ y_{I1} & y_{I2} & \cdots & y_{IJ} \end{bmatrix}_{I \times J} \tag{A.64}$$

Besides, we need to point out that there are two different notations, i.e., numerator layout notation and denominator layout notation. In the numerator layout notation, $\frac{d\vec{x}}{dy}$, $\frac{dx}{d\vec{y}}$, and $\frac{d\vec{x}}{d\vec{y}}$ have the shape of $I \times 1$, $1 \times J$, and $I \times J$, respectively. The denominator layout notation will generate the "opposite" shapes: $1 \times I$, $J \times 1$, and $J \times I$. As can be seen here, "numerator" layout notation means the shape of the generated scalar refers to the shape of the tensor in the numerator, while the shape of the tensor in the denominator is flipped. The use of these two notations can produce different results in some operations but not all. To avoid confusion, this subsection only presents operations and rules based on the numerator layout notation. Let us first take a look at $\frac{d\vec{x}}{dx}$, $\frac{dx}{d\vec{y}}$, $\frac{d\vec{x}}{d\vec{y}}$, $\frac{d\bar{X}}{dy}$, and $\frac{dx}{d\bar{Y}}$.

$$\frac{\partial \vec{x}}{\partial y} = \begin{bmatrix} \frac{\partial x_1}{\partial y} \\ \frac{\partial x_2}{\partial y} \\ \vdots \\ \frac{\partial x_I}{\partial y} \end{bmatrix}_{I \times 1} \tag{A.65}$$

$$\frac{\partial x}{\partial \vec{y}} = \begin{bmatrix} \frac{\partial x}{\partial y_1} & \frac{\partial x}{\partial y_2} & \cdots & \frac{\partial x}{\partial y_J} \end{bmatrix}_{1 \times J} \tag{A.66}$$

$$\frac{\partial \vec{x}}{\partial \vec{y}} = \begin{bmatrix} \frac{\partial x_1}{\partial y_1} & \frac{\partial x_1}{\partial y_2} & \cdots & \frac{\partial x_1}{\partial y_J} \\ \vdots & \ddots & & \vdots \\ \frac{\partial x_I}{\partial y_1} & \frac{\partial x_I}{\partial y_2} & \cdots & \frac{\partial x_I}{\partial y_J} \end{bmatrix}_{I \times J} \tag{A.67}$$

$$\frac{\partial \bar{X}}{\partial y} = \begin{bmatrix} \frac{\partial x_{11}}{\partial y} & \frac{\partial x_{12}}{\partial y} & \cdots & \frac{\partial x_{1J}}{\partial y} \\ \vdots & \ddots & & \vdots \\ \frac{\partial x_{I1}}{\partial y} & \frac{\partial x_{I2}}{\partial y} & \cdots & \frac{\partial x_{IJ}}{\partial y} \end{bmatrix}_{I \times J} \tag{A.68}$$

$$\frac{\partial x}{\partial \bar{Y}} = \begin{bmatrix} \frac{\partial x}{\partial y_{11}} & \frac{\partial x}{\partial x_{21}} & \cdots & \frac{\partial x}{\partial y_{I1}} \\ \vdots & \ddots & & \vdots \\ \frac{\partial x}{\partial y_{1J}} & \frac{\partial x}{\partial y_{2J}} & \cdots & \frac{\partial x}{\partial y_{JI}} \end{bmatrix}_{J \times I} \tag{A.69}$$

First, we have the following identity:

$$\frac{\partial \vec{x}}{\partial \vec{x}} = I_{I \times I} \tag{A.70}$$

Let us use two extra 1D arrays, $\vec{u}_{I \times 1}$ and $\vec{v}_{I \times 1}$, two 2D arrays, $\bar{U}_{I \times I}$ and $\bar{V}_{I \times I}$, and one 2D array, $\bar{A}_{I \times I}$. $\vec{u}_{I \times 1}$, $\vec{v}_{I \times 1}$, $\bar{U}_{I \times I}$, and $\bar{V}_{I \times I}$ are functions of $\vec{x}$, while $\bar{A}_{I \times I}$ is not a function of $\vec{x}$. Then, the common operations and rules associated with array calculus are listed as the following identities:

$$\frac{\partial (\vec{u}^T \cdot \vec{v})}{\partial \vec{x}} = \vec{u}^T \cdot \frac{\partial \vec{v}}{\partial \vec{x}} + \vec{v}^T \cdot \frac{\partial \vec{u}}{\partial \vec{x}} \tag{A.71}$$

$$\frac{\partial(\vec{u}+\vec{v})}{\partial\vec{x}} = \frac{\partial\vec{u}}{\partial\vec{x}} + \frac{\partial\vec{v}}{\partial\vec{x}} \tag{A.72}$$

$$\frac{\partial(\bar{U}\cdot\bar{V})}{\partial x} = \bar{U}\cdot\frac{\partial\bar{V}}{\partial x} + \bar{V}\cdot\frac{\partial\bar{U}}{\partial x} \tag{A.73}$$

$$\frac{\partial(\bar{U}\otimes\bar{V})}{\partial x} = \bar{U}\otimes\frac{\partial\bar{V}}{\partial x} + \bar{V}\otimes\frac{\partial\bar{U}}{\partial x} \tag{A.74}$$

$$\frac{\partial(\vec{u}^T\cdot\bar{V})}{\partial\vec{x}} = \vec{u}^T\cdot\frac{\partial\bar{V}}{\partial\vec{x}} + \bar{V}^T\cdot\frac{\partial\vec{u}}{\partial\vec{x}} \tag{A.75}$$

$$\frac{\partial(\vec{u}^T\cdot\bar{A}\otimes\vec{v})}{\partial\vec{x}} = \vec{u}^T\cdot\bar{A}\otimes\frac{\partial\vec{v}}{\partial\vec{x}} + \vec{v}^T\otimes\bar{A}^T\cdot\frac{\partial\vec{u}}{\partial\vec{x}} \tag{A.76}$$

In fact, $\vec{u}$, $\vec{v}$, and $\vec{x}$ in the above identities can be scalars. The above identities can be used to obtain the following identities:

$$\frac{\partial(a\vec{u})}{\partial\vec{x}} = a\frac{\partial\vec{u}}{\partial\vec{x}} \tag{A.77}$$

$$\frac{\partial(\bar{A}\cdot\vec{x})}{\partial\vec{x}} = \bar{A} \tag{A.78}$$

$$\frac{\partial(\vec{x}^T\cdot\bar{A})}{\partial\vec{x}} = \bar{A}^T \tag{A.79}$$

$$\frac{\partial(\vec{x}^T\cdot\bar{A}\cdot\vec{x})}{\partial\vec{x}} = \vec{x}^T(\bar{A}+\bar{A}^T) \tag{A.80}$$

$$\frac{\partial(\vec{x}^T\cdot\bar{A}\cdot\vec{x})}{\partial\vec{x}\cdot\partial\vec{x}^T} = \bar{A}+\bar{A}^T \tag{A.81}$$

$$\frac{\partial(\vec{x}^T\cdot\vec{x})}{\partial\vec{x}} = 2\vec{x}^T \tag{A.82}$$

$$\frac{\partial(\vec{a}^T\cdot\vec{x}\vec{x}^T\cdot\vec{b})}{\partial\vec{x}} = \vec{x}^T(\vec{a}\cdot\vec{b}^T+\vec{b}\cdot\vec{a}^T) \tag{A.83}$$

There are also identities with 2D arrays as the independent variables (in the denominator):

$$\frac{\partial(\vec{a}^T\cdot\bar{X}\cdot\vec{b})}{\partial\bar{X}} = \vec{b}\cdot\vec{a}^T \tag{A.84}$$

$$\frac{\partial(\vec{a}^T \cdot \bar{X}^T \cdot \vec{b})}{\partial \bar{X}} = \vec{a} \cdot \vec{b}^T \tag{A.85}$$

$$\frac{\partial tr(\bar{A} \cdot \bar{X})}{\partial \bar{X}} = \frac{\partial(\bar{X} \cdot \bar{A})}{\partial \bar{X}} = \bar{A} \tag{A.86}$$

$$\frac{\partial tr(\bar{A} \cdot \bar{X}^T)}{\partial \bar{X}} = \frac{\partial(\bar{X}^T \cdot \bar{A})}{\partial \bar{X}} = \bar{A}^T \tag{A.87}$$

In addition, the following identify involving the chain rule is also very useful:

$$\frac{\partial[\vec{u}(\vec{v})]}{\partial \vec{x}} = \frac{\partial[\vec{u}(\vec{v})]}{\partial \vec{u}} \cdot \frac{\partial \vec{u}}{\partial \vec{x}} = \frac{\partial[\vec{u}(\vec{v})]}{\partial \vec{u}} \cdot \frac{\partial \vec{u}}{\partial \vec{v}} \cdot \frac{\partial \vec{v}}{\partial \vec{x}} \tag{A.88}$$

A.3 Optimization

AI has a big overlap with the general area of optimization because the search for the desired model is a process of identifying the model that best meets the goal. In some algorithms, such models can be derived analytically. Such deduction can also be counted as optimization because we essentially use the exact optimization methods. Examples include the linear models and support vector machines, which involve the linear least squares method and the Lagrangian multiplier method, respectively. Though many people do not treat these efforts as optimization, most people would agree those gradient descent methods in deep learning, which involve an interactive process for identifying the optimal model, should touch the core of optimization. But no matter how we view this overlap, it seems to be consensed that optimization tools, at least some optimization methods, are an essential part of modern AI predominated by machine learning. In typical optimization problems, we try to maximize or minimize an objective function, $f(\vec{x})$, where $\vec{x}$ is a vector of continuous/discrete values.

The classification of optimization methods can be rather diverse, depending on classification criteria including but not limited to the consideration of uncertainty, continuity of the parameters to be optimized, continuity and differentiability of the objective functions, search for global vs. local optima, existence of constraints, and number of objective functions. As mentioned above, optimization methods in machine learning can be roughly classified as exact methods (linear programming, quadratic programming, dynamic programming, branch and bound), general approximate (heuristic) methods (hill climbing, Tabu search, simulated annealing, evolutionary algorithms (including genetic algorithm), swarming intelligence algorithms (ant colony, honeybee mating, particle swarm)), gradient-based methods (first-order gradient) (GD, BGD, MBGD, SGD, Nesterov, momentum, AdaGrad, Adam, AdaDelta, RMSProp), and Newton's and quasi-Newton's methods (second-order gradient) (including BGFS). The conjugate direction method and its nonlinear variants can be regarded as a hybrid between the gradient-based methods (first-order method that uses the steepest descent gradient) and Newton's method (second-order method that uses Hessian as well). Another major category of optimization methods in AI, expectation-maximization (EM) algorithms, can be roughly viewed as a category of more general approximate methods.

Exact methods, gradient-based methods, and EM are the three main categories of optimization methods in contemporary AI. In this book, linear least squares, Lagrangian multiplier, and gradient-based methods have been well discussed in the chapters for linear models, SVM, and deep learning. Newton's and quasi-Newton's methods gained popularity in deep learning but have not been discussed, so some information will be given in this section. Conjugate gradient methods, which are very useful but have not been that widely adopted in AI, i.e., in deep learning, due to a variety of reasons, will also be discussed considering their wide applicability and possible use in the future of AI. EM has lots of applications in machine learning including Gaussian mixtures, clustering algorithms, K-means, and HMM, etc., but it has not been systematically discussed in the previous chapters. Thus, EM will also be explained in the following.

A.3.1 Gradient-Based Methods

Gradient descent is a predominant category of optimization methods in deep learning, thus, a must-know. The basic algorithm is the batch gradient descent (BGD), which uses the gradients of the whole batch (all the training samples) for update. Mini-batch gradient descent (MGBD) is the one that we possibly most frequently use. Here, we draw a portion of samples, i.e., mini batch, from the whole batch and update the model parameters using the gradients calculated with this mini-batch of samples. Thus, this mini-batch actually corresponds to the "batch" in the parameter "batch size," which is used in most deep learning software. The use of such mini-batch introduces stochasticity, which sometimes helps jump out of local minimums. Following this direction, we can reach the extreme of introducing the utmost stochasticity: stochastic gradient descent (SGD). SGD uses one sample randomly drawn from the whole batch to compute the gradients for updating the model.

Most of the optimizers that we use in deep learning are variations of MGBD. The momentum method includes the gradients of the previous update step in addition to the use of the gradients of the current step, which helps the optimizer jump off the local optima or saddle points. The Nesterov or Nesterov momentum uses the gradients at a "future" point for update and thus can be viewed as a modification of the momentum method. AdaGrad is proposed to automatically adjust the learning rate. AdaDelta is proposed to address three issues with AdaGrad: (1) monotonically decreasing learning rate, which can lead to excessively small learning rates in late training stages, (2) inconsistent units on two sides of the update equation, and (3) the manually set initial learning rate. As a variation and a special case

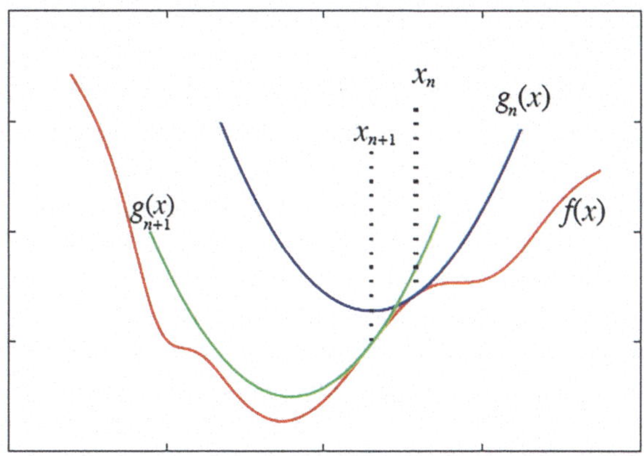

Fig. A.1 Approximation for optimization

of AdaDelta, RMSprop is between AdaGrad and AdaDelta. Adaptive Moment Estimation (Adam) is essentially RMSprop with momentum and can compute adaptive learning rates. Nadam can be viewed as an Adam variation with Nesterov momentum. More details are available in the chapter for deep learning.

Gradient-based methods are usually explained using the idea of hill climbing (in fact, "valley descending") as we move along the fastest descent direction to move toward the bottom of a valley. Alternatively, we can understand gradient-based methods as a process of approximating the true objective function using an approximate function and then using the minimal point of this approximate function as the next point to move to in an iterative process.

As illustrated in Fig. A.1, we use a function $g_n(x)$ to approximate the objective function $f(x)$, for which the exact function is unknown yet can be complex. Since we know the approximate function, its minimal value and the corresponding location can be conveniently identified. As we move to the next iterative step, we will use $g_{n+1}(x)$ for the approximation. $g_n(x)$ and $g_{n+1}(x)$ usually have the same mathematical function but include terms embedding local information, e.g., values and gradients at the current step. The approximate function in gradient-based methods is as follows:

$$g(x) = f(x_n) + f'(x_n)(x - x_n) + \frac{1}{2\alpha}(x - x_n)^2 \tag{A.89}$$

This curve depicts a second-order polynomial that is concave upward. So, we can easily identify the minimum, which happens at $g'(x) = 0$. This will give us the location for the minimum value:

$$x = x_n - \alpha f'(x_n) \tag{A.90}$$

where α is the well-known hyperparameter, learning rate, or learning step size, which is critical to optimization.

Therefore, we can obtain the following equation for update:

$$x_{n+1} = x_n - \alpha f'(x_n) \tag{A.91}$$

In high-order spaces, i.e., x has multiple entities (attributes), then this equation becomes the general update equation for gradient-based methods:

$$x_{n+1} = x_n - \alpha \nabla(x_n). \tag{A.92}$$

The same idea can be used for optimization but with different approximate functions. In the following, we will see that the above equation can be slightly modified by including a second-order gradient to reach Newton's method. In fact, the selection of more general approximate equations that do not necessarily include any gradients or even without an explicit math function could lead to the idea of expectation-maximization methods to be introduced in the final. The conjugate gradient method and its variations, as an intermediate method between gradient-based methods and Newton's method, use an approximate function with an implicit relationship between the $f'(x)$ and α, which needs to be obtained in an iterative way.

A.3.2 Newton's Method and Quasi-Newton's Methods

Using the same approximation concept, we adopt the following equation from Taylor's expansion to approximate the objective function:

$$g(x) = f(x_n) + f'(x_n)(x - x_n) + \frac{1}{2} f''(x_n)(x - x_n)^2. \tag{A.93}$$

For this equation, we can obtain the minimum objective function value at the following point:

$$x = x_n - \frac{f'(x_n)}{f''(x_n)}. \tag{A.94}$$

This will surrender the following equation for update:

$$x_{n+1} = x_n - \frac{f'(x_n)}{f''(x_n)}. \tag{A.95}$$

Newton's method can help us quickly find a global minimum if $f(x)$ is smooth enough. However, the actual objective function is usually much more complicated than that. If the computational demand is not a concern, Newton's method is, in general, much better than gradient-based methods. However, the calculation of the second-order gradients, i.e., the Hessian matrix, could be very computationally expensive. To address this issue, many quasi-Newton's methods such as BFGS, SR1 formula, and BHHH have been proposed to approximate the Hessian matrix.

It is worthwhile to mention Newton's method has also been widely used for solving equations. But equation solution and optimization can be correlated. The corresponding problem for the equation solution is $f'(x) = 0$. That is, the search for the minimum of $f(x)$ is the same problem as solving the equation $f'(x) = 0$.

A.3.3 Conjugate Gradient Methods

The conjugate gradient (CG) approach and its variations have been widely used for solving large-scale linear systems of equations and nonlinear optimization problems. First-order methods like gradient-based methods have a slow convergence speed, while second-order methods like Newton's methods are resource heavy. Conjugate gradient optimization is an intermediate algorithm; it integrates the advantages of only utilizing first-order information while ensuring the convergence speeds of high-order methods. Similar to gradient-based methods and quasi-Newton's methods, methods in the CG family are also among the approximate optimization methods, which use a function to approximate the objective function and search for the optimum using this approximate function. This can be proven, but the deduction is excluded here considering it is not essential in the understanding and use of this type of method.

Compared with gradient-based methods, CG can address one critical question: how to determine the learning step size. In gradient-based methods, such a step size is determined either manually based on experience or automatically using some algorithms. The determination of the step size is critical to the optimization speed and results in many cases and still needs to be carefully handled to ensure acceptable results. Newton's method and quasi-Newton's methods use the second-order gradient to address this issue, which offers a more accurate but computationally expensive solution. CG does not need to resort to second-order gradients, thus circumvents the trouble of computing them.

To reach a general deduction, we still resort to the following general equation for update:

$$\vec{x}_{i+1} = \vec{x}_i + \alpha_i \vec{r}_i \tag{A.96}$$

CG is mostly used for data spaces with two or more dimensions, i.e., the model coefficient $\vec{x}$ ($\vec{\theta}$ in other chapters; please be aware $\vec{x}$ is not a sample here) has two or more parameters. Thus, we use $\vec{x}$ rather than x as in those subsections for other optimization methods.

The essence of CG can be understood as follows. Like regular gradient-based methods, CG also moves along the fastest descent direction. Therefore, the step size vector $\alpha_i \vec{r}_i$ is proportional to the gradients: $\vec{r}_i = -\frac{df(\vec{x}_i)}{d\vec{x}_i}$ and $\vec{r}_{i+1} = -\frac{df(\vec{x}_{i+1})}{d\vec{x}_{i+1}}$. However, unlike the relatively "random" selection of α, CG determines the step size by searching for the lowest objective function value along this fastest descending direction. We know that the lowest objective function presents the following condition:

$$\frac{df(\vec{x}_{i+1})}{d\alpha_i} = 0 \tag{A.97}$$

where $\vec{x}_i$ is the state variable in the current step (or model parameters "$\vec{\theta}_i$" in the current optimization step) and thus $\vec{x}_{i+1}$ is the optimization state (model) that we will reach if we move along the fastest descending direction.

The following step is the key. We reformulate the above equation using the chain rule:

$$\frac{df(\vec{x}_{i+1})}{d\vec{x}_{i+1}} \cdot \frac{d\vec{x}_{i+1}}{d\alpha_i} = -\vec{r}_{i+1} \cdot \vec{r}_i = 0 \tag{A.98}$$

In the remaining deduction, we will use a quadratic optimization problem, which will minimize the following function if $\bar{A}$ is positive definite: $f(\vec{x}) = \frac{1}{2}\vec{x}^T \cdot \bar{A} \cdot \vec{x} - \vec{b}^T \cdot \vec{x} + c$.

It is noted the optimization problem is equivalent to the solution to the linear function: $\bar{A}^T \cdot \vec{x} = \vec{b}$.

Next, let us take a look at two basic parameters defined for this quadratic problem.

Error:

$$\vec{e}_i = \vec{x}_i - \vec{x} \tag{A.99}$$

Residual: $\bar{A}^T \cdot \vec{x}_i \neq \vec{b}$ because $\vec{x}_i \neq \vec{x}$

$$\vec{r}_i = \vec{b} - \bar{A}^T \cdot \vec{x}_i \tag{A.100}$$

We can derive

$$\vec{r}_i = (\vec{b} - \bar{A}^T \cdot \vec{x}_i) - (\vec{b} - \bar{A}^T \cdot \vec{x}) = -\bar{A}^T \cdot (\vec{x}_i - \vec{x}) = -\bar{A} \cdot \vec{e}_i \tag{A.101}$$

The above equation shows the relationship between $\vec{r}_i$ and $\vec{e}_i$, which is very useful for us to understand the problem. However, we use $-\frac{df(\vec{x}_i)}{d\vec{x}_i}$ to calculate $\vec{r}_i$, because $\vec{e}_i$ cannot be calculated as the true optimal point $\vec{x}_i$ is unknown.

Next, let us show how to derive the formulation for calculating α:

$$\begin{aligned}
\vec{r}_{i+1}^T \cdot \vec{r}_i &= (\vec{b} - \bar{A}^T \cdot \vec{x}_{i+1})^T \cdot \vec{r}_i \\
&= [\vec{b} - \bar{A}^T \cdot (\vec{x}_i + \alpha_i \vec{r}_i)]^T \cdot \vec{r}_i \\
&= (\vec{b} - \bar{A}^T \cdot \vec{x}_i)^T \cdot \vec{r}_i - \alpha_i (\bar{A}^T \cdot \vec{r}_i)^T \cdot \vec{r}_i \\
&= \vec{r}_i^T \cdot \vec{r}_i - \alpha_i (\bar{A}^T \cdot \vec{r}_i)^T \cdot \vec{r}_i = 0
\end{aligned} \tag{A.102}$$

Then, we can obtain the following equation for calculating α_i:

$$\alpha_i = \frac{\vec{r}_i^T \cdot \vec{r}_i}{\vec{r}_i^T \cdot \bar{A}^T \cdot \vec{r}_i} \tag{A.103}$$

CG can be implemented with different iterative procedures, leading to different CG variations. One intuitive way is to use the following three equations:

$$\vec{r}_i = \vec{b} - \bar{A}^T \cdot \vec{x}_i \tag{A.104}$$

$$\alpha_i = \frac{\vec{r}_i^T \cdot \vec{r}_i}{\vec{r}_i^T \cdot \bar{A}^T \cdot \vec{r}_i} \tag{A.105}$$

$$\vec{x}_{i+1} = \vec{x}_i + \alpha_i \vec{r}_i \tag{A.106}$$

Alternatively, we can multiply (dot product) the two sides of Eq. A.106 by $-\bar{A}$ and use the relationship $\vec{r}_i = -\bar{A}(\vec{x}_i - \vec{x})$, yielding

$$\vec{r}_{i+1} = \vec{r}_i - \alpha_i \bar{A} \cdot \vec{r}_i \tag{A.107}$$

This can be used with Eq. A.105 for the update. However, it is worthwhile to mention that this would involve more numerical errors. The use of $\vec{x}_i$ for calculating $\vec{r}_i$ frequently can help alleviate this issue.

The above deduction was presented based on a quadratic program problem (or linear equation solution). Notwithstanding, CG has been extended for handling nonlinear problems such as Fletcher-Reeves method, Hestenes-Stiefel method, and Dai-Yuan method [176]. However, it is noted that CG has rarely been used for large-scale deep learning problems. The nonlinear versions of CG require a line search to identify the lowest point along the direction of the fastest descent. This could cause difficulty in applying CG to nonlinear problems, such as additional computational effort, but this may not be the primary reason. One major concern lies in the use of batches for line search. The use of the whole batch can lead to high computing demands but, importantly, force the optimizer to converge to the nearest local minimum, which is not desired. Notwithstanding, such issues may be addressed in the future. Thus, CG is still introduced here considering its potential.

A.3.4 Expectation-Maximization Methods

As mentioned above, the expectation-maximization (EM) methods have a much broader definition. Due to the same reason, it is hard to define a general approximate function for them like what we did for gradient-based and Newton's methods. Also, the formulation and implementation of this type of method can be much different, depending on the applications and algorithms. Thus, the introduction to EM in this subsection will be divided into two parts. First, we will play with the math underneath one typical example to show what an approximate function could look like. Next, we will go through one simple example to get a more intuitive understanding and first-hand experience about the implementation of the basic EM method.

First, to peek into the approximate function, let us use cross-entropy to construct the objective function to set up an example:

$$\ell = -\sum_{x,y} \tilde{p}(x, y) \log p(x, y) \tag{A.108}$$

where x and y are two state variables, p is the probability, and $\tilde{p}$ is the prediction of p.

In a typical classification problem, let us use z to represent the class and then apply the chain rule. We can obtain

$$\ell = -\sum_{x,y} \tilde{p}(x, y) \log \sum_{z} p(x|z)p(z|y)p(y) \tag{A.109}$$

$p(y)$ is a fixed number. So, the above objective function can be modified by removing this term. So, we have

$$\ell = -\sum_{x,y} \tilde{p}(x, y) \log \sum_{z} p(x|z)p(z|y) \tag{A.110}$$

The following gradients can be obtained:

$$\frac{\partial \ell}{\partial p(x|z)} = -\sum_{y} \frac{\tilde{p}(x, y)p(z|y)}{\sum_z p(x|z)p(z|y)} \tag{A.111}$$

$$\frac{\partial \ell}{\partial p(z|y)} = -\sum_{x} \frac{\tilde{p}(x, y)p(x|z)}{\sum_z p(x|z)p(z|y)} \tag{A.112}$$

The application of gradient-based methods in this case is not easy because of extra constraints: $\sum_x p(x|z) = 1$, $\sum_y p(z|y) = 1$, $p(x|z) \geqslant 0$, and $p(z|y) \geqslant 0$. The position of log in the above loss function causes difficulty. Thus, we can introduce the following approximate function:

$$g = -\sum_{x,y} \tilde{p}(x, y) \sum_{z} \left[C_{x,y,z} \log \left(p(x|z)p(z|y)\right)\right] \tag{A.113}$$

The gradients of this appropriate objective function are

$$\frac{\partial g}{\partial p(x|z)} = -\sum_{y} \frac{\tilde{p}(x, y)C_{x,y,z}}{p(x|z)} \quad \text{(A.114)}$$

$$\frac{\partial g}{\partial p(z|y)} = -\sum_{x} \frac{\tilde{p}(x, y)C_{x,y,z}}{p(z|y)} \quad \text{(A.115)}$$

Comparing the gradients of this approximate function with those of the original function, e.g., $\nabla\ell$ vs ∇g, it is not difficult to obtain:

$$C_{x,y,z} = \frac{p(x|z)p(z|y)}{\sum_z [p(x|z)p(z|y)]} \quad \text{(A.116)}$$

Substituting the above equation into the g function will generate an approximate function for this specific example. As can be seen, this is much different from those in the gradient-based methods and Newton's method. Other approximate functions can be derived for other EM applications.

The above deduction is provided to show the similarity between EM and other optimization methods. However, implementations of EM do not need the above deduction and are different from the previous optimization methods. The key to most EM implementations is to construct and/or utilize hidden parameters. An iterative process involving alternately updating the major optimization parameters and the hidden parameters will lead to the optimal parameters and the model they define. Taking the above deduction as an example, we first can calculate $C_{x,y,z}$ using $p(x|z)$ and $p(z|y)$. Then, $C_{x,y,z}$ can be used to compute $p(x|z)$ and $p(z|y)$.

The above deduction is performed based on a meaningful yet abstract example. To gain a more intuitive understanding, let us consider one real-world example that is easy to understand.

In this example, we tossed two coins. Each time, we tossed one coin and recorded the number of times that the two sides of each coin appeared, i.e., heads and tails. Now, we will show how to compute the expected probabilities of getting the sides for each coin and guess which coin was used each time.

Table A.1 shows the results of the test. Please be aware that which coin was selected for each round of toss is unknown when solving the problem. However, this fact is recorded in the above table for validation purposes.

This problem can be solved using EM. The key is to identify a hidden parameter(s). Before that, we need to figure out the model to be sought or, in other words,

Table A.1 Results of coin toss

Round	Coin #A	Coin #B
1	–	5 heads, 5 tails
2	9 heads, 1 tail	–
3	8 heads, 2 tails	–
4	–	4 heads, 6 tails
5	7 heads, 3 tails	–

Table A.2 Results of EM learning process

Round	Coin A	Coin B
1	2.2 heads, 2.2 tails	2.8 heads, 2.8 tails
2	7.2 heads, 0.8 tails	1.8 heads, 0.2 tails
3	5.9 heads, 1.5 tails	2.1 heads, 0.5 tails
4	1.4 heads, 2.1 tails	2.6 heads, 3.9 tails
5	4.5 heads, 1.9 tails	2.5 heads, 1.1 tails
Total	21.3 heads, 8.6 tails	11.7 heads, 8.4 tails

the parameters to be optimized. These parameters can be the probabilities of getting the two sides for each coin. Let us use P_A and P_B to represent the probabilities of obtaining the heads for the two coins. Accordingly, the probabilities for getting the tail side for the two coins are $1 - P_A$ and $1 - P_B$. In the above problem description, we know there are other unknown parameters, that is, "which coin was tossed" in each of the five rounds. Intuitively, we can consider using these unknowns as the hidden parameters. Let us use $[\theta_1, \theta_2, \theta_3, \theta_4, \theta_5]$ to represent them, e.g., θ_1 is the probability of selecting Coin A in the first round, and likewise, $1 - \theta_1$ is the probability of selecting Coin B in the first round.

Let us start with random values $P_A = 0.6$ and $P_B = 0.5$ for initialization.

Iteration 1: $\theta_1 = 0.6$, $P_B = 0.5$

Round 1: $P_A = \frac{0.6^5 \cdot 0.4^5}{0.6^5 \cdot 0.4^5 + 0.5^5 \cdot 0.5^5} = 0.45$

For Coin A, we have the following numbers of times to get heads and tails: $5 * 0.45 = 2.2$, $5 * 0.45 = 2.2$.

For Coin B, we have the following numbers of times to get heads and tails: $5 * (1 - 0.45) = 2.8$, $5 * (1 - 0.45) = 2.8$.

Round 2: $P_A = \frac{0.6^9 \cdot 0.4^1}{0.6^9 \cdot 0.4^1 + 0.5^9 \cdot 0.5^1} = 0.8$

For Coin A, we have the following numbers of times to get heads and tails: $9 * 0.8 = 7.2$, $1 * 0.8 = 0.8$.

For Coin B, we have the following numbers of times to get heads and tails: $9 * (1 - 0.8) = 1.8$, $1 * (1 - 0.8) = 0.2$.

Finishing all the five rounds for the first iteration, as shown in Table A.2, we would get 21.3 heads and 8.6 tails for Coin A and 11.7 heads and 8.4 tails for Coin B. Then, we can use the following two equations to update P_A and P_B, which are needed for the next iteration:

$$P_A = \frac{21.3}{21.3 + 8.6} = 0.71 \tag{A.117}$$

$$P_B = \frac{11.7}{11.7 + 8.4} = 0.58 \tag{A.118}$$

The above iterative process can be repeated to approach the model that can best describe the data. This process illustrates a simple implementation of the maximum likelihood estimation.

A.4 Evaluation Metrics

A.4.1 Overview and Basics

Evaluation metrics are needed to quantitatively measure the performance of a model. Such performance indicators can be much more complicated than merely accuracy or success rate as we use in many daily-life applications. In fact, performance is usually a multifaceted and a relative term. That is, the measure of the model performance can be much different depending on what facet of the performance is more desirable to the model developer. In addition, the nature of the problem, i.e., classification (e.g., binary, multiclass, multi-label), regression, and clustering, also determines the available types and formulations of the evaluation metrics. Despite the discrepancy, there is a collection of metrics of good use. In the following, popular metrics for classification, regression, and clustering are presented.

A.4.2 Classification: Binary

For classification, some algorithms like SVM and KNN directly generate a class or label output. In a numeric setting, such output is usually either 0 or 1 in a binary classification problem. By contrast, many other algorithms directly output the probabilities of belonging to specific classifications/labels, such as logistic regression, random forest, and gradient boosting. These probability outputs can be easily converted to classification/labels using thresholds. It is noted that most of the following evaluation metrics are discussed using class/label output.

Confusion Matrix

The confusion matrix or error matrix is what we can start with to learn the common metrics for binary classification. A basic confusion matrix can be outlined as a table that contains the numbers of different types of samples classified according to their true labels and predicted labels (Fig. A.2). More comprehensive confusion matrices can contain parameters constructed with these numbers. Confusion matrices have been widely used in statistics, data mining, and machine learning, as well as other AI applications.

The core of this matrix/table is the numbers of the four cases corresponding to different combinations of predicted and actual samples. Listed below is an illustration of the four cases using a widely cited example of pregnancy diagnosis:

- True positive (TP): The person who was diagnosed to be pregnant is pregnant.
- True negative (TN): The person who was diagnosed not to be pregnant is not pregnant.

	Actual Positive (AP)	Actual Negative (AN)	Prevalence $=\frac{AP}{Total}$	Accuracy $=\frac{TP+TN}{Total}$	
Predicted Positive (PP)	True Positive (TP)	False Positive (FP)	Precision, Positive Predictive Value (PPV) $=\frac{TP}{PP}$	False Discovery Rate (FDR) $=\frac{FP}{PP}$	
Predicted Negative (PN)	False Negative (FN)	True Negative (TN)	False Omission Rate (FOR) $=\frac{FN}{PN}$	Negative Predictive Rate (NPV) $=\frac{TN}{PN}$	
	True Positive Rate (TPR), Recall, Sensitivity, Probability of Detection, Power $=\frac{TP}{AP}$	False Positive Rate (FPR), Fall-out, Probability of False Alarm $=\frac{FP}{AN}$	Positive Likelihood Ratio (LR+) $=\frac{TPR}{FPR}=\frac{TP\cdot AN}{AP\cdot FP}$	Diagnostic Odds Ratio (DOR) $=\frac{LR+}{LR-}=\frac{TP\cdot TN}{FP\cdot FN}$	F1 Score $=2\cdot\frac{Precision\cdot Recall}{Precision+Recall}$ $=2\cdot\frac{TP}{AP+PP}$
	False Negative Rate (FNR), Miss Rate $=\frac{FN}{AP}$	True Negative Rate (TNR), Specificity (SPC), Selectivity, $=\frac{TN}{AN}$	Negative Likelihood Ratio (LR-) $=\frac{FNR}{TNR}=\frac{FN\cdot AN}{AP\cdot TN}$		

Fig. A.2 Confusion matrix

- False positives (FP): The person who was diagnosed to be pregnant is not pregnant (also known as a "Type I error").
- False negatives (FN): The person who was diagnosed to be not pregnant is pregnant (also known as a "Type II error").

The above four cases including the two types of errors are illustrated in Fig. A.3. The relative significance of the two types of errors in this example is not that obvious. However, if we replace this example with the diagnosis of a detrimental disease like cancer, then Type II error "the person who was diagnosed to not have cancer has cancer" will cause a much more serious outcome than Type I error "the person who was diagnosed to have cancer does not have cancer." This is because the former threatens human lives. However, the condition may be the opposite if we apply it to the application of spam (email) detection. In this case, the Type II error "the email that was diagnosed not to be spam is spam" will cause a less serious outcome than the Type I error "the email that was diagnosed to be spam is not spam." This is because it is less acceptable to have important emails to be marked as spam and missed. Therefore, the meanings and significance of different cases can vary a lot in different applications.

As illustrated in Fig. A.2, more metrics have been proposed to describe the performance of AI models from different angles. Among them, the most widely adopted ones are accuracy, sensitivity, specificity, and precision.

As the most intuitive performance metric, accuracy tells the percentage of correct predictions:

$$\text{Accuracy} = \frac{\text{TP} + \text{TN}}{\text{Total}} \tag{A.119}$$

Sensitivity, which is also called recall, hit rate, or true positive rate, calculates the percentage of the positive samples that have been correctly detected ("recalled"). This metric measures how well the model recognizes a positive class:

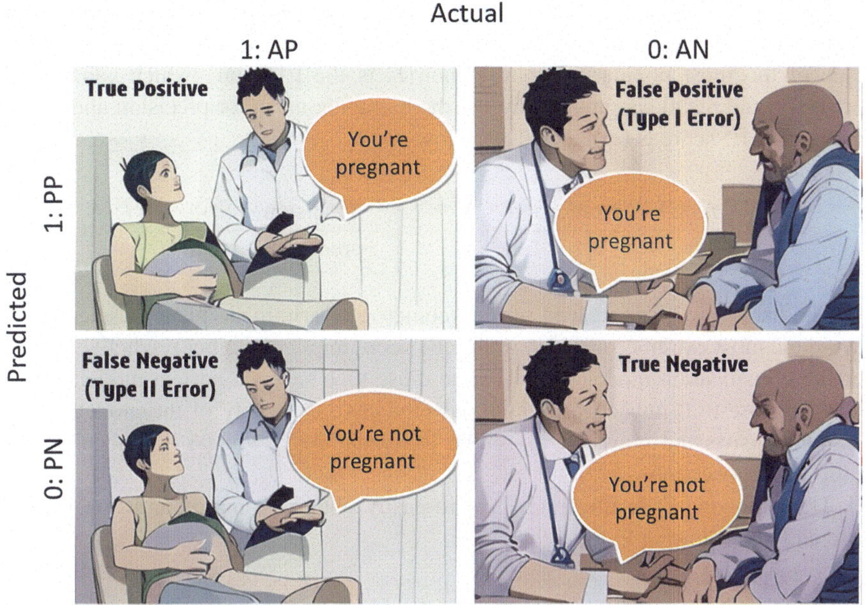

Fig. A.3 Two types of errors

$$\mathrm{TPR} = \frac{\mathrm{TP}}{\mathrm{AP}} = \frac{\mathrm{TP}}{\mathrm{TP} + \mathrm{FN}} = 1 - \mathrm{FNR} \tag{A.120}$$

Specificity, selectivity, or true negative rate shows the percentage of actual negative samples that have been correctly detected. Specificity is defined as follows:

$$\mathrm{TNR} = \frac{\mathrm{TN}}{\mathrm{AN}} = \frac{\mathrm{TN}}{\mathrm{TN} + \mathrm{FP}} = 1 - \mathrm{FPR} \tag{A.121}$$

Precision shows the probability that the predicted positive cases are correctly predicted. Precision or positive predictive value is formulated using the following equation:

$$\mathrm{PPV} = \frac{\mathrm{TP}}{\mathrm{TP} + \mathrm{FP}} = 1 - \mathrm{FDR}. \tag{A.122}$$

These different parameters and their combinations can convey specific information about the model's performance. For example, a situation of high recall and low precision implies that there are a few false-negative cases but lots of false positives. In other words, the model can identify most of the positive samples but is inclined to predict samples as positive. A combination of low recall and high precision indicates that we miss a lot of positive samples, i.e., high false negative. However, those we predict as positive are mostly positive predictions.

Besides the above quantitative descriptions using the basic metrics, composite metrics can also be constructed to consider more facets of the model's performance. One of the most popular composite metrics is the F1 score, which can assess precision and recall simultaneously. As the harmonic mean for precision and recall values, the F1 score is formulated as follows:

$$F_1 = \left(\frac{\text{recall}^{-1} + \text{precision}^{-1}}{2}\right)^{-1} = 2 \cdot \frac{\text{precision} \cdot \text{recall}}{\text{precision} + \text{recall}} \tag{A.123}$$

The harmonic mean instead of the geometric or arithmetic mean is selected to avoid the most extreme values. A higher F1 score implies a higher predictive power of the classification model. An F1 value close to 1 means a perfect model, while a score close to 0 indicates the minimal predictive capability of the model. More comprehensive and powerful tools like ROC and AUC will be described in the following subsection.

The following code illustrates how to easily obtain a simple confusion matrix and an F1 score using Python code:

```
from sklearn import datasets # Dateset
from sklearn.neighbors import KNeighborsClassifier
from sklearn.metrics import confusion_matrix, f1_score # Metric

# Loading the dataset
X, y_true = datasets.make_moons(n_samples=500,noise=0.3,
    random_state=10)  # Generate 500 samples

# Fitting using K-Nearest Neighbors Classifier
knnc = KNeighborsClassifier(n_neighbors=2).fit(X,y_true)
y_pred = knnc.predict(X)

# Print evaluation result using the metric
print(confusion_matrix(y_true,y_pred))
print(f1_score(y_true,y_pred))
```

In newer versions of Scikit-learn, the confusion matrix can also be conveniently plotted as follows:

```
from sklearn.metrics import plot_confusion_matrix
plot_confusion_matrix(y_true,y_pred)
```

ROC and AUC

An ROC curve, also called a receiver operating characteristic curve, employs a graph to assess the performance of a classification model. The ROC curve plots the relationship between the TPR (sensitivity) and the FPR. The ROC curve of a model can be generated by gradually changing the threshold for performing classification. Accordingly, each point in the curve corresponds to a specific decision threshold

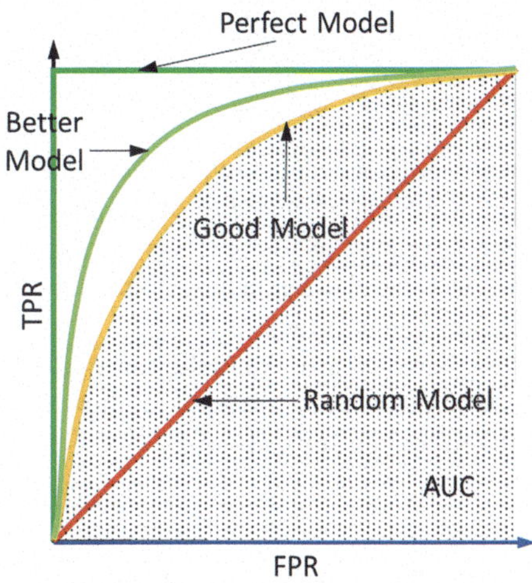

Fig. A.4 ROC and AUC

with its consequent TPR and FPR pair. Therefore, this metric applies to models with such thresholds, such as those using a logistic function for computing the final class. For example, if we have a low classification threshold, then we can classify more items as positive, leading to an increase in both false positives and true positives. Figure A.4 illustrates a typical ROC curve.

As shown in the figure, the best model, or a perfect one, delivers the dark green ROC curve. The performance deteriorates as the ROC curve moves away from the green curve. The deterioration continues until the red curve, which represents the performance of a random model.

The area under the ROC curve, which is termed AUC-ROC (area under the curve) or just AUC (area under the ROC curve), provides a quantitative way of describing the model performance. AUC is used to evaluate the quality of a model's predictive ability regardless of the selected threshold.

Logarithmic Loss

AUC-ROC quantifies the model's ability to discriminate between two different classes. However, it does not provide a way to update the predicted probabilities. By contrast, logarithmic loss or log loss, as a probabilistic metric, measures the difference between the predicted probabilities and the actual class labels. Therefore, log loss is particularly useful for handling models that output probabilities, e.g., logistic regression or ANNs. The use of this function provides feedback for model evaluation during training and thus helps improve the predicted probabilities.

The following equation for the log loss, ℓ, is presented here for general classification tasks. A similar math formulation was introduced in the subsection for cross-entropy:

$$\ell = -\frac{1}{I}\sum_{i=1}^{I} \cdot [y_i \ln(P_i) + (1 - y_i)\ln(1 - P_i)] \tag{A.124}$$

where y_i tells whether sample i belongs to class 1 (e.g., value 1 for "belong to") and P_i indicates the probability of sample i belonging to class 1, i.e., "positive." Log loss has a range of $[0, \infty)$. That is, this metric is positive and has no upper bound limit. In general, a value near 0 indicates higher accuracy, whereas a high loss value indicates low accuracy.

Log loss is also suitable for multiclass classification problems. The above equation needs to be adjusted to allow for multiple classes. This will be introduced in the following subsection.

A.4.3 Classification: Multiclass

Indirect Methods

Indirect methods for extending the above evaluation metrics from binary classification to multiclass classification involve the decomposition of one multiclass classification problem into multiple binary classification problems in a one-vs-all or a one-vs-one fashion. Taking one-vs-all, for example, one class of interest is viewed as positive, and the other classes are labeled as negative. In this way, the binary classification metrics such as precision, recall, and F1 score can be applied to each binary classification problem and then averaged to yield the metric value for the multiclass classification problem. Depending on the average approach, we can have at least three indirect methods:

- Macroaverage: First calculate the metric value for each binary classification problem, and then obtain the arithmetic average. For example, one classification problem with K classes can be transformed into $K - 1$ binary classification problems. Then, we can calculate the macro average metric as

$$\text{Precision}_k = \frac{\text{TP}_k}{\text{TP}_k + \text{FP}_k} \tag{A.125}$$

$$\text{Precision}_{\text{macro}} = \frac{\sum_{k=1}^{K-1} \text{Precision}_k}{K - 1} \tag{A.126}$$

- Microaverage: Gather the results from all binary classification problems and then use these results to calculate the metric for the multiclass problem:

$$\text{Precision}_{\text{micro}} = \frac{\sum_{k=1}^{K-1} \text{TP}_k}{\sum_{k=1}^{K-1} \text{TP}_k + \sum_{k=1}^{K-1} \text{FP}_k} \tag{A.127}$$

- Weighted average: Modify macroaverage with different weights for the metrics from the different binary classification problems to consider possible data imbalance issues:

$$\text{Precision}_{\text{macro}} = \frac{\sum_{k=1}^{K-1} \text{Precision}_k \times w_k}{K-1} \tag{A.128}$$

where w_k is the weight for the kth binary classification problem.

The macromethod considers all classes equally. Therefore, classes with few samples may be unnecessarily emphasized. By contrast, the micromethod can better consider cases with imbalanced classes. In fact, the selection of these indirect methods may not matter too much for tasks with balanced classes. In addition, such selection also relies on how we consider the significance of classes with fewer samples. If we want to emphasize them, the macromethod could be a better choice. When a more delicate consideration is needed, we can also resort to the weighted average method.

Confusion Matrix

The confusion matrix can be applied to multiclass classification without making major changes. We just need to replace the positive and negative labels with multiclass labels such as "1,2,3,...." As a result, the core of the confusion will have a size much greater than 2×2. We can get a good understanding of the confusion matrix by implementing the following code to generate a confusion matrix for a classification problem with 10 classes:

```
from sklearn import datasets # Dateset
from sklearn.neighbors import KNeighborsClassifier
from sklearn.metrics import confusion_matrix, roc_auc_score,
    f1_score # Metric

# Loading the dataset
X, y_true = datasets.make_blobs(n_samples=500, centers=10,
    n_features=2, random_state=0) # Generate 500 samples from 10
    classes

# Fitting using K-Nearest Neighbors Classifier
knnc = KNeighborsClassifier(n_neighbors=2).fit(X,y_true)
y_pred = knnc.predict(X)

# Print evaluation result
print(confusion_matrix(y_true,y_pred))
from sklearn.metrics import ConfusionMatrixDisplay
ConfusionMatrixDisplay.from_predictions(y_true,y_pred,cmap = '
    plasma')
```

The above code generates Fig. A.5 for the confusion matrix.

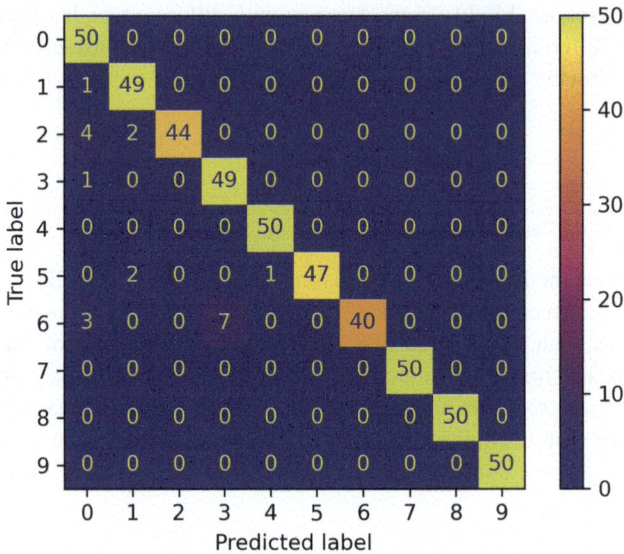

Fig. A.5 Confusion matrix for multiclass classification problem

The diagonal cells in light colors are the correctly predicted samples. The numbers in these cells show the corresponding numbers of samples. Off the diagonal, for example, the cell in the first column and second row tells that four samples, which have a true label of 2, were predicted to be 0.

Logarithmic Loss

The log loss can also be applied to multiclass classification. Based on the same cross-entropy concept, a general mathematical formula for the log loss in multiclass classification tasks is usually constructed as follows:

$$\ell = -\frac{1}{I} \sum_{i=1}^{I} \sum_{k=1}^{K} y_{ik} \cdot \ln(P_{ik}) \tag{A.129}$$

where y_{ik} tells whether sample i belongs to class k (e.g., value 1 for "belong to") and P_{ik} indicates the probability of sample i belonging to class k. Log loss has a range of $[0, \infty)$. This way of labeling the samples is the widely adopted "one-hot" encoding. Attributed to the adoption of this labeling method, the use of the above equation can be fairly simple because only the term containing P_{ik}, which the sample belongs to, needs to be considered while the other terms are zeros.

Kappa Coefficient

The kappa coefficient, also known as Cohen's kappa coefficient, quantifies the agreement between two sets of multiclass labels. If one set of the labels is true, then the other set can be the predictions from the classification task to be evaluated. The kappa coefficient has the following formula:

$$\kappa = \frac{p - p_e}{1 - p_e} = \frac{p - \sum_{i=1}^{K}(I_k \cdot \tilde{I}_k)/I^2}{1 - \sum_{i=1}^{K}(I_k \cdot \tilde{I}_k)/I^2} \tag{A.130}$$

where p is the precision, I is the number of all the samples, I_k is the number of the samples in Class k, and $\tilde{I}_k$ is the number of samples with a predicted label of k.

The values of the kappa coefficient can vary from -1 to 1 in theory. But the range in most applications is [0, 1]. A value of [0,0.2] implies very low similarity, [0.2,0.4] is low, [0.4,0.6] is medium, [0.6,0.8] is good, and [0.8,1.0] is excellent agreement.

Hinge Loss

The hinge loss was originally proposed for the use of SVM for solving binary classification problems. A general formulation for this purpose is as follows:

$$\ell_{\text{hinge}} = \max(0, 1 - y_i \cdot \tilde{y}_i) \tag{A.131}$$

where y is the actual sample label, e.g., 1 and -1, and $\tilde{y}$ is the prediction. It is noted that $\tilde{y}$ here is not necessarily the predicted label. It can be the raw calculation results like $f(\vec{x})$. For example, this $f(\vec{x})$ can be a linear model ($y = \vec{w}^T \cdot \vec{x} + b$) or an ANN. To use this loss as an evaluation metric, we can obtain the total or average hinge loss for all the samples.

The hinge loss can be extended to multiclass classification in two different ways. The first way is to use the indirect methods by breaking down the multiclass classification problem into multiple binary classification problems in a one-vs-all or a one-vs-one fashion. The second way is to modify the formula directly. There are multiple variations for this purpose. The following is a popular one:

$$\ell_{\text{hinge}} = \max\left(0, 1 - \max_{i \neq k}(\tilde{y}_{ij}) - y_{ik}\right) \tag{A.132}$$

where sample i has a true label of k. y_{ik} can be the probability or score for the kth class.

A.4.4 Classification: Multi-Label

Hamming Distance

Hamming distance is suitable for multi-label classification problems. It compares two classification results, which are usually the predicted results and the true labels. As a “distance” measure, it quantifies the difference or dissimilarity between the two classification results:

$$D_{\text{hamming}}(\tilde{y}_i, y_i) = \frac{1}{L}\sum_{l=1}^{L} \text{Sign}(\tilde{y}_{il} \neq y_{il}) \tag{A.133}$$

where the above Hamming distance is for sample i, which has L labels.

Jaccard Similarity Coefficient

Jaccard similarity coefficient measures the similarity by checking the numbers of samples predicted by two models, e.g., the predictions made by the model of interest and the true labels:

$$J(\tilde{y}_i, y_i) = \frac{|\tilde{y}_i \cap y_i|}{|\tilde{y}_i \cup y_i|} \tag{A.134}$$

where $|\tilde{y}_i \cap y_i|$ and $|\tilde{y}_i \cup y_i|$ are the numbers of samples in both sets and in only one of them, respectively.

A.4.5 Regression

Root Mean Squared Error

Root mean squared error (RMSE) is among the most popular metrics, if not the most popular one, for regression problems. RMSE is defined by the standard deviation of the prediction errors. These prediction errors, which are also called residuals in some places, measure the distance of data points from the regression line:

$$\text{RMSE} = \sqrt{\frac{1}{I}\sum_{i=1}^{I}(y_i - \tilde{y}_i)^2} \tag{A.135}$$

where $(y_i - \tilde{y}_i)^2$ is the square of the difference between the predicted and actual values for sample i and I is the number of samples.

RMSE tells us how well the data points are clustered around the regression line. In particular, RMSE is efficient at handling a large number of data points, leading to more reliable error construction. However, it is noted that RMSE is heavily influenced by outliers (data points that differ significantly from others). Therefore, the exclusion of outliers can be critical before the use of RMSE. RMSE can be easily obtained with many machine learning packages. Taking Scikit-learn, for example, the following code gives out RMSE if we know the true labels and predicted labels.

```
sklearn.metrics.root_mean_squared_error(y_true, y_pred)
```

Most of the regression metrics to be introduced next can be obtained in a similar way.

Mean Absolute Error

Mean absolute error (MAE) is the average of the differences between the predicted and actual labels. MAE also measures the average magnitude of errors, i.e., how far the predictions are from the actual values. However, MAE does not inform the direction of errors, i.e., whether we are overfitting or underfitting the data:

$$\mathrm{MAE} = \frac{1}{I}\sum_{i=1}^{I}|y_i - \tilde{y}_i| \tag{A.136}$$

Mean Squared Error

Mean squared error (MSE) is defined as the average of the squares of the differences between the predicted and actual values:

$$\mathrm{MSE} = \frac{1}{I}\sum_{i=1}^{I}(y_i - \tilde{y}_i)^2 \tag{A.137}$$

MSE, together with MAE and other metrics, is commonly employed to construct the loss function in the optimization/solution of machine learning problems. In such optimization applications, e.g., loss reduction in backpropagation, the computation of gradients in MSE is easier than in MAE, which requires computational tools to compute gradients.

Also, speaking of sensitivity to outliers, the order is MSE > RMSE > MAE. RMSE, which is more sensitive to outliers than MAE but less sensitive than MSE. Therefore, RMSE can be considered if a compromise between MAE and MSE is desired. By contrast, MAE is preferred if we care more about small errors than large ones or if the data contains many outliers. MSE stands out when we want to focus on large errors or outliers.

Root Mean Squared Logarithmic Error

The root mean squared logarithmic error (RMSLE) adopts the log of the predicted and actual values. RMSLE can be considered if we do not want to penalize big differences (errors) between the predicted and the actual values:

$$\mathrm{RMSLE} = \sqrt{\frac{1}{I}\sum_{i=1}^{I}[\ln(y_i + 1) - \ln(\tilde{y}_i + 1)]^2} \tag{A.138}$$

R^2 and Adjusted R^2

R^2, also known as the coefficient of determination, is a statistical measure of how closely the data points are fitted to the regression line. R^2 values always lie between 0% and 100%. A R^2 value of 0% indicates that the model predicts no (0%) of the relationship between the input and output, e.g., a constant-only regression, while 100% marks that all of the data points are located on the regression line. Thus, a higher R^2 value corresponds to a better model:

$$R_a^2 = 1 - \frac{\mathrm{MSE}}{\mathrm{Var}} = 1 - \frac{\frac{1}{I}\sum_{i=1}^{I}(y_i - \tilde{y}_i)^2}{\frac{1}{I}\sum_{i=1}^{I}(y_i - \bar{y}_i)^2} \tag{A.139}$$

However, R^2 cannot determine whether the coefficient estimates and predictions are biased (toward certain independent variables or attributes). The adjusted R^2 or R_a^2 was proposed to address this issue. R_a^2 tells the percentage of the variance for a dependent variable (label) that is explained by an independent variable:

$$R_a^2 = 1 - (1 - R^2)\left[\frac{I - 1}{I - (J + 1)}\right] \tag{A.140}$$

where J is the number of independent variables.

The adjusted R_a^2 value increases only if the new variables/attributes improve the model. In other words, R_a^2 decreases if we add useless/unnecessary/irrelevant variables to a model and increases if newly added variables are useful.

A.4.6 Clustering

The evaluation of clustering models or tasks is more challenging in many ways than classification. In particular, true labels of the observations are usually not available in most clustering tasks. As a result, many evaluation metrics rely only on the goodness of splitting the samples but not on the labels anymore. This fact leads to the wide use of internal (or called intrinsic metrics/measures/techniques/indices) in clustering, which is applicable to all unsupervised learning applications. In such measures without labels, a general way of telling good clustering is to find one that generates clusters with small intra-cluster (within-cluster) variance, i.e., data points in the same cluster are similar/close to each other, and large intercluster (between-cluster) variance, i.e., clusters are dissimilar to other clusters.

Many metrics are available for clustering: inertia (i.e., within-cluster sum of squares), Dunn index (DI), silhouette coefficient, Davies-Bouldin index (DB or DBI), Rand index (RI), Adjusted Rand Index (ARI), Adjusted Mutual Information (AMI), and Calinski-Harabasz index (variance ratio criterion). As mentioned above, internal metrics have been widely used in clustering, though external metrics are still available by comparing one clustering against another reference one, which can be "true" labels in many cases. Among the above metrics, RI, ARI, and NMI are external metrics, while the others are internal measures. In the following, let us check the two most simple internal metrics first, i.e., inertia and Dunn index. Next, we will go through more complicated internal metrics, followed by the introduction to external metrics. A slightly different introduction to some of these metrics was presented in the chapter for clustering.

Inertia and Dunn Index

Inertial evaluates intra-cluster distance as the sum of distances of all the points within a cluster from the centroid of that cluster. Usually, we use Euclidean distance for numeric attributes (regression) and Manhattan distance for categorical attributes:

$$\text{Inertia} = \text{Intracluster Distance} = \sum_{i=1, X_i \in C_k}^{I} (X_i - \mu_k)^2 \tag{A.141}$$

where X_i is the attribute values (coordinates) of sample i and μ_k is the center (coordinates) of cluster k, i.e., C_k. The above equation calculates the inertia of one cluster. In many software packages like Scikit-learn, the inertia value is the sum of the inertia values for all the clusters.

The following example shows how inertia can be obtained in a Scikit-learn example:

```
from sklearn import datasets # Dateset
from sklearn.cluster import KMeans # Algorithm
from sklearn.metrics import silhouette_score # Metric

# Loading the dataset
X, y_true = datasets.make_blobs(n_samples=300, centers=4,
    cluster_std=0.50, random_state=0)

# Fitting using K-Means
kmeans = KMeans(n_clusters=4, random_state=1).fit(X)
y_pred = kmeans.labels_
```

Inertia only considers the intra-cluster distance. The Dunn index (DI) is proposed to additionally consider the intercluster distance. For this purpose, the Dunn index is defined as the ratio between the minimum intercluster distances and the maximum intra-cluster distances:

$$\mathrm{DI} = \frac{\min\text{ (Intercluster Distance)}}{\max\text{ (Intracluster Distance)}} = \frac{\min_{l \neq k} \Delta_{k,l}}{\max \delta_k} \tag{A.142}$$

where δ and Δ are intra- and intercluster distances, respectively; l is the cluster number other than cluster k.

Silhouette

The silhouette score or coefficient measures the similarity of each data point to its own cluster compared to other clusters. This metric value is calculated with the mean intra-cluster distance and the mean nearest-cluster distance for each point as follows:

$$S_i = \frac{b_i - a_i}{\max(a_i, b_i)} \tag{A.143}$$

where a_i is the mean distance between Point i and all the other data points in the same cluster and b_i is the smallest mean distance of Point i to the points in any other cluster.

The values of silhouette score vary between -1 and +1. Higher silhouette values correspond to better clustering. Following the above code (for inertia and Dunn index), the silhouette score can be obtained as

```
# Print evaluation result using the metric
print(silhouette_score(X, y_pred))
```

Davies-Bouldin Index

DB index or DBI is defined as the average of the maximum ratio of the intra-cluster distance and the intercluster distance for each cluster:

$$\mathrm{DB} = \frac{1}{K} \sum_{k=1, l \neq k}^{K} \max\left(\frac{\delta_k + \delta_l}{\Delta_{k,l}}\right) \tag{A.144}$$

The values of DB index fall into $[0, +\infty]$. Smaller values indicate better clustering results. DBI can be easily obtained using the following code:

```
from sklearn.metrics import davies_bouldin_score # Metric
```

```
# Print evaluation result using the metric
print(davies_bouldin_score(X, y_pred))
```

Calinski-Harabasz Index

Unlike the above internal metrics, the Calinski-Harabasz index (CHI or CH) adopts the intra-cluster and intercluster variances. To calculate CH, we will need to calculate the intercluster covariance matrices, Σ_B, and intra-cluster covariance matrices, Σ_W first. Then, CH can be calculated using the traces of these two matrices, the total sample number I, and the cluster number K as

$$\text{CH} = \frac{\text{tr}(\Sigma_B)(I - K)}{\text{tr}(\Sigma_W)(K - 1)} \tag{A.145}$$

The following code shows how to obtain the CH value for the example given above:

```
from sklearn.metrics import calinski_harabasz_score # Metric

# Print evaluation result using the metric
print(calinski_harabasz_score(X, y_pred))
```

Adjusted Rand Index

The Adjusted Rand Index (ARI) measures the similarity between two data clusterings instead of evaluating the quality of a single clustering. Accordingly, distinct from the internal metrics introduced above, ARI and the following external metrics compare the labels (predicted cluster numbers) of one clustering with those of another clustering, which is a reference cluster and are viewed as "true labels" in many cases:

$$\text{RI} = \frac{(a + b)}{\binom{I}{2}} = \frac{2(a + b)}{I(I - 1)} \tag{A.146}$$

where I is the number of observations in a sample, a is the number of observation pairs that belong to the same cluster in both clusterings, and b is the number of observations that are assigned to different clusters in the two clusterings.

RI has a value range of [0, 1]. Higher values indicate the agreement between the two clusterings.

The raw RI score can be "adjusted for chance" into the ARI score as follows so that ARI is not sensitive to chance:

$$\text{ARI} = \frac{\text{RI} - \mathbb{E}(\text{RI})}{\max(\text{RI}) - \mathbb{E}(\text{RI})} \tag{A.147}$$

where $\mathbb{E}_{(RI)}$ is the expected RI.

RI and ARI can be obtained using Scikit-learn code in the above example. As can be seen, we now compare the cluster labels of two clusterings instead of assessing the data points and labels of one clustering. Within a range of $[-1, 1]$, higher values of this metric imply higher similarity between the two clusterings.

```
from sklearn.metrics import rand_score,adjusted_rand_score #
    Metric

# Print evaluation result using the metric
print(rand_score(y_true,y_pred))
print(adjusted_rand_score(y_true,y_pred))
```

Adjusted Mutual Information

The Adjusted Mutual Information (AMI), together with the related mutual information and normalized mutual information, forms a category of metrics developed with information theory concepts. MI is defined with the entropy function to allow for the likelihood of assigning a cluster. NMI applies normalization to the MI. AMI is similar to the ARI in that both of them are independent of the permutation of labels.

MI is formulated using the following equation:

$$\text{MI}(U, V) = \sum \sum \left(P_{k,l} \times \log \frac{P_{k,l}}{P_k \times P_l} \right) \tag{A.148}$$

where P_k is the percentage of the cluster k samples in all the samples and $P_{k,l}$ is the percentage of the samples marked as cluster k in both clusterings in all the samples. For example, for $U = [1, 2, 2]$, then $P_1 = 1/3$ and $P_2 = 2/3$. The log function can have a base of 2 or e.

MI can be normalized to obtain the normalized mutual information (NMI):

$$\text{NMI}(U, V) = \frac{\text{MI}(U, V)}{F(H(U), H(V))} \tag{A.149}$$

where $\text{MI}(U, V)$ is the mutual information score between U and V clusterings, $H()$ is the information entropy function ($H(U) = -\sum(P_i \log(P_i))$), and F can be an average function (geometric or arithmetic) or a comparison function (max or min).

AMI uses an equation similar to ARI to adjust for chance: it corrects the effect of the agreement solely due to chance between clusterings:

$$\text{AMI}(U, V) = \frac{\text{MI}(U, V) - \mathbb{E}[\text{MI}(U, V)]}{F(H(U), H(V)) - \mathbb{E}[\text{MI}(U, V)]} \tag{A.150}$$

AMI lies between 0 and 1. A value close to 0 means splits are independent, while a value close to 1 means they are similar.

These three metrics can be easily obtained in Scikit-learn as follows:

```
from sklearn.metrics import mutual_info_score,
    normalized_mutual_info_score, adjusted_mutual_info_score
# Print evaluation result using the metric
print(mutual_info_score(y_true,y_pred))
print(normalized_mutual_info_score(y_true,y_pred))
print(adjusted_mutual_info_score(y_true,y_pred))
```

Bibliography

1. Accenture, How AI boosts industry profits and innovation. Technical report, Accenture (2017)
2. The AI Index Report 2022 (2022). https://aiindex.stanford.edu/ai-index-report-2022/. Accessed 07 July 2024
3. The AI Index Report 2024 (2024). https://aiindex.stanford.edu/report/. Accessed 07 July 2024
4. Artificial Intelligence (AI) Software Market Size, Report 2032 — precedenceresearch.com. https://www.precedenceresearch.com/artificial-intelligence-software-market. Accessed 07 Feb 2024
5. Critical reasons for crashes investigated in the national motor vehicle crash causation survey. Technical report (2023)
6. S. Singh, Critical reasons for crashes investigated in the national motor vehicle crash causation survey. Technical report (2015)
7. K. Abraham, D. Schwarcz, Courting disaster: the underappreciated risk of a cyber insurance catastrophe. Connecticut Insur. Law J. **27**(2), 407–473 (2021)
8. M. Colagrossi, 10 golden age science fiction novels (2019). https://bigthink.com/high-culture/10-golden-age-science-fiction-novels/. Accessed 25 July 2024
9. W.S. McCulloch, W. Pitts, A logical calculus of the ideas immanent in nervous activity. Bull. Math. Biophys. **5**, 115–133 (1943)
10. A.M. Turing, *Computing Machinery and Intelligence* (Springer, Berlin, 2009)
11. S.H. Lavington, The Manchester Mark I and Atlas: a historical perspective. Commun. ACM **21**(1), 4–12 (1978)
12. A.L. Samuel, Some studies in machine learning using the game of checkers. IBM J. Res. Dev. **3**(3), 210–229 (1959)
13. P. McCorduck, C. Cfe, *Machines Who Think: A Personal Inquiry into the History and Prospects of Artificial Intelligence* (AK Peters/CRC Press, Natick/Boca Raton, 2004)
14. J. Moor, The Dartmouth College artificial intelligence conference: the next fifty years. AI Mag. **27**(4), 87–87 (2006)
15. M. Zemčík, A brief history of chatbots. DEStech Trans. Comput. Sci. Eng. **10**, 14–18 (2019)
16. R. Ciesla, *The Book of Chatbots: From ELIZA to ChatGPT* (Springer Nature, Berlin, 2024)
17. M. Minsky, S.A. Papert, *Perceptrons, Reissue of the 1988 Expanded Edition with a New Foreword by Léon Bottou: An Introduction to Computational Geometry* (MIT Press, Cambridge, 2017)
18. D.A. Waterman, *A Guide to Expert Systems* (Addison-Wesley Longman Publishing Co., Inc., Boston, 1985)
19. E.A. Feigenbaum, P. McCorduck, *The Fifth Generation* (Pan Books, London, 1984)
20. D.A. Medler, A brief history of connectionism. Neural Comput. Surv. **1**, 18–72 (1998)

Z. "L." Liu, *Artificial Intelligence for Engineers*,
https://doi.org/10.1007/978-3-031-75953-6

21. R. Brooks, *Flesh and Machines: How Robots Will Change Us* (Vintage, New York City, 2003)
22. G.E. Hinton, S. Osindero, Y.-W. Teh, A fast learning algorithm for deep belief nets. Neural Comput. **18**(7), 1527–1554 (2006)
23. R. Raina, A. Madhavan, A.Y. Ng, Large-scale deep unsupervised learning using graphics processors, in *Proceedings of the 26th Annual International Conference on Machine Learning* (2009), pp. 873–880
24. J. Deng, W. Dong, R. Socher, L.-J. Li, K. Li, L. Fei-Fei, ImageNet: a large-scale hierarchical image database, in *2009 IEEE Conference on Computer Vision and Pattern Recognition* (IEEE, Piscataway, 2009), pp. 248–255
25. X. Glorot, A. Bordes, Y. Bengio, Deep sparse rectifier neural networks, in *Proceedings of the Fourteenth International Conference on Artificial Intelligence and Statistics*. JMLR Workshop and Conference Proceedings (2011), pp. 315–323
26. A. Krizhevsky, I. Sutskever, G.E. Hinton, ImageNet classification with deep convolutional neural networks. Adv. Neural Inf. Proces. Syst. **25**, 1–9 (2012)
27. I. Goodfellow, J. Pouget-Abadie, M. Mirza, B. Xu, D. Warde-Farley, S. Ozair, A. Courville, Y. Bengio, Generative adversarial nets. Adv. Neural Inf. Proces. Syst. **27**, 1–9 (2014)
28. P.P. Ray, ChatGPT: a comprehensive review on background, applications, key challenges, bias, ethics, limitations and future scope. Internet Things Cyber-Phys. Syst. **3**, 121–154 (2023)
29. TIOBE Index - TIOBE — tiobe.com. https://www.tiobe.com/tiobe-index/. Accessed 17 July 2024
30. T.E. Oliphant et al., *Guide to NumPy*, vol. 1 (Trelgol Publishing, USA, 2006)
31. S. Van Der Walt, S.C. Colbert, G. Varoquaux, The NumPy array: a structure for efficient numerical computation. Comput. Sci. Eng. **13**(2), 22–30 (2011)
32. P. Virtanen, R. Gommers, T.E. Oliphant, M. Haberland, T. Reddy, D. Cournapeau, E. Burovski, P. Peterson, W. Weckesser, J. Bright et al., SciPy 1.0: fundamental algorithms for scientific computing in Python. Nat. Methods **17**(3), 261–272 (2020)
33. W. McKinney et al., pandas: a foundational python library for data analysis and statistics. Python High Perform. Sci. Comput. **14**(9), 1–9 (2011)
34. J.D. Hunter, Matplotlib: a 2d graphics environment. Comput. Sci. Eng. **9**(03), 90–95 (2007)
35. M.L. Waskom, Seaborn: statistical data visualization. J. Open Source Software **6**(60), 3021 (2021)
36. F. Pedregosa, G. Varoquaux, A. Gramfort, V. Michel, B. Thirion, O. Grisel, M. Blondel, P. Prettenhofer, R. Weiss, V. Dubourg et al., Scikit-learn: machine learning in Python. J. Mach. Learn. Res. **12**, 2825–2830 (2011)
37. M. Abadi, P. Barham, J. Chen, Z. Chen, A. Davis, J. Dean, M. Devin, S. Ghemawat, G. Irving, M. Isard et al., {TensorFlow}: a system for {Large-Scale} machine learning, in *12th USENIX Symposium on Operating Systems Design and Implementation (OSDI 16)* (2016), pp. 265–283
38. N. Ketkar, J. Moolayil, N. Ketkar, J. Moolayil, Introduction to PyTorch, in *Deep Learning with Python: Learn Best Practices of Deep Learning Models with PyTorch* (2021), pp. 27–91
39. N. Ketkar, N. Ketkar, Introduction to Keras, in *Deep Learning with Python: A Hands-on Introduction* (2017), pp. 97–111
40. G. Brockman, V. Cheung, L. Pettersson, J. Schneider, J. Schulman, J. Tang, W. Zaremba, OpenAI Gym. Preprint. arXiv:1606.01540 (2016)
41. Y. Jia, E. Shelhamer, J. Donahue, S. Karayev, J. Long, R. Girshick, S. Guadarrama, T. Darrell, Caffe: convolutional architecture for fast feature embedding, in *Proceedings of the 22nd ACM International Conference on Multimedia* (2014), pp. 675–678
42. D.J. Higham, N.J. Higham, *MATLAB Guide* (SIAM, Philadelphia, 2016)
43. A.E. Hoerl, R.W. Kennard, Ridge regression: biased estimation for nonorthogonal problems. Technometrics **12**(1), 55–67 (1970)
44. R. Tibshirani, Regression shrinkage and selection via the Lasso. J. R. Stat. Soc. Ser. B: Stat. Methodol. **58**(1), 267–288 (1996)
45. D.W. Hosmer Jr, S. Lemeshow, R.X. Sturdivant, *Applied Logistic Regression* (John Wiley & Sons, Hoboken, 2013)

46. J.W. Hardin, J.M. Hilbe, *Generalized Linear Models and Extensions* (Stata Press, College Station, 2007)
47. A.J. Dobson, A.G. Barnett, *An Introduction to Generalized Linear Models* (Chapman and Hall/CRC, Boca Raton, 2018)
48. T. Hofmann, B. Schölkopf, A.J. Smola, Kernel methods in machine learning (2008). https://projecteuclid.org/journals/annals-of-statistics/volume-36/issue-3/Kernel-methods-in-machine-learning/10.1214/009053607000000677.full
49. O.Z. Maimon, L. Rokach, *Data Mining with Decision Trees: Theory and Applications*, vol. 81 (World Scientific, Singapore, 2014)
50. J.R. Quinlan, Induction of decision trees. Mach. Learn. **1**, 81–106 (1986)
51. J.R. Quinlan, *C4.5: Programs for Machine Learning* (Elsevier, Amsterdam, 2014)
52. L. Breiman, *Classification and Regression Trees* (Routledge, Milton Park, 2017)
53. X. Ying, An overview of overfitting and its solutions, in *Journal of Physics: Conference Series*, vol. 1168 (IOP Publishing, Bristol, 2019), p. 022022
54. L.A. Breslow, D.W. Aha et al., Simplifying decision trees: a survey. Knowl. Eng. Rev. **12**(1), 1–40 (1997)
55. C. Cortes, V. Vapnik, Support-vector networks. Mach. Learn. **20**, 273–297 (1995)
56. J.-P. Vert, K. Tsuda, B. Schölkopf, A primer on kernel methods (2004). https://direct.mit.edu/books/edited-volume/3898/Kernel-Methods-in-Computational-Biology
57. V. Jakkula, Tutorial on support vector machine (SVM). School of EECS, Washington State University **37**(2.5), 3 (2006)
58. J.C. Platt, Fast training of support vector machines using sequential minimal optimization (1998). https://www.researchgate.net/publication/234786663_Fast_Training_of_Support_Vector_Machines_Using_Sequential_Minimal_Optimization
59. M.A. Tanner, *Tools for Statistical Inference*, vol. 3 (Springer, Berlin, 1993)
60. N. Friedman, D. Geiger, M. Goldszmidt, Bayesian network classifiers. Mach. Learn. **29**, 131–163 (1997)
61. G.I. Webb, J.R. Boughton, Z. Wang, Not so naive bayes: aggregating one-dependence estimators. Mach. Learn. **58**, 5–24 (2005)
62. N. Friedman, D. Geiger, M. Goldszmidt, Bayesian network classifiers. Mach. Learn. **29**, 131–163 (1997)
63. A. Ankan, A. Panda, pgmpy: Probabilistic graphical models using Python, in *SciPy* (Citeseer, 2015), pp. 6–11
64. S. Ghosal, A.W. van der Vaart, *Fundamentals of Nonparametric Bayesian Inference*, vol. 44 (Cambridge University Press, Cambridge, 2017)
65. E. Schulz, M. Speekenbrink, A. Krause, A tutorial on gaussian process regression: modelling, exploring, and exploiting functions. J. Math. Psychol. **85**, 1–16 (2018)
66. D.J. Livingstone, *Artificial Neural Networks: Methods and Applications*, vol. 458 (Springer, Berlin, 2008)
67. T. Trappenberg, *Fundamentals of Computational Neuroscience* (OUP Oxford, Oxford, 2009)
68. D.A. Medler, A brief history of connectionism. Neural Comput. Surv. **1**, 18–72 (1998)
69. F. Rosenblatt, The perceptron: a probabilistic model for information storage and organization in the brain. Psychol. Rev. **65**(6), 386 (1958)
70. H.J. Kelley, Gradient theory of optimal flight paths. ARS J. **30**(10), 947–954 (1960)
71. S. Dreyfus, The numerical solution of variational problems. J. Math. Anal. Appl. **5**(1), 30–45 (1962)
72. A.G. Ivakhnenko, V.G. Lapa et al., Cybernetic predicting devices, in *Joint Publications Research Service* (1966)
73. S. Linnainmaa, The representation of the cumulative rounding error of an algorithm as a Taylor expansion of the local rounding errors. PhD thesis, Master's Thesis (in Finnish). University of Helsinki (1970)
74. A.G. Ivakhnenko, Polynomial theory of complex systems. IEEE Trans. Syst. Man Cybernet. **4**, 364–378 (1971)

75. K. Fukushima, Neocognitron: a self-organizing neural network model for a mechanism of pattern recognition unaffected by shift in position. Biol. Cybern. **36**(4), 193–202 (1980)
76. J.J. Hopfield, Neural networks and physical systems with emergent collective computational abilities. Proc. Natl. Acad. Sci. **79**(8), 2554–2558 (1982)
77. P.J. Werbos, Applications of advances in nonlinear sensitivity analysis, in *System Modeling and Optimization: Proceedings of the 10th IFIP Conference New York City, August 31–September 4, 1981* (Springer, Berlin, 2005), pp. 762–770
78. D.H. Ackley, G.E. Hinton, T.J. Sejnowski, A learning algorithm for boltzmann machines. Cognit. Sci. **9**(1), 147–169 (1985)
79. D.E. Rumelhart, G.E. Hinton, R.J. Williams, Learning representations by back-propagating errors. Nature **323**(6088), 533–536 (1986)
80. Y. LeCun, B. Boser, J. Denker, D. Henderson, R. Howard, W. Hubbard, L. Jackel, Handwritten digit recognition with a back-propagation network. Adv. Neural Inf. Proces. Syst. **2**, 396–404 (1989)
81. G. Cybenko, Approximation by superpositions of a sigmoidal function. Math. Control Sig. Syst. **2**(4), 303–314 (1989)
82. S. Hochreiter, Untersuchungen zu dynamischen neuronalen netzen. Diploma, Technische Universität München **91**(1), 31 (1991)
83. S. Hochreiter, J. Schmidhuber, Long short-term memory. Neural Comput. **9**(8), 1735–1780 (1997)
84. N. Srivastava, G. Hinton, A. Krizhevsky, I. Sutskever, R. Salakhutdinov, Dropout: a simple way to prevent neural networks from overfitting. J. Mach. Learn. Res. **15**(1), 1929–1958 (2014)
85. D. Silver, J. Schrittwieser, K. Simonyan, I. Antonoglou, A. Huang, A. Guez, T. Hubert, L. Baker, M. Lai, A. Bolton et al., Mastering the game of Go without human knowledge. Nature **550**(7676), 354–359 (2017)
86. X. Glorot, Y. Bengio, Understanding the difficulty of training deep feedforward neural networks, in *Proceedings of the Thirteenth International Conference on Artificial Intelligence and Statistics*. JMLR Workshop and Conference Proceedings (2010), pp. 249–256
87. K. Simonyan, A. Zisserman, Very deep convolutional networks for large-scale image recognition. Preprint. arXiv:1409.1556 (2014)
88. C. Szegedy, W. Liu, Y. Jia, P. Sermanet, S. Reed, D. Anguelov, D. Erhan, V. Vanhoucke, A. Rabinovich, Going deeper with convolutions, in *Proceedings of the IEEE Conference on Computer Vision and Pattern Recognition* (2015), pp. 1–9
89. K. He, X. Zhang, S. Ren, J. Sun, Deep residual learning for image recognition, in *Proceedings of the IEEE Conference on Computer Vision and Pattern Recognition* (2016), pp. 770–778
90. X. Zhang, X. Zhou, M. Lin, J. Sun, Shufflenet: an extremely efficient convolutional neural network for mobile devices, in *Proceedings of the IEEE Conference on Computer Vision and Pattern Recognition* (2018), pp. 6848–6856
91. R. Girshick, Fast R-CNN, in *Proceedings of the IEEE International Conference on Computer Vision* (2015), pp. 1440–1448
92. S. Ren, K. He, R. Girshick, J. Sun, Faster R-CNN: towards real-time object detection with region proposal networks. IEEE Trans. Pattern Anal. Mach. Intel. **39**(6), 1137–1149 (2016)
93. J. Redmon, S. Divvala, R. Girshick, A. Farhadi, You only look once: unified, real-time object detection, in *Proceedings of the IEEE Conference on Computer Vision and Pattern Recognition* (2016), pp. 779–788
94. T.N. Kipf, M. Welling, Semi-supervised classification with graph convolutional networks. Preprint. arXiv:1609.02907 (2016)
95. F. Scarselli, M. Gori, A.C. Tsoi, M. Hagenbuchner, G. Monfardini, The graph neural network model. IEEE Trans. Neural Networks **20**(1), 61–80 (2008)
96. Z. Wu, S. Pan, F. Chen, G. Long, C. Zhang, S.Y. Philip, A comprehensive survey on graph neural networks. IEEE Trans. Neural Networks Learn. Syst. **32**(1), 4–24 (2020)
97. A. Vaswani, N. Shazeer, N. Parmar, J. Uszkoreit, L. Jones, A.N. Gomez, Ł. Kaiser, I. Polosukhin, Attention is all you need. Adv. Neural Inf. Proces. Syst. **30**, 1–15 (2017)

98. O. Sagi, L. Rokach, Ensemble learning: a survey. Wiley Interdiscip. Rev.: Data Min. Knowl. Discovery **8**(4), e1249 (2018)
99. R. Polikar, Ensemble learning, in *Ensemble Machine Learning: Methods and Applications* (2012), pp. 1–34
100. B.V. Dasarathy, B.V. Sheela, A composite classifier system design: concepts and methodology. Proc. IEEE **67**(5), 708–713 (1979)
101. B. Efron, Bootstrap methods: another look at the jackknife, in *Breakthroughs in Statistics: Methodology and Distribution* (Springer, Berlin, 1992), pp. 569–593
102. M. Kearns, Thoughts on hypothesis boosting. Unpublished Manuscript **45**, 105 (1988)
103. M. Kearns, L. Valiant, Cryptographic limitations on learning boolean formulae and finite automata. J. ACM **41**(1), 67–95 (1994)
104. R.E. Schapire, The strength of weak learnability. Mach. Learn. **5**, 197–227 (1990)
105. Y. Freund, R.E. Schapire, A decision-theoretic generalization of on-line learning and an application to boosting. J. Comput. Syst. Sci. **55**(1), 119–139 (1997)
106. L.K. Hansen, P. Salamon, Neural network ensembles. IEEE Trans. Pattern Anal. Mach. Intel. **12**(10), 993–1001 (1990)
107. D.H. Wolpert, Stacked generalization. Neural Networks **5**(2), 241–259 (1992)
108. L. Breiman, Bagging predictors. Mach. Learn. **24**, 123–140 (1996)
109. Z. Kalal, J. Matas, K. Mikolajczyk, P-N learning: Bootstrapping binary classifiers by structural constraints, in *2010 IEEE Computer Society Conference on Computer Vision and Pattern Recognition* (IEEE, Piscataway, 2010), pp. 49–56
110. L. Deng, D. Yu, J. Platt, Scalable stacking and learning for building deep architectures, in *2012 IEEE International Conference on Acoustics, Speech and Signal Processing (ICASSP)* (IEEE, Piscataway, 2012), pp. 2133–2136
111. T. Chen, C. Guestrin, XGboost: a scalable tree boosting system, in *Proceedings of the 22nd ACM SIGKDD International Conference on Knowledge Discovery and Data Mining* (2016), pp. 785–794
112. T. Hastie, R. Tibshirani, J.H. Friedman, J.H. Friedman, *The Elements of Statistical Learning: Data Mining, Inference, and Prediction*, vol. 2 (Springer, Berlin, 2009)
113. Y. Freund, R.E. Schapire et al., Experiments with a new boosting algorithm, in *Machine Learning: Proceedings of the Thirteenth International Conference*, vol. 96 (Citeseer, 1996), pp. 148–156
114. J.H. Friedman, Greedy function approximation: a gradient boosting machine. Ann. Stat. **29**(5), 1189–1232 (2001). https://www.researchgate.net/publication/2424824_Greedy_Function_Approximation_A_Gradient_Boosting_Machine
115. Z. Ghahramani, Unsupervised learning, in *Summer School on Machine Learning* (Springer, Berlin, 2003), pp. 72–112
116. T. Hastie, R. Tibshirani, J. Friedman, T. Hastie, R. Tibshirani, J. Friedman, Unsupervised learning, in *The Elements of Statistical Learning: Data Mining, Inference, and Prediction* (2009), pp. 485–585
117. K. Kameshwaran, K. Malarvizhi, Survey on clustering techniques in data mining. Int. J. Comput. Sci. Inf. Technol. **5**(2), 2272–2276 (2014)
118. J. MacQueen et al., Some methods for classification and analysis of multivariate observations, in *Proceedings of the Fifth Berkeley Symposium on Mathematical Statistics and Probability*, Oakland, CA, vol. 1 (1967), pp. 281–297
119. D. Arthur, S. Vassilvitskii, k-Means++: The advantages of careful seeding. Technical report. Stanford (2006)
120. Y. Cheng, Mean shift, mode seeking, and clustering. IEEE Trans. Pattern Anal. Mach. Intel. **17**(8), 790–799 (1995)
121. M. Ester, H.-P. Kriegel, J. Sander, X. Xu et al., A density-based algorithm for discovering clusters in large spatial databases with noise, in *KDD'96: Proceedings of the Second International Conference on Knowledge Discovery and Data Mining*, vol. 96 (1996), pp. 226–231
122. R.O. Duda, P.E. Hart et al., *Pattern Classification and Scene Analysis*, vol. 3 (Wiley, New York, 1973)

123. F. Murtagh, P. Contreras, Algorithms for hierarchical clustering: an overview. Wiley Interdiscip. Rev.: Data Min. Knowl. Discovery **2**(1), 86–97 (2012)
124. L. Van Der Maaten, E.O. Postma, H.J. Van Den Herik et al., Dimensionality reduction: a comparative review. J. Mach. Learn. Res. **10**(66–71), 13 (2009)
125. B. Venkatesh, J. Anuradha, A review of feature selection and its methods. Cybernet. Inf. Technol. **19**(1), 3–26 (2019)
126. R. Zebari, A. Abdulazeez, D. Zeebaree, D. Zebari, J. Saeed, A comprehensive review of dimensionality reduction techniques for feature selection and feature extraction. J. Appl. Sci. Technol. Trends **1**(1), 56–70 (2020)
127. K. Pearson, LIII. on lines and planes of closest fit to systems of points in space. London, Edinburgh, Dublin Philos. Mag. J. Sci. **2**(11), 559–572 (1901)
128. R.A. Fisher, The use of multiple measurements in taxonomic problems. Ann. Eugenics **7**(2), 179–188 (1936)
129. J. Hérault, Détection de grandeurs primitives dans un message composite par une architecture de calcul neuromimétique en apprentissage non supervisé, in *Proceedings of GRETSI* (1985), pp. 1017–1020
130. J.B. Tenenbaum, V.d. Silva, J.C. Langford, A global geometric framework for nonlinear dimensionality reduction. Science **290**(5500), 2319–2323 (2000)
131. M.D. Ritchie, L.W. Hahn, N. Roodi, L.R. Bailey, W.D. Dupont, F.F. Parl, J.H. Moore, Multifactor-dimensionality reduction reveals high-order interactions among estrogen-metabolism genes in sporadic breast cancer. Am. J. Hum. Genet. **69**(1), 138–147 (2001)
132. S.T. Roweis, L.K. Saul, Nonlinear dimensionality reduction by locally linear embedding. Science **290**(5500), 2323–2326 (2000)
133. B.H. Menze, B.M. Kelm, R. Masuch, U. Himmelreich, P. Bachert, W. Petrich, F.A. Hamprecht, A comparison of random forest and its Gini importance with standard chemometric methods for the feature selection and classification of spectral data. BMC Bioinf. **10**, 1–16 (2009)
134. B.N. Parlett, *The Symmetric Eigenvalue Problem* (SIAM, Philadelphia, 1998)
135. V. Chandola, A. Banerjee, V. Kumar, Anomaly detection: a survey. ACM Comput. Surv. **41**(3), 1–58 (2009)
136. G. Pang, C. Shen, L. Cao, A.V.D. Hengel, Deep learning for anomaly detection: a review. ACM Comput. Surv. **54**(2), 1–38 (2021)
137. S. Omar, A. Ngadi, H.H. Jebur, Machine learning techniques for anomaly detection: an overview. Int. J. Comput. Appl. **79**(2), 33–41 (2013)
138. D. Samariya, A. Thakkar, A comprehensive survey of anomaly detection algorithms. Ann. Data Sci. **10**(3), 829–850 (2023)
139. F.E. Grubbs, *Sample Criteria for Testing Outlying Observations* (University of Michigan, Ann Arbor, 1949)
140. M. Goldstein, A. Dengel, Histogram-based outlier score (HBOS): a fast unsupervised anomaly detection algorithm. KI-2012: Poster and Demo Track **1**, 59–63 (2012)
141. O. Alghushairy, R. Alsini, T. Soule, X. Ma, A review of local outlier factor algorithms for outlier detection in big data streams. Big Data Cognit. Comput. **5**(1), 1 (2020)
142. F.T. Liu, K.M. Ting, Z.-H. Zhou, Isolation-based anomaly detection. ACM Trans. Knowl. Discovery Data **6**(1), 1–39 (2012)
143. M.E. Villa-Pérez, M.A. Alvarez-Carmona, O. Loyola-González, M.A. Medina-Pérez, J.C. Velazco-Rossell, K.-K.R. Choo, Semi-supervised anomaly detection algorithms: a comparative summary and future research directions. Knowl.-Based Syst. **218**, 106878 (2021)
144. J. Zhai, S. Zhang, J. Chen, Q. He, Autoencoder and its various variants, in *2018 IEEE International Conference on Systems, Man, and Cybernetics (SMC)* (IEEE, Piscataway, 2018), pp. 415–419
145. Q. Zhao, S.S. Bhowmick, Association rule mining: a survey. Nanyang Technol. Univ. Singapore **135**, 18 (2003)
146. R. Agrawal, T. Imieliński, A. Swami, Mining association rules between sets of items in large databases, in *Proceedings of the 1993 ACM SIGMOD International Conference on Management of Data* (1993), pp. 207–216

147. D.T. Larose, C.D. Larose, *Discovering Knowledge in Data: An Introduction to Data Mining*, vol. 4 (John Wiley & Sons, Hoboken, 2014)
148. P.C. Wong, P. Whitney, J. Thomas, Visualizing association rules for text mining, in *Proceedings 1999 IEEE Symposium on Information Visualization (InfoVis' 99)* (IEEE, Piscataway, 1999), pp. 120–123
149. S. Brin, R. Motwani, J.D. Ullman, S. Tsur, Dynamic itemset counting and implication rules for market basket data, in *Proceedings of the 1997 ACM SIGMOD International Conference on Management of Data* (1997), pp. 255–264
150. R. Agrawal, R. Srikant et al., Fast algorithms for mining association rules, in *Procedings 20th International Conference on Very Large Data Bases, VLDB*, Santiago, vol. 1215 (1994), pp. 487–499
151. J. Han, J. Pei, Y. Yin, Mining frequent patterns without candidate generation. ACM Sigmod Record **29**(2), 1–12 (2000)
152. M.J. Zaki, S. Parthasarathy, W. Li, A localized algorithm for parallel association mining, in *Proceedings of the Ninth Annual ACM Symposium on Parallel Algorithms and Architectures* (1997), pp. 321–330
153. M.L. Puterman, Markov decision processes. Handbooks Oper. Res. Manage. Sci. **2**, 331–434 (1990)
154. E.A. Feinberg, A. Shwartz, *Handbook of Markov Decision Processes: Methods and Applications*, vol. 40 (Springer Science & Business Media, Berlin, 2012)
155. R. Bellman, A Markovian decision process. J. Math. Mech. **6**(5), 679–684 (1957)
156. R.A. Howard, *Dynamic Programming and Markov Processes* (John Wiley, Hoboken, 1960)
157. D. Blackwell, Discrete dynamic programming. Ann. Math. Stat. **33**(2), 719–726 (1962). https://celebratio.org/Blackwell_DH/article/248/
158. G. Brockman, V. Cheung, L. Pettersson, J. Schneider, J. Schulman, J. Tang, W. Zaremba, OpenAI Gym. Preprint. arXiv:1606.01540 (2016)
159. C.J. Watkins, P. Dayan, Q-learning. Mach. Learn. **8**, 279–292 (1992)
160. K. Arulkumaran, M.P. Deisenroth, M. Brundage, A.A. Bharath, Deep reinforcement learning: a brief survey. IEEE Sig. Proces. Mag. **34**(6), 26–38 (2017)
161. Y. Li, Deep reinforcement learning: an overview. Preprint. arXiv:1701.07274 (2017)
162. G.A. Rummery, M. Niranjan, *On-line Q-learning using Connectionist Systems*, vol. 37 (University of Cambridge, Department of Engineering Cambridge, Cambridge, 1994)
163. V. Mnih, K. Kavukcuoglu, D. Silver, A. Graves, I. Antonoglou, D. Wierstra, M. Riedmiller, Playing Atari with deep reinforcement learning. Preprint. arXiv:1312.5602 (2013)
164. R.S. Sutton, A.G. Barto, *Reinforcement Learning: An Introduction* (MIT Press, Cambridge, 2018)
165. L.P. Kaelbling, M.L. Littman, A.W. Moore, Reinforcement learning: a survey. J. Artif. Intell. Res. **4**, 237–285 (1996)
166. R. Williams, A class of gradient-estimation algorithms for reinforcement learning in neural networks, in *Proceedings of the International Conference on Neural Networks* (1987), pp. II–601
167. L. Engstrom, A. Ilyas, S. Santurkar, D. Tsipras, F. Janoos, L. Rudolph, A. Madry, Implementation matters in deep RL: a case study on ppo and trpo, in *International Conference on Learning Representations* (2019)
168. Y. Li, Deep reinforcement learning: an overview. Preprint. arXiv:1701.07274 (2017)
169. S. Kumar, Balancing a CartPole system with reinforcement learning–a tutorial. Preprint. arXiv:2006.04938 (2020)
170. A.G. Barto, R.S. Sutton, C.W. Anderson, Neuronlike adaptive elements that can solve difficult learning control problems. IEEE Trans. Syst. Man Cybernet. **13**(5), 834–846 (1983). http://incompleteideas.net/papers/barto-sutton-anderson-83.pdf
171. L. Weaver, N. Tao, The optimal reward baseline for gradient-based reinforcement learning. Preprint. arXiv:1301.2315 (2013)
172. V. Mnih, A.P. Badia, M. Mirza, A. Graves, T. Lillicrap, T. Harley, D. Silver, K. Kavukcuoglu, Asynchronous methods for deep reinforcement learning, in *International Conference on Machine Learning* (PMLR, 2016), pp. 1928–1937

173. R.S. Sutton, D. McAllester, S. Singh, Y. Mansour, Policy gradient methods for reinforcement learning with function approximation. Adv. Neural Inf. Proces. Syst. **12**, 1057–1063 (1999)
174. Z. Liu, *Multiphysics in Porous Materials* (Springer, Berlin, 2018)
175. K.B. Petersen, M.S. Pedersen, The matrix cookbook (2012). http://www2.compute.dtu.dk/pubdb/pubs/3274-full.html
176. L. Lucambio Pérez, L. Prudente, Nonlinear conjugate gradient methods for vector optimization. SIAM J. Optim. **28**(3), 2690–2720 (2018)

Index

Z. "L." Liu, *Artificial Intelligence for Engineers*,
https://doi.org/10.1007/978-3-031-75953-6

The manufacturer's authorised representative in the EU is Springer Nature Customer Service Centre GmbH, Europaplatz 3, 69115 Heidelberg, Germany. If you have any concerns regarding our products, please contact ProductSafety@springernature.com

Printed and bound by CPI Group (UK) Ltd, Croydon, CR0 4YY
07/07/2026
02160911-0004